Two-Point Boundary Value Problems: Shooting Methods

Modern Analytic *and* Computational Methods *in* Science *and* Mathematics

A GROUP OF MONOGRAPHS AND ADVANCED TEXTBOOKS

Richard Bellman, EDITOR
University of Southern California

Published

1. R. E. Bellman, R. E. Kalaba, and Marcia C. Prestrud, Invariant Imbedding and Radiative Transfer in Slabs of Finite Thickness, 1963

2. R. E. Bellman, Harriet H. Kagiwada, R. E. Kalaba, and Marcia C. Prestrud, Invariant Imbedding and Time-Dependent Transport Processes, 1964

3. R. E. Bellman and R. E. Kalaba, Quasilinearization and Nonlinear Boundary-Value Problems, 1965

4. R. E. Bellman, R. E. Kalaba, and Jo Ann Lockett, Numerical Inversion of the Laplace Transform: Applications to Biology, Economics, Engineering, and Physics, 1966

5. S. G. Mikhlin and K. L. Smolitskiy, Approximate Methods for Solution of Differential and Integral Equations, 1967

6. R. N. Adams and E. D. Denman, Wave Propagation and Turbulent Media, 1966

7. R. L. Stratonovich, Conditional Markov Processes and Their Application to the Theory of Optimal Control, 1968

8. A. G. Ivakhnenko and V. G. Lapa, Cybernetics and Forecasting Techniques, 1967

9. G. A. Chebotarev, Analytical and Numerical Methods of Celestial Mechanics, 1967

10. S. F. Feshchenko, N. I. Shkil', and L. D. Nikolenko, Asymptopic Methods in the Theory of Linear Differential Equations, 1967

11. A. G. Butkovskiy, Distributed Control Systems, 1969

12. R. E. Larson, State Increment Dynamic Programming, 1968

13. J. Kowalik and M. R. Osborne, Methods for Unconstrained Optimization Problems, 1968

14. S. J. Yakowitz, Mathematics of Adaptive Control Processes, 1969

15. S. K. Srinivasan, Stochastic Theory and Cascade Processes, 1969

16. D. U. von Rosenberg, Methods for the Numerical Solution of Partial Differential Equations, 1969

17. R. B. Banerji, Theory of Problem Solving: An Approach to Artificial Intelligence, 1969

18. R. Lattès and J.-L. Lions, The Method of Quasi-Reversibility: Applications to Partial Differential Equations. Translated from the French edition and edited by Richard Bellman, 1969

19. D. G. B. Edelen, Nonlocal Variations and Local Invariance of Fields, 1969

20. J. R. Radbill and G. A. McCue, Quasilinearization and Nonlinear Problems in Fluid and Orbital Mechanics, 1970

21. W. Squire, Integration for Engineers and Scientists, 1970

22. T. Parthasarathy and T. E. S. Raghavan, Some Topics in Two-Person Games, 1971

23. T. Hacker, Flight Stability and Control, 1970

24. D. H. Jacobson and D. Q. Mayne, Differential Dynamic Programming, 1970

25. H. Mine and S. Osaki, Markovian Decision Processes, 1970

26. W. Sierpinski, 250 Problems in Elementary Number Theory, 1970

27. E. D. Denman, Coupled Modes in Plasmas, Elastic Media, and Parametric Amplifiers, 1970

28. F. H. Northover, Applied Diffraction Theory, 1971

29. G. A. Phillipson, Identification of Distributed Systems, 1971

30. D. H. Moore, Heaviside Operational Calculus: An Elementary Foundation, 1971

31. S. M. Roberts and J. S. Shipman, Two-Point Boundary Value Problems: Shooting Methods

32. V. F. Demyanov and A. M. Rubinov, Approximate Methods in Optimization Problems, 1970

33. S. K. Srinivasan and R. Vasudevan, Introduction to Random Differential Equations and Their Applications, 1971

34. C. J. Mode, Multitype Branching Processes: Theory and Applications, 1971

35. R. Tomovic and M. Vukobratovic, General Sensitivity Theory

37. W. T. Tutte, Introduction to the Theory of Matroids, 1971

In Preparation

36. J. G. Krzyz, Problems in Complex Variable Theory

38. B. W. Rust and W. R. Burrus, Mathematical Programming and the Numerical Solution of Linear Equations

39. J. O. Mingle, The Invariant Imbedding Theory of Nuclear Transport.

40. H. M. Lieberstein, Mathematical Physiology.

Two-Point Boundary Value Problems: Shooting Methods

Sanford M. Roberts
International Business Machines Corporation
Basic Scientific Development
Palo Alto, California

Jerome S. Shipman
International Business Machines Corporation
Federal Systems Division
Gaithersburg, Maryland

American Elsevier
Publishing Company, Inc.
NEW YORK

AMERICAN ELSEVIER PUBLISHING COMPANY, INC.
52 Vanderbilt Avenue, New York, N.Y. 10017

ELSEVIER PUBLISHING COMPANY
335 Jan Van Galenstraat, P.O. Box 211
Amsterdam, The Netherlands

International Standard Book Number 0-444-00102-6

Library of Congress Card Number 72-153420

Printed in the United States of America

To our wives, NANCY and BETTE

and to MAX SHIPMAN

who were pleased when we began the book
and were pleased when we finished it
—but for different reasons

CONTENTS

Chapter 4

Method of Complementary Functions

Chapter 5

Quasilinearization

Chapter 6

Newton-Raphson Method and Two-Point Boundary Value Problems

Chapter 7

Continuation

Chapter 8

Finite Difference Methods and Related Topics

Chapter 9

Second-Order Newton-Raphson Methods

PREFACE

Two-point boundary value problems associated with systems of linear and nonlinear ordinary differential equations occur in many branches of mathematics, engineering, and the various sciences. In these problems, conditions are specified at the endpoints of an interval, and a solution of the differential equations over the interval is sought which satisfies the given endpoint conditions. Generally the equations cannot be solved analytically, so recourse must be made to a numerical approach. In this book we describe, develop, and exploit one such class of methods; namely, shooting methods. In a shooting method, a set of values of the unspecified conditions at the initial point of the interval ("missing initial conditions") is assumed, and the differential equations are numerically integrated to the terminal point ("shooting" at the target terminal points). If the computed terminal values satisfy the specified terminal conditions, the problem has been solved. If they do not (the normal course of events), the differences between the computed and specified terminal conditions (the "miss distances") are used to adjust the missing initial conditions. If the differential equations and boundary conditions are linear, the adjustment need only be made once, but if the differential equations or the boundary conditions are nonlinear, the adjustment of the missing initial conditions is an iterative process.

In the past, shooting methods were regarded with a certain amount of suspicion, chiefly for two reasons. The first had to do with the lack of a theoretical foundation for the iterative process just described, with the result that conditions under which the iterative process converged and estimates of the rate of convergence were not available. The second reason concerns the apparent inability of shooting methods to handle numerically sensitive (unstable) problems.

We first became aware of the lack of an adequate theory when we tried to reconcile some anomalous results produced by the Goodman-Lance method of adjoints, one of the best known and widely used shooting methods. Once we discovered that the Goodman-Lance procedure was a concrete realization of the abstract Newton-Kantorovich method, we had in hand a satisfactory theory of convergence and error estimates for it, and, subsequently, for other shooting methods as well. Then we were able to devise extensions of shooting methods, such as continuation, which enabled them to handle numerically sensitive problems. In this book we develop the basic methods and their extensions, along with the appropriate theory, and illustrate the techniques by applying them to a variety of problems drawn from practice.

Our book is intended for applied mathematicians and engineers, students and professionals, who are familiar with the numerical integration of initial value problems, and who want to solve two-point boundary value problems with a minimum of problem preparation. Since we believe that computing is the art of the possible, we give the methods in sufficient detail that the reader can generate his own programs. A well designed shooting method program will be independent of the particular problem to be solved, with the exception of the subroutine into which the right hand side of the set of ordinary differential equations to be solved is inserted. Our methods work just as well on small machines as on large ones.

While our main concern is to produce working tools for the applied mathematician and engineer, we do not think that these tools can be used properly without an understanding of the underlying theory. We have chosen to develop the necessary theory in the framework of functional analysis, following Kantorovich (who first exploited the application of functional analysis to applied mathematics and numerical analysis), Collatz, and others. We have found that functional analysis unifies seemingly disparate results, and that it often furnishes valuable geometrical insight through which a known procedure can be better understood, and new techniques developed.

Since we realize that many of our readers may not be acquainted with the concepts and terminology of functional analysis we have attempted throughout the text to restate key phrases and ideas in more familiar terms. In addition, we have included an Appendix in which the various concepts and terms are defined rigorously. Occupying, as we do, the middle ground between the mathematician and the engineer, we hope our stance of generating and explaining practical methods with sufficient theoretical underpinning will bridge the gap between their disciplines.

While our emphasis is on shooting methods, we also devote space to other methods, in particular, quasilinearization and finite difference methods, which have wide applicability in the computer solution of two-point boundary value problems. We regret not having been able to include still other approaches (invariant imbedding, for example), but we believe that the methods in this book will give the analyst a repertoire of techniques which should enable him to solve numerically the majority of two-point boundary value problems he is likely to encounter. This, in fact, has been our experience for we ourselves have solved almost every one of the numerical examples presented in this text by at least one, and often several, of the methods discussed.

SANFORD M. ROBERTS
JEROME S. SHIPMAN

Houston, Texas and
Gaithersburg, Maryland
Spring 1971

Chapter 1

INTRODUCTION

The purpose of this book is to describe in some detail various "shooting" and related methods for the numerical solution of two-point boundary value problems for linear and nonlinear ordinary differential equations. Two-point boundary value problems occur in a number of areas of applied mathematics, theoretical physics, and engineering, among them boundary layer theory, the study of stellar interiors, and control and optimization theory. Since it is usually impossible to obtain analytic (closed-form) solutions to the two-point boundary value problems met in practice, these problems must be attacked by numerical methods. The methods treated in this book have enabled applied mathematicians, programmers, and engineers who are not specialists in two-point boundary value problems to obtain numerical solutions to a wide variety of problems, within the time limits so often placed on their work.

In contrast to initial value problems for ordinary differential equations in which all the conditions are specified for one value of the independent variable (the initial point), two-point boundary value problems, as the name implies, have the property that conditions are specified at two values of the independent variable (the initial point and the final point; collectively, the boundary points). *

This apparently minor change can lead to profound changes in the behavior of the solution of the differential equations. It is not hard to give examples of linear differential equations that possess unique solutions as initial value problems, but which may have no solution, a unique solution, or an infinite number of solutions as two-point boundary value problems. For example, the initial value problem

$$\ddot{y}+y = 0, \qquad y(0) = c_1, \qquad \dot{y}(0) = c_2$$

has the unique solution $y(x) = c_1 \cos x + c_2 \sin x$ for any set of values c_1, c_2.

---

* Multipoint boundary value problems, in which data are specified at more than two values of the independent variable, are sometimes encountered. Methods for handling such problems are not treated at length, but it is pointed out how certain of the methods can be generalized to handle multipoint boundary value problems.

However, the boundary value problem

$$\ddot{y}+y = 0, \qquad y(0) = 1, \qquad y(\pi) = 0$$

has no solution; the problem

$$\ddot{y}+y = 0, \qquad y(0) = 1, \qquad y(2) = 0$$

has the unique solution $y(x) = \cos x - (\text{cotan } 2) \sin x$; while the problem

$$\ddot{y}+y = 0, \qquad y(0) = 0, \qquad y(\pi) = 0$$

has an infinite number of solutions $y(x) = B \sin x$, where B may have any value.

In the examples above, values of the solution at the two ends of the interval were specified, and different combinations of end points and values of the solution led to different conditions of existence and uniqueness of solutions. The specification of the derivative of the solution, rather than the value of the solution itself, may also lead to different conclusions with regard to the existence and uniqueness of solutions of two-point boundary value problems. Consider, for example, the differential equation

$$\ddot{y}+f(y, \dot{y}, x) = 0,$$

where $f(y, \dot{y}, x)$ is continuous in the strip $D: -\infty \leqq y \leqq \infty, -\infty \leqq \dot{y} \leqq \infty$, $a \leqq x \leqq b$, and $f(y, \dot{y}, x)$, satisfies the Lipschitz condition

$$|f(y, \dot{y}, x)-f(u, \dot{u}, x)| \leqq K|y - u| + L|\dot{y} - \dot{u}|,$$

where K and L are positive constants and y, $\dot{y}$, u, $\dot{u}$ are points in D. Then Bailey *et al.* [1] show that the existence of a unique solution to the two-point boundary value problem $y(a) = A$, $y(b) = B$ can be guaranteed over an interval $[a, b]$ *twice as long* as the interval for which existence and uniqueness obtain for the problem $y(a) = A$, $\dot{y}(b) = m$.

In view of the complicated behavior that solutions of two-point boundary value problems can exhibit, it should not be surprising that the theory of the existence and uniqueness of solutions of these problems is in a less satisfactory condition than the corresponding theory for initial value problems. And it should be expected that the numerical solution of a two-point boundary problem for a given ordinary differential equation will in general be a more difficult matter than the numerical solution of the corresponding initial value problem.

There now exist a number of efficient methods for the step-by-step numerical integration of initial value problems, and it is assumed that the reader is familiar with the use of standard one-step methods such as Runge-Kutta and multistep methods such as Hamming's modification of Milne's method.

These methods have in common that the solution is computed at a succession of values of the independent variable, say $x_1, x_2, x_3, \ldots$, where x_0 is the initial point. The initial conditions at x_0 contain sufficient information for the solution to be computed at x_1; and so on. (The progression of the solution from x_0 to x_1 to x_2, etc., explains why initial value problems are sometimes called "marching" problems.) Iteration at the points x_i is sometimes used to improve the numerical accuracy, but no "guessing" is involved because the method has already furnished a good first approximation. In two-point boundary value problems, on the other hand, there is not sufficient information at the initial point to start a step-by-step solution; hence a way must be found to determine the missing initial conditions, or an approach other than step-by-step integration must be used. Also, iteration is more likely to be an essential feature of a method for the solution of two-point boundary value problems, and it is usual that missing initial conditions or even solution profiles must be guessed, with no other *a priori* knowledge.

Two-point boundary value problems have been attacked by a variety of techniques, among them:

1. Interpolation methods. Solutions of the differential equations are found by numerical integration for sets of values of the missing initial conditions. These solutions will not in general satisfy the prescribed boundary conditions. The correct values of the missing initial conditions are then found by inverse interpolation [2, 3].

2. Variational methods. The two-point boundary value problem is replaced by the variational problem of minimizing a certain integral, and the resulting variational problem is solved by the Rayleigh-Ritz methods [2, 4].

3. Method of collocation. The solution of the two-point boundary value problem is represented by a function of several parameters which satisfies the boundary conditions for any set of values of the parameters. The approximate solution is substituted in the differential equations and the parameters are determined by the satisfaction of some error criterion [2].

4. Picard's method. The two-point boundary value problem is put in a form symbolically represented by $x = F(x)$. A sequence of approximate solutions $x^{(n)}$ is developed by the process $x^{(n)} = F(x^{(n-1)})$ which converges to the solution of the original problem under certain conditions [1–3].

5. Discrete methods. The derivatives occurring in the differential equations are replaced by appropriate finite differences, and the solution to the two-point boundary value is sought at discrete values of the independent variable. The effect is to replace the original problem by the problem of solving a finite number of algebraic or transcendental equations [2–6].

6. Quasilinearization. In this method, applicable only to two-point boundary value problems for systems of nonlinear ordinary differential equations, the original nonlinear problem is replaced by a sequence of more easily solved linear problems whose solutions converge under appropriate conditions to the solution of the original problem [7, 8].

7. Shooting methods. They take their name from the situation in the two-point boundary value problem for a single second-order differential equation with initial and final values of the solution prescribed. Varying the initial slope gives rise to a set of profiles which suggest the trajectory of a projectile "shot" from the initial point. That initial slope is sought which results in the trajectory "hitting" the target; that is, the final value [1, 6, 8–11].

This hit-or-miss method is of course unsuitable for the solution of two-point boundary value problems on high-speed digital computers. What is needed is a systematic way to vary the missing conditions based on the amount by which the final values are missed. The shooting methods we are concerned with have this property. In fact, linear problems can be solved by shooting methods without iteration, and the iterations necessary for nonlinear problems can be shown to converge under appropriate conditions.

For nonlinear differential equations, shooting methods have certain advantages for the problem solver. First of all, the methods are quite general and are applicable to a wide variety of differential equations. It is not necessary for the applicability of shooting methods that the equations be of special types such as even-order self-adjoint. Second, shooting methods require a minimum of problem analysis and preparation. It is relatively easy to implement shooting methods on digital computers using standard subroutines for the numerical integration of ordinary differential equations, solutions of linear algebraic equations, etc. With a properly written code, only one subroutine need be altered from problem to problem, the one in which the right-hand side of the system of differential equations written in a standard form is entered. All other parts of the code will handle automatically any problem from a broad class.

Despite their advantages, shooting methods, like all methods, have their limitations. Shooting methods sometimes fail to converge for problems which are sensitive to the initial conditions. In some problems modest changes in the initial conditions result in numerical difficulties such as machine overflow. Procedures such as continuation and reorthogonalization have been developed which extend the usefulness of shooting methods, and they are discussed.

However, problems are sometimes encountered in practice which cannot be solved even by the extended shooting methods. For this reason we have included a brief treatment of finite difference methods, one of whose virtues is their capability of handling numerically sensitive problems. We also discuss quasilinearization, which has been successfully applied to the practical solution of two-point boundary value problems.

Several apparently different shooting methods have been presented in the literature, often as purely formal manipulations with little attention to the conditions under which the methods work. Our emphasis here is on the derivation of the various methods, their interrelationships, and the demonstration that they are all realizations of a generalization of the familiar Newton-Raphson method for the solution of equations. Once the shooting methods are shown to be a kind of Newton-Raphson method, conditions for convergence, rates of convergence, and error estimates can be derived. In addition, we are concerned with the practical computer implementation of the techniques, and the solution of problems that arise in scientific and engineering applications.

REFERENCES

1. P. B. Bailey, L. F. Shampine, and P. E. Waltman, *Nonlinear Two Point Boundary Value Problems*, Academic, New York, 1968.
2. L. Collatz, *The Numerical Treatment of Differential Equations*, 3rd ed., Springer-Verlag, Berlin, 1960.
3. L. Fox, *Numerical Solution of Two-Point Boundary Problems in Ordinary Differential Equations*, Oxford, London, 1957.
4. R. S. Varga, *Matrix Iterative Analysis*, Prentice-Hall, Englewood Cliffs, N. J., 1962.
5. P. Henrici, *Discrete Variable Methods in Ordinary Differential Equations*, Wiley, New York, 1962.
6. H. B. Keller, *Numerical Methods for Two-Point Boundary Value Problems*, Blaisdell, Waltham, Mass., 1968.
7. R. Bellman and R. Kalaba, *Quasilinearization and Nonlinear Boundary Value Problems*, American Elsevier, New York, 1965.
8. S. M. Roberts, J. S. Shipman, and C. V. Roth, Continuation in Quasilinearization, *J. Optimization Theory and Appl.* (3) **2** (May 1968), 164–178.
9. T. R. Goodman and G. N. Lance, The Numerical Solution of Two Point Boundary Value Problems, *MTAC*, **10** (1956), 82–86.
10. S. M. Roberts and J. S. Shipman, The Kantorovich Theorem and Two-Point Boundary Value Problems, *IBM J. Res. Develop.* (5) **10** (Sept. 1966), 402–406.
11. S. M. Roberts and J. S. Shipman, Continuation in Shooting methods for Two-Point Boundary Value Problems, *J. Math. Anal. Appl.* (1) **18** (April 1967), 45–58.

General References

L. Fox, *Numerical Solution of Ordinary and Partial Differential Equations*, Addison-Wesley, Reading, Mass., 1962.

R. H. Moore, Newton's Method and Variations, in *Nonlinear Integral Equations*, P. M. Anselone, ed., Univ. of Wisconsin Press, 1964.

Chapter 2

SHOOTING METHODS

2.1. INTRODUCTION

In this chapter we discuss the statement of the two-point boundary value problems as dealt with in this book; namely, in terms of a system of first-order ordinary differential equations. The reduction of higher-order systems of differential equations to a system of first-order differential equations is described. A brief description of each of the shooting methods is given.

2.2. TWO-POINT BOUNDARY VALUE PROBLEM STATEMENT

Two point-boundary value problems are problems in which, for a set of possibly nonlinear ordinary differential equations, some boundary conditions are specified at the initial value of the independent variable, while the remainder of the boundary conditions are specified at the terminal value of the independent variable. The boundary conditions are therefore split between the two points, the initial and terminal values of the independent variable.

In this book, by two-point boundary value problems we mean problems with the following characteristics:

1. n first-order ordinary differential equations to be solved over the interval $[t_0, t_f]$, where t is the independent variable, t_0 is the initial point, and t_f is the final point.

2. r boundary conditions specified at the initial value of the independent variable.

3. $(n - r)$ boundary conditions specified at the terminal value of the independent variable.

As a rule, the differential equations will be nonlinear, but linear equations will play an important role in the numerical methods to be developed.

Problems originally expressed as higher-order nonlinear ordinary differential equations can be reduced to a system of first-order nonlinear ordinary differential equations as described in Section 2.3.

In this book the two-point boundary value problem is written as follows: The set of n nonlinear ordinary differential equations is

$$\dot{y}_i = g_i(y_1, y_2, \ldots, y_n, t), \qquad i = 1, 2, \ldots, n, \tag{2.2.1}$$

where the differential equations can be explicitly solved for the derivative, the g_i functions are assumed to be twice differentiable with respect to each of the dependent variables y_j, t is the independent variable, and $\dot{y}_i = dy_i/dt$.

The initial boundary conditions at the initial independent variable t_0 are

$$y_i(t_0) = c_i, \qquad i = 1, 2, \ldots, r. \tag{2.2.2}$$

The terminal boundary conditions at the terminal value of the independent variable t_f are

$$y_{i_m}(t_f) = c_{i_m}, \qquad m = 1, 2, \ldots, n-r. \tag{2.2.3}$$

More complicated boundary conditions can be given, but they usually can be reduced to the statement above or can be handled with modest changes in the shooting procedures. See Chapter 3, Section 3.8. Moreover, problems are sometimes encountered in which conditions are specified at more than two points (multipoint boundary value problems), but they are not treated here at length. See Chapter 8, Sections 8.7 and 8.8.

This statement of the problem assumes that the equations and variables can be renamed or reordered so the first r variables will have boundary conditions specified at the initial independent variable.

The subscripts on the specified terminal conditions are written i_m to allow for the possibility that the set of variables specified at the initial value of the independent variable and the set of variables specified at the final value of the independent variable may not be disjoint. For example, let the number of equations $n = 6$; let the initial boundary conditions be specified for $y_1(t_0), y_2(t_0)$, and $y_3(t_0)$; and let the terminal conditions be given for $y_2(t_f), y_4(t_f)$, and $y_6(t_f)$. Under these circumstances, y_2 is fixed at both the initial and final points; therefore the sets of variables specified at the initial and terminal points are not disjoint. Note that y_5 is not given at either the initial or terminal points. The indexing for the terminal conditions is therefore $i_1 = 2$, $i_2 = 4$, $i_3 = 6$, so $y_{i_1}(t_f)$ is $y_2(t_f)$, $y_{i_2}(t_f)$ is $y_4(t_f)$, and $y_{i_3}(t_f)$ is $y_6(t_f)$.

It should be mentioned here that the indexing for both the initial and terminal conditions is a convenient device to state the two-point boundary value problem concisely for purposes of discussion. In practice we do not rename or reorder the equations every time the boundary conditions are changed. For a computer solution of the two-point boundary value problem, we indicate in the input data to the program which variables are specified at the initial and final values of the independent variable and which variables are not specified. The computer program then tests each variable against the input data and treats each variable accordingly. Thus the "bookkeeping" of the variables specified at the initial and terminal values of the independent variable is handled automatically by the code.

2.3. REDUCTION OF *n*th-ORDER EQUATION

It is convenient to reduce all higher-order differential equations to a system of first-order equations. From an analysis point of view and from a computer program point of view, we can then focus our attention on essentially one class of problem.

The method for doing this is standard: Given a single *n*th-order nonlinear ordinary differential equation

$$y^{(n)}(t) = g(y, y^{(1)}, y^{(2)}, \ldots, y^{(n-1)}, t),$$

where

$$y^{(k)} = \frac{d^k y}{dt^k},$$

we define a system of n first-order nonlinear differential equations as follows:

$$\begin{aligned} y_1 &= y, \\ y_2 &= \dot{y}_1 = y^{(1)}, \\ y_3 &= \dot{y}_2 = \ddot{y}_1 = y^{(2)}, \\ &\vdots \\ y_n &= \dot{y}_{n-1} = \ldots = y^{(n-1)}, \end{aligned}$$

where

$$\dot{y}_k = dy_k/dt.$$

The $y^{(n)}(t)$ equation is therefore replaced by the system of n first-order nonlinear ordinary differential equations

$$\begin{aligned} \dot{y}_n &= g(y_1, y_2, \ldots, y_n, t), \\ \dot{y}_{n-1} &= y_n, \\ \dot{y}_{n-2} &= y_{n-1}, \\ &\vdots \\ \dot{y}_1 &= y_2. \end{aligned}$$

Thus any system of higher-order ordinary differential equations can be reduced to a system of first-order ordinary differential equations.

2.4. ANALYTICAL SOLUTION

If the differential equation can be solved analytically, then the two-point boundary value problem can generally be solved without difficulty. For linear differential equations the solution of a two-point boundary value problem reduces to determining the values of the constants from the given boundary conditions as the solution to a set of linear algebraic equations.

Consider the following example which, because of the simplicity of the problem, we do not bother to put in the form (2.2.1):

$$\frac{d^2y}{dt^2} = y+t, \qquad y(0) = 0, \qquad y(1) = 1.$$

The solution of the homogeneous differential equations is

$$y_h(t) = c_1 e^t + c_2 e^{-t},$$

while the particular solution is

$$y_p(t) = -t.$$

The general solution is therefore

$$y(t) = y_h(t) + y_p(t) = c_1 e^t + c_2 e^{-t} - t.$$

By the first boundary condition,

$$y(0) = 0 = c_1 + c_2$$

and, by the second boundary condition,

$$y(1) = 1 = c_1 e + c_2 e^{-1} - 1.$$

Solving this set of linear algebraic equations for c_1 and c_2 gives

$$c_1 = -c_2 = \frac{2}{e-e^{-1}} = \frac{1}{\sinh 1}.$$

The general solution is therefore

$$y(t) = \frac{2\sinh t}{\sinh 1} - t.$$

The fact that the solution to the boundary value problem could ultimately be made to depend on the solution of a set of algebraic equations will turn out to be characteristic of shooting methods.

2.5. NUMERICAL SOLUTION

If the two-point boundary value problem cannot be solved analytically, which is the usual case, then recourse must be made to numerical methods. Since no single numerical method works all the time for all two-point boundary value problems, a number of special techniques have been developed, each of which has its area of applicability. Part of the art of the numerical analyst lies in selecting the technique best suited for the problem at hand.

Since we are primarily interested in shooting techniques, we want to characterize their special approach to two-point boundary value problems. We said in our description in Chapter 1 that in the application of a shooting method the values of the missing initial conditions are incremented by an amount which depends on the amount by which the calculated final values miss the prescribed values. There are $n - r$ final values to be met, and the miss distance ϕ_k, that is, the prescribed final value minus the calculated final value, is a function of the r initial conditions $c_1, c_2, \ldots, c_r$, and the $(n - r)$ missing initial conditions.

What is wanted is the set of initial values $y_{r+1}(t_0), y_{r+2}(t_0), \ldots, y_n(t_0)$ which make the miss distance zero. Thus the problem has been reduced to the solution of $n - r$ algebraic equations:

$$\phi_k(y_{r+1}(t_0), y_{r+2}(t_0), \ldots, y_n(t_0)) = 0, k = 1, 2, \ldots, n - r,$$

the numerical integration of the differential equations being necessary to evaluate the functions ϕ_k.

This way of looking at shooting methods is quite useful. The various shooting methods are simply different ways of solving the system of equations $\phi_k = 0$, and known theorems about the convergence of methods of their solution will give us convergence theorems for the solution of two-point boundary value problems. In particular, shooting methods for *linear* two-point boundary value problems will give rise to a set of linear algebraic equations to solved, so the essential features of our example in Section 2.4 will be preserved even for systems for which no analytic solution is known. Furthermore, shooting methods for nonlinear problems will be shown to reduce to a sequence of linear problems; thus shooting methods for nonlinear problems are characterized by *iteration*. The various methods differ in the way the coefficients in the linear algebraic equations are obtained. Of the two most important methods, the method of adjoints employs backward integration of the adjoint equations, and the method of complementary functions employs forward integration of the variational equations.

The various techniques for solving two-point boundary value problems may be compared by such criteria as: amount of problem preparation; initial information or data necessary to commence the problem solution; computer storage requirements; speed of solution; numerical stability; accuracy achievable; and the type of problem for which the technique is best suited. As various methods are developed and discussed, we will touch on certain aspects of these criteria.

2.6. STABILITY

As we have seen, shooting methods depend on assuming the missing initial conditions and integrating the differential equations forward over the interval. It is of crucial importance, therefore, that the corresponding initial value problem have a solution over the interval, not only for the correct initial conditions but also for initial conditions near the correct values. In practice, it is usually not enough that the initial value problems have a solution in these circumstances: the solution must be close to the true solution.

What is required, then, is that small changes to the initial values should result in small changes in the solution. Problems which have this property are said to be *well-posed, well-conditioned,* or *stable.* Naturally, problems which do not have this property are said to be *not well-posed, ill-conditioned, unstable,* or sometimes *sensitive.* Quite innocuous looking problems can turn out to be not well-posed. A striking example due to B. A. Troesch [1] as given by Keller [2] is the problem: $\ddot{y} = 16 \sinh 16\, y$ with the boundary conditions $y(0) = y(1) = 0$. Because this problem has the unique solution $y(x) = 0$, the true missing initial condition is $\dot{y}(0) = 0$. An attempt to solve this problem numerically by shooting methods would require the forward integration of the differential equation with the initial conditions $y(0) = 0$, $\dot{y}(0) = s$, where s must be assumed or guessed. But it can be proved that, if $s > 10^{-7}$ (approximately), then the solution to the corresponding initial value problem is *singular* in the interval $[0, 1]$; that is, it actually goes to infinity at some point x_∞, $0 < x_\infty < 1$. (As a matter of fact, x_∞ is approximately $(1/16) \ln(8/s)$; this means that the numerical solution for the reasonable initial guess $s = 0.01$ would "blow up" before the independent variable x reached 0.42.)

Techniques have been developed for dealing with problems which are not well-posed. For example, we may supplement the shooting method by orthogonalization (Sections 4.9–4.12,Chapter 4) or by continuation methods (Chapter 7). Other possibilities are quasilinearization (Chapter 5), and finite difference methods and multiple shooting (both discussed in Chapter 8).

The type of instability exemplified by Troesch's problem is inherent in the problem and is quite independent of the numerical method used to integrate the differential equation. But the numerical method itself may be the source of the instability a particular problem exhibits, and in practice it is often difficult if not impossible to tell whether the observed instability is inherent or numerical. It is important, therefore, to choose numerical integration methods which are stable or, at the very least, whose stability properties are well understood.

A thorough discussion of numerical stability belongs properly to a book on the numerical integration of initial value problems, and we have assumed that the reader is familiar with this material. However, because of the importance of stability considerations in the practical solution of two-point boundary value problems, we now give a brief review of the subject [2–13].

Recall that the *truncation error* ε_j of a method of numerical integration at the point x_j is the difference between the solution y_j of the equations of the method and the true solution $y(x_j)$ of the differential equation, and that the *round-off error* r_j, which arises because machine computations are performed with only a finite number of digits, is the difference between the computed solution $\tilde{y}_j$ and y_j. All methods for numerical integration contain a parameter h, the step size or the distance between the points x_j at which the solution is computed. Loosely speaking, a method is said to be *convergent* if $y_j \to y(x_j)$ as $h \to 0$ for all starting values of the method which approach the exact initial conditions as $h \to 0$.

Most *multistep* methods of numerical integration are linear k-step of the form

$$\sum_{s=0}^{k} \alpha_s y_{n+1-s} = h \sum_{s=0}^{k} \beta_s f(x_{n+1-s}, y_{n+1-s}), \tag{2.6.1}$$

where α_j, β_j, $j = 1, 2, \ldots, k$, are real constants; $\alpha_k \neq 0$; $|\alpha_0| + |\beta_0| > 0$. (The term "linear" refers to the formula and not to $f(x, y)$, which may be, and usually is, nonlinear in its arguments.) Because of round-off error r_n, the equations (2.6.1) are perturbed, and the equations actually solved are of the form

$$\sum_{s=0}^{k} \alpha_s \tilde{y}_{n+1-s} = h \sum_{s=0}^{k} \beta_s f(x_{n+1-s}, \tilde{y}_{n+1-s}) + r_n, \tag{2.6.2}$$

where $\tilde{y}_{n+1-s}$ = numerical approximation to y_{n+1-s}.

It is usually assumed that $f(x, y)$ is continuous in the domain D: $-\infty \leqq y \leqq \infty$, $a \leqq x \leqq b$, and satisfies a Lipschitz condition with respect to y there. Roughly speaking, a method is *strongly stable* if it is insensitive to small perturbations r_n. Strong stability is the numerical analog of well-posedness, and strongly stable methods have the desirable property that small round-off errors do not cause divergence of the computed solution. It can be shown that a necessary and sufficient condition for convergence is that the method be strongly stable and possess another property called *consistency*, that is, $\varepsilon_j \to 0$ as $h \to 0$. It turns out that strong stability and consistency (and therefore convergence) depend on the roots of the polynomials

$$\rho(\zeta) = \alpha_k \zeta^k + \alpha_{k-1} \zeta^{k-1} + \ldots + \alpha_0, \tag{2.6.3a}$$

$$\sigma(\zeta) = \beta_k \zeta^k + \beta_{k-1} \zeta^{k-1} + \ldots + \beta_0. \tag{2.6.3b}$$

In particular, a method is strongly stable if: the modulus of every root of (2.6.3a) ≤ 1; and the roots of modulus 1 are simple. A method is consistent if: $\rho(1) = 0$ and $\dot{\rho}(1) = \sigma(1)$. It can be assumed that any method of numerical integration of practical value is strongly stable.

Strong stability refers to the behavior of solutions of (2.6.1) at a fixed point x_k as $h \to 0$ and, when it obtains, does so for all $f(x, y)$ in a wide class of functions. Another type of stability is *weak stability*, sometimes called *parasitic* or *stepwise stability*, which is concerned with the growth of solutions at x_k as $k \to \infty$ for fixed h, and is thus an asymptotic phenomenon. Weak stability depends on both the numerical method and the specific form of $f(x, y)$. Moreover, a method may be strongly stable and yet not be weakly stable. As an example, consider the problem

$$\dot{y} = -y - 1, \qquad y(0) = a_0, \tag{2.6.4}$$

with the analytic solution $y(x) = (a_0 + 1)\, e^{-x} - 1$, to be solved by the method

$$y_{k+1} - y_{k-1} = -2h\,(y_k + 1), \qquad y_0 = a_0. \tag{2.6.5}$$

The method is strongly stable and consistent, hence convergent. However, the solution of the difference equation (2.6.5) is

$$y_k = c_1 e^{-kh} + c_2 (-1)^k e^{kh} - 1 \tag{2.6.6}$$

and, since c_2 is, in general, not zero the term, $c_2(-1)^k e^{kh}$ oscillates unboundedly as $k \to \infty$, so that as the computation proceeds the error in the numerical solution of (2.6.4) becomes arbitrarily large. The reason for this behavior is that the *first-order* differential equation has been approximated by a *second-order* difference equation, which of necessity has two solutions. One, of the form e^{-kh}, approximates the solution of the differential equation which is of the form e^{-x}, but the other, the extraneous or parasitic solution e^{kh}, does not, and it is the parasitic solution which ultimately dominates the numerical solution.

Now consider the problem

$$\dot{y} = y - 1, \qquad y(0) = a_0 \tag{2.6.7}$$

(which differs from (2.6.4) only in the sign of y) whose solution is $y(x) = (a_0 - 1)e^x + 1$. Approximate (2.6.7) by the difference scheme analogous to (2.6.5); that is,

$$y_{k+1} - y_{k-1} = 2h(y_k - 1). \tag{2.6.8}$$

The method is strongly stable but is not consistent. In this case the difference equation (2.6.8) has the solution

$$y_k = c_1 e^{kh} + c_2(-1)^k e^{-kh} + 1, \tag{2.6.9}$$

but here the parasitic solution e^{-kh} decreases as $k \to \infty$ and so does not cause any numerical difficulties. However (2.6.7) is inherently unstable, somewhat like Troesch's problem. When the initial condition is $a_0 = 1$, the solution is $y(x) = 1$, while for a small perturbation of the initial condition $a_0 = 1 + \delta$ the solution is $y(x) = \delta e^x + 1$ which grows without bound as x increases. Since any method of numerical integration will, because of round-off error, in effect perturb the initial condition $a_0 = 1$ to $a_0 = 1 + \delta$, the numerical solution will grow without bound; that is, the error will become arbitrarily large as k increases.

Thus we see that, if the two problems (2.6.4) and (2.6.7) are solved numerically by essentially the same method, the computed solution will "blow up" in each case, but for different reasons. In the case of (2.6.4) the cause can be traced to the numerical method, which is *weakly unstable*, while in the case of (2.6.7) the problem is *inherently unstable*. The examples illustrate the difficulty confronting the analyst when a practical problem blows up, for he has available neither the solution of the differential equation nor the solution of the difference equation.

No single approach seems adequate to cope with all cases of weak stability. Since weak stability depends on the approximating formula as well as on the form of $f(x, y)$, we can always try a different approximating formula (that is, another method of numerical integration) whose roots corresponding to parasitic solutions have more favorable growth properties. Sometimes, backward integration of the equations curtails the growth of parasitic solutions. Finally, a reformulation of the problem may be attempted.

A final type of stability to be mentioned is *partial stability*, encountered in methods like Runge-Kutta, whose characteristic behavior is that the numerical solution approximates the true solution for values of the step size h below a certain value, but fails to approximate the true solution when

h is greater than this value. As an illustration of partial stability, consider the example given by Fox [6]:

$$\dot{y} = -10y+9-10x, \qquad (2.6.10)$$

whose general solution is

$$y = Ae^{-10x}-x+1. \qquad (2.6.11)$$

Solving this equation numerically by a second-order Runge-Kutta method is equivalent to the recurrence formula

$$y_{r+1} = (1-10h+50h^2)y_r+h(9-10x_r)+50h^2(x_r-1), \qquad (2.6.12)$$

which has a solution of the form $C\lambda^r-x_r+1$, where $\lambda = 1-10h+50h^2$. As $h \to 0$, the term $C\lambda^r$ at a fixed value of $x = rh$ is $C(1-10h+50h^2)^{x/h}$, which tends to e^{-10x} and therefore approximates the term Ae^{-10x} in the true solution. However, if $h > 0.2$, then $\lambda > 1$, and the term $C\lambda^r$ grows without bound, so it cannot possibly approximate Ae^{-10x}.

The obvious way around the problem of partial instability is to reduce h. However, the reduction of the magnitude of h often brings about other difficulties, particularly in the case of so-called "stiff" differential equations. Special methods of numerical integration have accordingly been devised for these problems [14].

In our brief discussion of stability we have confined ourselves to first-order differential equations. It can be expected that higher-order equations, or systems of first-order equations, will exhibit all the difficulties described above, and that the corresponding treatment will be a great deal more complicated than the one given here.

REFERENCES

1. B. A. Troesch, "Intrinsic Difficulties in the Numerical Solution of a Boundary Value Problem," Internal Report NN–142, Jan. 29, 1960, TRW Inc., Redondo Beach, Calif.
2. H. B. Keller, *Numerical Methods for Two-Point Boundary Value Problems*, Blaisdell, Waltham, Mass., 1968.
3. L. Collatz, *The Numerical Treatment of Differential Equations*, 3rd ed., Springer-Verlag, Berlin, 1960.
4. R. L. Crane and R. W. Klopfenstein, A Predictor-Corrector Algorithm with an Increased Range of Absolute Stability, *J. Assoc. Comput. Mach.*, (2) **12** (April 1965), 227–241.
5. L. Fox, A Note on the Numerical Integration of First Order Differential Equations, *Quart. J. Mech. Appl. Math.*, **7**, pt. 3 (1954), 367–378.
6. L. Fox, *Numerical Solution of Ordinary and Partial Differential Equations*, Addison-Wesley, Reading, Mass., 1962.
7. R. W. Hamming, *Numerical Methods for Scientists and Engineers*, McGraw-Hill, New York, 1962.

8. P. Henrici, *Discrete Variable Methods in Ordinary Differential Equations*, Wiley, New York, 1962.
9. P. Henrici, "The Propagation of Error in Digital Integration of Ordinary Differential Equations", in *Error in Digital Computation*, Vol. 1, L. B. Rall, ed., Wiley, New York, 1965, 185–205.
10. F. B. Hildebrand, *Finite-Difference Equations and Simulation*, Prentice-Hall, Englewood Cliffs, N. J., 1968.
11. E. Issacson and H. B. Keller, *Analysis of Numerical Methods*, Wiley, New York, 1966.
12. B. Noble, *Numerical Methods*, Vol. 2, Oliver and Boyd, or Interscience, New York, 1964.
13. J. Todd, Solution of Differential Equations by Recurrence Equations, *MTAC* (29–32) **29** (1950), 39–44.
14. J. Certaine, "The Solution of Ordinary Differential Equations with Large time Constants," in Ralston and Wilf, eds., *Mathematical Methods for Digital Computers*, Wiley, New York, 1960, pp. 128–132.

General References

P. B. Bailey, L. F. Shampine, and P. E. Waltman, *Nonlinear Two Point Boundary Value Problems*, Academic, New York, 1968.

P. E. Chase, Stability Properties of Predictor-Corrector Methods for Ordinary Differential Equations, *J. Assoc. Comput. Mach.* (4), **9** (Oct. 1962), 457–468.

H. P. Decell, L. F. Guseman, and R. N. Lea, Concerning the Numerical Solution of Differential Equations, *Math. of Comp.* **20** (1966), 431–434.

G. E. Forsythe and W. R. Wasow, *Finite Difference Methods for Partial Differential Equations*, Wiley, New York, 1960.

L. Fox, *Numerical Solution of Two-Point Boundary Problems in Ordinary Differential Equations*, Oxford, London, 1957.

L. Fox and E. T. Goodwin, Some New Methods for the Numerical Integration of Ordinary Differential Equations, *Proc. Cambridge Phil. Soc.* **45** (1949), 373–388.

L. Fox and A. R. Mitchell, Boundary Value Techniques for the Numerical Solution of Initial-Value Problems in Ordinary Differential Equations, *Quart. J. Mech. Appl. Math.*, **10** (1957), 232–238.

T. R. Goodman and G. N. Lance, The Numerical Solution of Two Point Boundary Value Problems, *MTAC* **10** (1956), 82–86.

F. B. Hildebrand, *Introduction to Numerical Analysis*, Prentice-Hall, Englewood Cliffs, N.J., 1956.

T. E. Hull and A. C. R. Newberry, Integration Procedures Which Minimize Propagated Errors, *J. Soc. Indust. Appl. Math.* (1) **9** (March 1961), 31–47.

E. L. Ince, *Ordinary Differential Equations*, reprinted by Dover, New York,

W. C. Martin, K. C. Paulson, and L. Sashkin, A General Method of Systematic Interval Computation for Numerical Integration of Initial Value Problems, *Comm. ACM* (10) **9** (Oct. 1966), 754–757.

A. R. Mitchell, The Influence of Critical Boundary Conditions on Finite Difference Solutions of Two Point Boundary Value Problems, *MTAC* (65–68) **13** (1959), 252–260.

R. A. Usmani, Boundary Value Techniques for the Numerical Solution of Certain Initial Value Problems in Ordinary Differential Equations, *J. Assoc. Comput. Mach.* (2) **13** (April 1966), 287–295.

Chapter 3

METHOD OF ADJOINTS

3.1. INTRODUCTION

The method of adjoints is an important practical shooting method. For the linear two-point boundary value problem the method of adjoints finds the set of missing initial conditions in one pass through the process. For the nonlinear two-point boundary value problem it generates automatically by an iterative process a sequence of corrections to the trial values of the missing initial conditions. The method of adjoints is a general technique applicable to any problem of the form (2.2.1)–(2.2.3) of Chapter 2. First we discuss the method of adjoints for linear problems, and then generalize it to apply to nonlinear problems [1] *.

3.2. LINEAR TWO-POINT BOUNDARY VALUE PROBLEM

Consider the set of n linear ordinary differential equations with variable coefficients

$$\dot{\mathbf{y}} = \mathbf{A}(t)\mathbf{y}+\mathbf{f}(t), \tag{3.2.1}$$

where

$\mathbf{A}(t) = n \times n$ matrix with elements $a_{ij}(t)$, $i, j = 1, 2, \ldots, n$,

$\mathbf{y}(t) = n \times 1$ vector with components $y_1(t), y_2(t), \ldots, y_n(t)$,

$\dot{\mathbf{y}}(t) = n \times 1$ vector, derivative of $\mathbf{y}$ with respect to t, with components $\dot{y}_1, \dot{y}_2, \ldots, \dot{y}_n$,

$\mathbf{f}(t) = n \times 1$ vector with components $f_1(t), f_2(t), \ldots, f_n(t)$.

---

* The terms "method of adjoints" and the "Goodman-Lance method" will be used interchangeably.

The initial conditions are

$$y_i(t_0) = c_i, \qquad i = 1, 2, \ldots, r, \tag{3.2.2}$$

and the terminal conditions are

$$y_{i_m}(t_f) = c_{i_m}, \qquad m = 1, 2, \ldots, n-r. \tag{3.2.3}$$

Referring to (2.2.1) in Chapter 2, we see that here

$$g_i(y_1, y_2, \ldots, y_n, t) = \sum_{j=1}^{n} a_{ij}(t) y_j + f_i(t), \qquad i = 1, 2, \ldots, n.$$

If we assume that a solution to the linear two-point boundary value problem exists, then the data in (3.2.2) and (3.2.3) contain all the information required to produce the solution. Unfortunately for numerical work, the boundary value data are not in the appropriate form because numerical integration is suitable only for the initial value problem. We therefore need other relationships which can be extracted from (3.2.1), (3.2.2), and (3.2.3) to produce numerical results. One such relationship involves the adjoint equations. With every set of linear ordinary differential equations is associated a companion set of equations called the adjoint equations, defined as the set of homogeneous linear ordinary differential equations whose matrix of coefficients is the negative transpose of the matrix of the original set of linear ordinary differential equations. The adjoint equations are important for our purposes because the initial and terminal conditions of the adjoint equations are related to the initial and terminal boundary conditions of the original system by a certain identity, which we exploit for the numerical solution of two-point boundary value problems. We now develop this identity.

The adjoint system described in the preceding paragraph has the form

$$\dot{\mathbf{x}} = -\mathbf{A}^T(t)\mathbf{x}, \tag{3.2.4}$$

where

$\mathbf{x}$ = the adjoint variable vector, an $n \times 1$ vector with components $x_1(t), x_2(t), \ldots, x_n(t)$,

$\mathbf{A}^T(t)$ = $n \times n$ matrix, the transpose of the matrix $\mathbf{A}(t)$ in (3.2.1).

The adjoint equations, like the original system (3.2.1), are a set of n linear ordinary differential equations with variable coefficients. Note that the adjoint equations do not contain the forcing function $\mathbf{f}(t)$ which appears in (3.2.1).

If we multiply the ith equation of (3.2.1) by $x_i(t)$, we have

$$x_i(t)\dot{y}_i = x_i(t)\,[a_{i1}(t)y_1(t)+a_{i2}(t)y_2(t)+\ \dots\ +a_{in}(t)y_n(t)]+x_i(t)f_i(t). \quad (3.2.5)$$

Summing (3.2.5) over all n equations yields

$$\sum_{i=1}^{n} x_i(t)\dot{y}_i = \sum_{i=1}^{n} x_i(t)\,[a_{i1}(t)y_1(t)+a_{i2}(t)y_2(t)+\ \dots\ +a_{in}(t)y_n(t)]+ + \sum_{i=1}^{n} x_i(t)f_i(t). \quad (3.2.6)$$

Similiarly we multiply the ith equation of the adjoint equations by $y_i(t)$:

$$\dot{x}_i y_i(t) = -y_i(t)\,[a_{1i}(t)x_1(t)+a_{2i}(t)x_2(t)+\ \dots\ +a_{ni}(t)x_n(t)], \quad (3.2.7)$$

and sum over all n equations (3.2.7):

$$\sum_{i=1}^{n} \dot{x}_i y_i(t) = -\sum_{i=1}^{n} y_i(t)\,[a_{1i}(t)x_1(t)+a_{2i}(t)x_2(t)+\ \dots\ +a_{ni}(t)x_n(t)]. \quad (3.2.8)$$

On adding (3.2.6) and (3.2.8), we find

$$\sum_{i=1}^{n} (x_i(t)\dot{y}_i+\dot{x}_i y_i(t)) = \sum_{i=1}^{n} x_i(t)f_i(t). \quad (3.2.9)$$

Equation (3.2.9) may also be written as

$$\sum_{i=1}^{n} \frac{d}{dt} x_i(t)y_i(t) = \sum_{i=1}^{n} x_i(t)f_i(t) \quad (3.2.10)$$

or as

$$\frac{d}{dt}\sum_{i=1}^{n} x_i(t)y_i(t) = \sum_{i=1}^{n} x_i(t)f_i(t). \quad (3.2.11)$$

On integrating (3.2.11) over $[t_0, t_f]$, we have

$$\int_{t_0}^{t_f} \frac{d}{dt}\left(\sum_{i=1}^{n} x_i(t)y_i(t)\right)dt = \int_{t_0}^{t_f} \sum_{i=1}^{n} x_i(t)f_i(t)\,dt \quad (3.2.12)$$

or

$$\sum_{i=1}^{n} x_i(t_f)y_i(t_f)-\sum_{i=1}^{n} x_i(t_0)y_i(t_0) = \int_{t_0}^{t_f} \sum_{i=1}^{n} x_i(t)f_i(t)\,dt. \quad (3.2.13)$$

Equation (3.2.13) is the fundamental identity for the method of adjoints. It gives the relationship between the y_i variables of the original equation and the adjoint variables x_i at the initial and terminal points. As a matter of fact, the identity is true for any interval (t_1, t_2) over which (3.2.11) may be integrated. For any consistent set of specified initial and terminal conditions on the y_i variables and for a choice of either the initial or terminal conditions for the x_i variables, the identity is a linear algebraic relationship among the unspecified $y_i(t_0)$ variables.

To utilize the fundamental identity of the method of adjoints to solve linear two-point boundary value problems, integrate backward the adjoint equations $(n-r)$ times with the terminal boundary conditions

$$x_i^{(m)}(t_f) = \left.\begin{matrix} 1, & i = i_m \\ 0, & i \neq i_m \end{matrix}\right\}, \qquad m = 1, 2, \ldots, n-r, \tag{3.2.14}$$

where the superscript m refers to the mth backward integration of the adjoint equations and i_m refers to the subscripts on the specified terminal conditions $y_{i_m}(t_f)$ in (3.2.3). With these terminal conditions, the first summation in (3.2.13) reduces to

$$x_{i_m}^{(m)}(t_f)y_{i_m}(t_f) = 1 \cdot y_{i_m}(t_f) = c_{i_m}.$$

Since by (3.2.2) the initial conditions are given for $y_i(t_0)$, $i = 1, 2, \ldots, r$, (3.2.13) can be written as

$$\sum_{i=r+1}^{n} x_i^{(m)}(t_0)y_i(t_0) = y_{i_m}(t_f) - \sum_{i=1}^{r} x_i^{(m)}(t_0)y_i(t_0) - \int_{t_0}^{t_f} \sum_{i=1}^{n} x_i^{(m)}(t)f_i(t)\,dt. \tag{3.2.15}$$

In the application to the solution of (3.2.1), everything on the right side of (3.2.15) is known. The $y_{i_m}(t_f)$, $m = 1, 2, \ldots, n-r$ are specified in (3.2.3). The $x_i^{(m)}(t)$ and the $x_i^{(m)}(t_0)$ are known from the backward integrations of the adjoint equations. The $f_i(t)$ are known functions of t. On the left side of (3.2.15), the $x_i^{(m)}(t_0)$ are known from the backward integrations of the adjoint equations. The generation of (3.2.15), for the $(n-r)$ specified terminal conditions, $y_{i_m}(t_f)$, and the corresponding $x_{i_m}(t_f)$, yields a set of $(n-r)$ linear algebraic equations in the $(n-r)$ unknowns $y_{r+1}(t_0), y_{r+2}(t_0), \ldots, y_n(t_0)$.

In matrix form we have, for (3.2.15),

$$\begin{bmatrix} x_{r+1}^{(1)}(t_0) & x_{r+2}^{(1)}(t_0) & \dots & x_n^{(1)}(t_0) \\ x_{r+1}^{(2)}(t_0) & x_{r+2}^{(2)}(t_0) & \dots & x_n^{(2)}(t_0) \\ \vdots & \vdots & & \vdots \\ x_{r+1}^{(n-r)}(t_0) & x_{r+2}^{(n-r)}(t_0) & \dots & x_n^{(n-r)}(t_0) \end{bmatrix} \begin{bmatrix} y_{r+1}(t_0) \\ y_{r+2}(t_0) \\ \vdots \\ y_n(t_0) \end{bmatrix}$$

$$= \begin{bmatrix} y_{i_1}(t_f) - \sum_{i=1}^{r} x_i^{(1)}(t_0) y_i(t_0) - \int_{t_0}^{t_f} \sum_{i=1}^{n} x_i^{(1)}(t) f_i(t)\, dt \\ y_{i_2}(t_f) - \sum_{i=1}^{r} x_i^{(2)}(t_0) y_i(t_0) - \int_{t_0}^{t_f} \sum_{i=1}^{n} x_i^{(2)}(t) f_i(t)\, dt \\ \vdots \\ y_{i_{n-r}}(t_f) - \sum_{i=1}^{r} x_i^{(n-r)}(t_0) y_i(t_0) - \int_{t_0}^{t_f} \sum_{i=1}^{n} x_i^{(n-r)}(t) f_i(t)\, dt \end{bmatrix}. \tag{3.2.16}$$

The set of missing initial conditions is found by solving (3.2.16):

$$\begin{bmatrix} y_{r+1}(t_0) \\ y_{r+2}(t_0) \\ \vdots \\ y_n(t_0) \end{bmatrix} = \begin{bmatrix} x_{r+1}^{(1)}(t_0) & x_{r+2}^{(1)}(t_0) & \dots & x_n^{(1)}(t_0) \\ x_{r+1}^{(2)}(t_0) & x_{r+2}^{(2)}(t_0) & \dots & x_n^{(2)}(t_0) \\ \vdots & \vdots & & \vdots \\ x_{r+1}^{(n-r)}(t_0) & x_{r+2}^{(n-r)}(t_0) & \dots & x_n^{(n-r)}(t_0) \end{bmatrix}^{-1}$$

$$\times \begin{bmatrix} y_{i_1}(t_f) - \sum_{i=1}^{r} x_i^{(1)}(t_0) y_i(t_0) - \int_{t_0}^{t_f} \sum_{i=1}^{n} x_i^{(1)}(t) f_i(t)\, dt \\ y_{i_2}(t_f) - \sum_{i=1}^{r} x_i^{(2)}(t_0) y_i(t_0) - \int_{t_0}^{t_f} \sum_{i=1}^{n} x_i^{(2)}(t) f_i(t)\, dt \\ \vdots \\ y_{i_{n-r}}(t_f) - \sum_{i=1}^{r} x_i^{(n-r)}(t_0) y_i(t_0) - \int_{t_0}^{t_f} \sum_{i=1}^{n} x_i^{(n-r)}(t) f_i(t)\, dt \end{bmatrix}. \tag{3.2.17}$$

provided, of course, the inverse exist.

To recapitulate, the method of adjoints for two-point boundary value problems with linear ordinary differential equations is carried out as follows:

1. Set the counter $m = 1$.

2. Integrate the adjoint equations (3.2.4) backward from t_f to t_0 for the mth set of boundary conditions (3.2.14).

3. Evaluate the mth row in (3.2.16).

4. If m is equal to $(n - r)$, solve (3.2.17) for the missing initial conditions $y_i(t_0)$, $i = r + 1, \ldots, n$. Go to item 6.

5. If m is less than $(n - r)$, set $m = m+1$ and return to item 2.

6. Using the specified initial conditions (3.2.2) and the calculated missing initial conditions from item 4, integrate (3.2.1) forward from t_0 to t_f to obtain the solution of the boundary value problem (3.2.1), (3.2.2), (3.2.3).

We observe that the process is noniterative, since the solution of the set of algebraic equations (3.2.16) gives the missing initial conditions directly. The differential equations (3.2.1) are not integrated in the course of finding the missing initial conditions. Once the missing initial conditions are found, the differential equations (3.2.1) are integrated to produce the $y_i(t)$ profiles, $i = 1, 2, \ldots, n$; $t_0 \leq t \leq t_f$.

3.3. DISCUSSSION

From the theory of linear ordinary differential equations, it is known that solutions which are linearly independent at the initial point remain independent for all values of the independent variable for which a solution exists [2]. The linear independence of the set of terminal vectors of the adjoint variables (specified by (3.2.14), which are not only linearly independent but also orthogonal) assures theoretically the linear independence of the set of initial vectors of the adjoint variables. This, of course, is necessary in order to solve for the set of missing initial conditions.

In practice we find that sometimes, even though the adjoint equations are integrated backward with a set of $(n - r)$ linearly independent terminal vectors, the initial vectors are not numerically independent. In particular, this can happen when the matrix $\mathbf{A}(t)$ in (3.2.1) has eigenvalues widely separated in value. An orthonormalization procedure is described in Sections 4.9–4.12, Chapter 4, to cope with these numerical problems.

Although in theory only linear independence for the set of terminal conditions for the adjoint variables is required, the use of the Kronecker delta terminal conditions reduces to the minimum, $n - r$, the number of integrations of the adjoint equations necessary. If we were to choose merely linear independence for the set of terminal conditions $x_i^{(p)}(t_f)$, then the adjoint equations would need to be solved n times to develop a system of n equations in n unknowns, the $(n - r)$ missing initial conditions and the r missing terminal conditions. To see this let us choose a set of linearly independent terminal conditions for the adjoint equations.

$$x_i^{(p)}(t_f) = a_i^{(p)}; \qquad i, p = 1, 2, \ldots, n, \tag{3.3.1}$$

where $a_i^{(p)} \neq 0$, $\quad i, p = 1, 2, \ldots, n$.

Equation (3.2.13) appears as

$$\sum_{i=1}^{n} x_i^{(p)}(t_f)y_i(t_f) - \sum_{i=1}^{n} x_i^{(p)}(t_0)y_i(t_0)$$
$$= \int_{t_0}^{t_f} \sum_{i=1}^{n} x_i^{(p)}(t)f_i(t)\,dt, \qquad p = 1,2,\ldots,n. \tag{3.3.2}$$

Since $y_{i_m}(t_f)$, $m = 1,2,\ldots,n-r$ are known and, since $y_i(t_0)$, $i = 1,2,\ldots,r$ are known, (3.3.2) may be rearranged:

$$\sum_{i \neq i_1, i_2, \ldots, i_{n-r}} x_i^{(p)}(t_f)y_i(t_f) - \sum_{i=r+1}^{n} x_i^{(p)}(t_0)y_i(t_0) = -\sum_{m=1}^{n-r} x_{i_m}^{(p)}(t_f)y_{i_m}(t_f)$$
$$+ \sum_{i=1}^{r} x_i^{(p)}(t_0)y_i(t_0) + \int_{t_0}^{t_f} \sum_{i=1}^{n} x_i^{(p)}(t)f_i(t)\,dt, \qquad p = 1,2,\ldots,n. \tag{3.3.3}$$

The first summation on the left side of (3.3.3) is taken over the variables not specified at the terminal point. The right side is a known vector. Equations (3.3.3) are a set of n linear algebraic equations in the $(n-r)$ unknown initial conditions $y_{r+i}(t_0)$, $i = 1,2,\ldots,n-r$, and the r unspecified terminal conditions $y_i(t_f)$, $i \neq i_1, i_2, \ldots, i_{n-r}$.

It is therefore a considerable disadvantage (in the way of r additional backward integrations of the adjoint equations and the solution of n algebraic equations rather than $n-r$ algebraic equations) to use terminal conditions on the adjoint equations other than the Kronecker delta conditions, when the boundary conditions are given by (3.2.2) and (3.2.3). A similiar analysis shows that the specification of linearly independent initial conditions for the adjoint equations also gives rise to n algebraic equations in n unknowns. Thus the choice of the Kronecker delta terminal conditions and backward integration of the adjoint equations is the most efficient one to find the missing initial conditions, when the boundary conditions are given by (3.2.2) and (3.2.3).

Finally, it is worth noting that it can always be assumed that $n - r \leq r$ for, if this is not the case, the roles of the initial point t_0 and the final point t_f can simply be reversed.

3.4. EXAMPLE: LINEAR SYSTEM

To illustrate in detail the method of adjoints in Section 3.2, let us consider a system of five linear ordinary differential equations:

$$\dot{\mathbf{y}} = \mathbf{A}(t)\mathbf{y} + \mathbf{f}(t),$$

where

$$\mathbf{y}(t) = 5\times 1 \text{ vector},$$
$$\mathbf{f}(t) = 5\times 1 \text{ vector},$$
$$\mathbf{A}(t) = 5\times 5 \text{ matrix},$$

with the initial boundary conditions

$$y_1(t_0) = c_1, \qquad y_2(t_0) = c_2, \qquad y_3(t_0) = c_3,$$

and the terminal conditions

$$y_3(t_f) = c_{i_1}, \qquad y_4(t_f) = c_{i_2}.$$

In this problem there are five linear ordinary differential equations, so $n = 5$; three initial conditions, $r = 3$; and two terminal conditions, $n-r = 2$. Note that $n-r < r$.

In view of the two terminal conditions the adjoint equations $\dot{\mathbf{x}} = -\mathbf{A}(t)^T\mathbf{x}$ are integrated backward for two linearly independent sets of terminal conditions. For the first backward integration, $m = 1, i_1 = 3$, so the terminal conditions are

$$x_1^{(1)}(t_f) = 0,$$
$$x_2^{(1)}(t_f) = 0,$$
$$x_3^{(1)}(t_f) = 1,$$
$$x_4^{(1)}(t_f) = 0,$$
$$x_5^{(1)}(t_f) = 0.$$

For the second backward integration, $m = 2, i_2 = 4$, so the terminal conditions are

$$x_1^{(2)}(t_f) = 0,$$
$$x_2^{(2)}(t_f) = 0,$$
$$x_3^{(2)}(t_f) = 0,$$
$$x_4^{(2)}(t_f) = 1,$$
$$x_5^{(2)}(t_f) = 0.$$

For $m = 1$, (3.2.13) appears as

$$y_3(t_f) - \sum_{i=1}^{5} x_i^{(1)}(t_0)y_i(t_0) = \int_{t_0}^{t_f} \sum_{i=1}^{5} x_i^{(1)}(t) f_i(t)\,dt$$

and, for $m = 2$, as

$$y_4(t_f) - \sum_{i=1}^{5} x_i^{(2)}(t_0) y_i(t_0) = \int_{t_0}^{t_f} \sum_{i=1}^{5} x_i^{(2)}(t) f_i(t)\, dt.$$

In matrix form we have, by (3.2.15) or (3.2.16),

$$\begin{bmatrix} x_4^{(1)}(t_0) & x_5^{(1)}(t_0) \\ x_4^{(2)}(t_0) & x_5^{(2)}(t_0) \end{bmatrix} \begin{bmatrix} y_4(t_0) \\ y_5(t_0) \end{bmatrix}$$

$$= \begin{bmatrix} y_3(t_f) - \sum_{i=1}^{3} x_i^{(1)}(t_0) y_i(t_0) - \int_{t_0}^{t_f} \sum_{i=1}^{5} x_i^{(1)}(t) f_i(t)\, dt \\ y_4(t_f) - \sum_{i=1}^{3} x_i^{(2)}(t_0) y_i(t_0) - \int_{t_0}^{t_f} \sum_{i=1}^{5} x_i^{(2)}(t) f_i(t)\, dt \end{bmatrix}.$$

As mentioned before, all the terms on the right side are known; the elements of the matrix on the left side are known; hence the solution of this set of equations yields the missing initial conditions $y_4(t_0)$ and $y_5(t_0)$. In combination with the prescribed $y_1(t_0), y_2(t_0), y_3(t_0)$, they would then give the complete set of initial conditions sufficient for the numerical integration of the system of equations.

3.5. EXAMPLE: PROBLEM OF SECTION 2.4

As another illustration, let us consider the example given in Section 2.4, Chapter 2:

$$\frac{d^2y}{dt^2} = y + t$$

with the boundary conditions

$$y(0) = 0, \qquad y(1) = 1.$$

We have previously solved this problem analytically, but we now attack it by the method of adjoints. Let

$$y_1 = y, \qquad y_2 = \dot{y}_1.$$

The second-order system can be written as two first-order equations:

$$\begin{pmatrix} \dot{y}_1 \\ \dot{y}_2 \end{pmatrix} = \begin{pmatrix} 0 & 1 \\ 1 & 0 \end{pmatrix} \begin{pmatrix} y_1 \\ y_2 \end{pmatrix} + \begin{pmatrix} 0 \\ t \end{pmatrix}$$

with the boundary conditions

$$y_1(0) = 0, \qquad y_1(1) = 1.$$

The adjoint equations are

$$\begin{pmatrix} \dot{x}_1 \\ \dot{x}_2 \end{pmatrix} = -\begin{pmatrix} 0 & 1 \\ 1 & 0 \end{pmatrix} \begin{pmatrix} x_1 \\ x_2 \end{pmatrix}.$$

The analytical solution to the adjoint equations is

$$x_1(t) = a_1 e^t + a_2 e^{-t}, \qquad x_2(t) = -a_1 e^t + a_2 e^{-t},$$

where a_1 and a_2 are constants to be determined. Since there is only one specified terminal boundary condition, namely, $y_1(1) = 1$, the terminal conditions for $m = 1$ for the adjoint equations are

$$\begin{pmatrix} x_1^{(1)}(t_f) \\ x_2^{(1)}(t_f) \end{pmatrix} = \begin{pmatrix} x_1^{(1)}(1) \\ x_2^{(1)}(1) \end{pmatrix} = \begin{pmatrix} 1 \\ 0 \end{pmatrix}$$

The constants a_1 and a_2 are evaluated by solving

$$a_1 e + a_2 e^{-1} = 1, \qquad -a_1 e + a_2 e^{-1} = 0,$$

which yields $a_1 = \frac{1}{2}e^{-1}$, $a_2 = \frac{1}{2}e$. The solution to the adjoint equation is

$$x_1(t) = \tfrac{1}{2}e^{t-1} + \tfrac{1}{2}e^{1-t}, \qquad x_2(t) = -\tfrac{1}{2}e^{t-1} + \tfrac{1}{2}e^{1-t}.$$

By the fundamental identity of the method of adjoints we have

$$(x_1^{(1)}(1)y_1(1) + x_2^{(1)}(1)y_2(1)) - (x_1^{(1)}(0)y_1(0) + x_2^{(1)}(0)y_2(0))$$

$$= \int_0^1 x_1^{(1)}(t)f_1(t) + x_2^{(1)}(t)f_2(t)\,dt,$$

which by the boundary conditions and the solution to the adjoint equations yields

$$y_2(0) = \frac{2}{\sinh 1} - 1.$$

The solution to the system of first-order equations can be found therefore by integrating forward with the initial conditions $y_1 = 0$ and $y_2 = (2/\sinh 1) - 1$.

As a check on the $y_2(0)$ found by the method of adjoints, we recall from Section 2.4, Chapter 2, that the analytical solution for the second-order equation is

$$y(t) = \frac{2\sinh t}{\sinh 1} - t.$$

Since $y_2 = \dot{y}_1 = \dot{y}$,

$$\dot{y}(t) = \frac{2\cosh t}{\sinh 1} - 1.$$

At $t = 0$,

$$\dot{y}(0) = \frac{2}{\sinh 1} - 1,$$

which agrees with answer obtained by the method of adjoints, as it should.

3.6. NONLINEAR TWO POINT BOUNDARY VALUE PROBLEMS

We now show how the method of adjoints for linear problems can be applied in an iterative fashion to solve nonlinear problems. The derivation given here is heuristic; a proof that the method actually converges under suitable conditions, as well as estimates of the speed of convergence and error of the computed solution is given in Chapter 6.

Consider the set of n nonlinear ordinary differential equations

$$\dot{y}_i = g_i(y_1, y_2, \ldots, y_n, t), \qquad i = 1, 2, \ldots, n, \tag{3.6.1}$$

with the initial conditions

$$y_i(t_0) = c_i, \qquad i = 1, 2, \ldots, r, \tag{3.6.2}$$

and the terminal conditions

$$y_{i_m}(t_f) = c_{i_m}, \qquad m = 1, 2, \ldots, n-r. \tag{3.6.3}$$

If $y_i(t)$, $i = 1, 2, \ldots, n$; $t_0 \le t \le t_f$, is the solution of (3.6.1), let us consider a nearby solution, $y_i(t) + \delta y_i(t)$, $i = 1, 2, \ldots, n$, where $\delta y_i(t)$ is often called the variation, a first-order correction to $y_i(t)$. For our purposes the $y_i(t)$ may be thought of as solutions corresponding to guessed values of the missing initial conditions and $\delta y_i(t)$ as corrections necessary to produce the actual solution of the boundary value problem (3.6.1)–(3.6.3).

The differential equations of the "nearby" equations are

$$\dot{y}_i(t)+\delta\dot{y}_i(t) = g_i(y_1(t)+\delta y_1(t), y_2(t)+\delta y_2(t), \ldots, y_n(t)+\delta y_n(t), t),$$
$$i = 1,2,\ldots,n. \qquad (3.6.4)$$

Expanding the right side of (3.6.4) in a Taylor's series up to and including first-order terms, we obtain

$$\dot{y}_i(t)+\delta\dot{y}_i(t) = g_i(y_1, y_2, \ldots, y_n, t)+\sum_{j=1}^{n} \frac{\partial g_i}{\partial y_j}\delta y_j(t), \qquad i = 1,2,\ldots,n, \qquad (3.6.5)$$

where the partial derivatives $\partial g_i/\partial y_j$ are evaluated at $(y_1, y_2, \ldots, y_n, t)$. If (3.6.1) is subtracted from (3.6.5), we have

$$\delta\dot{y}_i(t) = \sum_{j=1}^{n} \frac{\partial g_i}{\partial y_j}\delta y_j(t), \qquad i = 1,2,\ldots,n. \qquad (3.6.6)$$

These equations, called the variational equations, are linear ordinary differential equations with variable coefficients.

Following now a development comparable to that for the two-point boundary value problems with linear ordinary differential equations in Section 3.2, we form the equations which are adjoint to the variational equations:

$$\dot{x}_i = -\sum_{j=1}^{n} \frac{\partial g_j}{\partial y_i} x_j(t), \qquad i = 1,2,\ldots,n. \qquad (3.6.7)$$

As before, the matrix of coefficients of the adjoint equations is the negative transpose of the matrix of coefficients in (3.6.6). The adjoint equations are again linear ordinary differential equations in $x_i, i = 1,2,\ldots,n$.

The fundamental identity (3.2.13) of the method of adjoints for the linear ordinary differential equations $\delta\dot{y}_i(t)$ in (3.6.6) and the adjoint equations $\dot{x}_i(t)$ in (3.6.7) reduces to

$$\sum_{i=1}^{n} x_i(t_f)\delta y_i(t_f)-\sum_{i=1}^{n} x_i(t_0)\delta y_i(t_0) = 0. \qquad (3.6.8)$$

The integral term, which occurred on the right side of (3.2.13), does not appear here because the variational equations (3.6.6) do not have a forcing function $f_i(t)$.

We have interpreted the variation $\delta y_i(t)$ to be the difference between the true, but unknown, profile and the calculated profile; that is,

$$\delta y_i(t) = y_{i_{\text{true}}}(t)-y_{i_{\text{calc.}}}(t), \qquad i = 1,2,\ldots,n, \quad t_0 \leqq t \leqq t_f, \qquad (3.6.9)$$

but, since Eqs. (3.6.5) for $\delta y_i(t)$ are only approximate systems, the process of finding the true profiles will be an iterative process which terminates when $\delta y_i(t)$, $i = 1, 2, \ldots, n$; $t_0 \leq t \leq t_f$, are sufficiently small. Equation (3.6.9) is better written as

$$\delta y_i^{(k)}(t) = y_{i_{\text{true}}}(t) - y_i^{(k)}(t), \qquad i = 1, 2, \ldots, n, \ t_0 \leqq t \leqq t_f, \quad (3.6.10)$$

where

$$y_i^{(k)}(t) = y_{i_{\text{calc.}}}(t), \quad \text{the profile for the } k\text{th iteration.}$$

With this interpretation for the variation, we observe that, for the variables specified at the initial time, $y_i(t_0)$, $i = 1, 2, \ldots, r$, that

$$\delta y_i^{(k)}(t_0) = 0, \qquad i = 1, 2, \ldots, r, \quad k = 0, 1, \ldots, \quad (3.6.11)$$

since the $y_{i_{\text{calc.}}}(t_0)$ is always taken equal to the given initial condition, which, of course, is $y_{i_{\text{true}}}(t_0)$. Similarly, for the variables specified at the final time $y_{i_m}(t_f)$, $m = 1, 2, \ldots, n - r$:

$$\delta y_{i_m}^{(k)}(t_f) = 0, \qquad m = 1, 2, \ldots, n - r, \qquad k = 0, 1, \ldots. \quad (3.6.12)$$

The fundamental identity (3.6.8) can be used to find the corrections $\delta y_i(t_0)$, $i = r + 1, \ldots, n$ to the set of missing initial conditions $y_i(t_0)$, $i = r + 1, \ldots, n$. As in Section 3.2, we choose the Kronecker delta terminal conditions for the adjoint variables $x_i(t_f)$:

$$x_i^{(m)}(t_f) = \left.\begin{matrix} 1, & i = i_m \\ 0, & i \neq i_m \end{matrix}\right\}, \qquad m = 1, 2, \ldots, n - r, \quad (3.6.13)$$

where

$x_i^{(m)}(t_f)$ = the terminal conditions for the mth backward integration of the adjoint equations,

i_m = the set of indices specified in (3.6.3);

and we integrate the adjoint equations backward.

The fundamental identity for the $(n - r)$ backward integrations of the adjoint equations provides a set of $(n - r)$ linear algebraic equations in the $(n - r)$ unknowns $\delta y_i^{(k)}(t_0)$, $i = r + 1, \ldots, n$:

$$\begin{bmatrix} x_{r+1}^{(1)}(t_0) & x_{r+2}^{(1)}(t_0) & \cdots & x_n^{(1)}(t_0) \\ x_{r+1}^{(2)}(t_0) & x_{r+2}^{(2)}(t_0) & \cdots & x_n^{(2)}(t_0) \\ \vdots & \vdots & & \vdots \\ x_{r+1}^{(n-r)}(t_0) & x_{r+2}^{(n-r)}(t_0) & \cdots & x_n^{(n-r)}(t_0) \end{bmatrix} \begin{bmatrix} \delta y_{r+1}^{(k)}(t_0) \\ \delta y_{r+2}^{(k)}(t_0) \\ \vdots \\ \delta y_n^{(k)}(t_0) \end{bmatrix} = \begin{bmatrix} \delta y_{i_1}^{(k)}(t_f) \\ \delta y_{i_2}^{(k)}(t_f) \\ \vdots \\ \delta y_{i_{n-r}}^{(k)}(t_f) \end{bmatrix}. \quad (3.6.14)$$

In order to integrate the adjoint equations we must first develop analytically the partial derivatives $\partial g_i/\partial y_j$, $i, j = 1, 2, \ldots, n$. To evaluate numerically these analytical partial derivatives requires assuming trial values for the missing initial conditions $y_i(t_0)$, $i = r+1, r+2, \ldots, n$, and integrating (3.6.1). Once the $y_i(t)$, $i = 1, 2, \ldots, n$, profiles are known, the partial derivatives can be evaluated.

The integration of (3.6.1) with the kth iterate of the initial conditions

$$y_1^{(k)}(t_0), y_2^{(k)}(t_0), \ldots, y_r^{(k)}(t_0), y_{r+1}^{(k)}(t_0), \ldots, y_n^{(k)}(t_0)$$

(where $y_i^{(k)}(t_0) = y_i(t_0) = c_i$, $i = 1, 2, \ldots, r$) generates the terminal conditions $y_i^{(k)}(t_f)$, $i = 1, 2, \ldots, n$. The right-hand side of (3.6.14) is found by forming

$$\delta y_{i_m}^{(k)}(t_f) = y_{i_m}(t_f) - y_{i_m}^{(k)}(t_f), \qquad m = 1, 2, \ldots, n-r, \qquad (3.6.15)$$

where the first term on the right side is the specified terminal condition, given in (3.6.3), and the second term on the right side is the calculated terminal value for y_{i_m} for the kth iteration through the process.

The solution to (3.6.12) may be expressed as

$$\begin{bmatrix} \delta y_{r+1}^{(k)}(t_0) \\ \delta y_{r+2}^{(k)}(t_0) \\ \vdots \\ \delta y_n^{(k)}(t_0) \end{bmatrix} = \begin{bmatrix} x_{r+1}^{(1)}(t_0) & x_{r+2}^{(1)}(t_0) & \cdots & x_n^{(1)}(t_0) \\ x_{r+1}^{(2)}(t_0) & x_{r+2}^{(2)}(t_0) & \cdots & x_n^{(2)}(t_0) \\ \vdots & \vdots & & \vdots \\ x_{r+1}^{(n-r)}(t_0) & x_{r+2}^{(n-r)}(t_0) & \cdots & x_n^{(n-r)}(t_0) \end{bmatrix}^{-1} \begin{bmatrix} \delta y_{i_1}^{(k)}(t_f) \\ \delta y_{i_2}^{(k)}(t_f) \\ \vdots \\ \delta y_{i_{n-r}}^{(k)}(t_f) \end{bmatrix} \qquad (3.6.16)$$

provided the inverse exists.

For the next iteration through the process the new initial conditions are found from

$$y_i^{(k+1)}(t_0) = y_i(t_0) = c_i, \qquad i = 1, 2, \ldots, r,$$

$$y_i^{(k+1)}(t_0) = y_i^{(k)}(t_0) + \delta y_i^{(k)}(t_0), \qquad i = r+1, \ldots, n. \qquad (3.6.17)$$

To recapitulate, the method of adjoints for nonlinear ordinary differential equations is carried out as follows:

1. Determine analytically the partial derivatives $\partial g_i/\partial y_j$, $i, j = 1, 2, \ldots, n$.
2. Initialize the counter on the iterative process. Set $k = 0$.
3. For $k = 0$, guess the missing initial conditions, $y_i^{(0)}(t_0)$, $i = r+1, \ldots, n$.

4. Integrate (3.6.1) with the initial conditions

$$y_i^{(k)}(t_0) = c_i, \qquad i = 1,2,\ldots,r,$$

$$y_i^{(k)}(t_0), \qquad i = r+1,\ldots,n,$$

and store the profiles.

5. Set the counter on the integrations of the adjoint equations, $m = 1$.

6. For the mth row of (3.6.14) determine the coefficients of the left hand side of the equation. For the mth set of Kronecker delta terminal boundary conditions (3.6.13), integrate backward the adjoint equations from t_f to t_0. Save $x_i^{(m)}(t_0)$, $i = r+1,\ldots,n$. Note that, for each of the backward integrations of the adjoint equations, the stored profiles $y_i^{(k)}(t)$, $i = 1,2,\ldots,n$; $t_0 \le t \le t_f$, are used to evaluate the partial derivatives $\partial g_i/\partial y_j$, $i,j = 1,2,\ldots,n$.

7. For the mth row of (3.6.14) form the right side of (3.6.14) by taking the difference between the specified terminal value $y_{i_m}(t_f) = c_{i_m}$ in (3.6.3) and the calculated value $y_{i_m}^{(k)}(t_f)$ found in item 4.

8. If $m < n-r$, set $m = m+1$. Go to item 6.

9. Form the set of $(n-r)$ linear algebraic equations (3.6.14) and solve for $\delta y_i^{(k)}(t_0)$, $i = r+1,\ldots,n$.

10. Form the next set of trial initial values by

$$y_i^{(k+1)}(t_0) = y_i^{(k)}(t_0) + \delta y_i^{(k)}(t_0), \qquad i = r+1,\ldots,n.$$

11. Set $k = k+1$. Return to item 4.

12. Terminate the calculation whenever $\max\{\delta y_{i_m}^{(k)}(t_f), m = 1,2,\ldots,n-r\}$ is less than a preassigned tolerance or whenever k exceeds a maximum iteration count.

3.7. DISCUSSION

In contrast to the solution of two-point boundary value problems with linear ordinary differential equations, the solution of the nonlinear differential equations by the method of adjoints is an iterative process. The method for nonlinear problems does not compute the missing initial conditions, but rather computes corrections to the trial values for the missing initial conditions. The nonlinear process, however, does deal with linear equations, namely, the variational equations and the adjoint equations associated with the variational equations. In contrast to the linear problem, the nonlinear process requires integration of the $\dot{y}_i$ equations in order to compute corrections to the set of missing initial conditions. Per iteration, the nonlinear process requires one integration of the $\dot{y}_i$ equations plus $(n-r)$ integrations of the adjoint equations for a total of $(n-r+1)$ integrations. The entire linear process requires $(n-r)$ integrations of the adjoint equations, $(n-r)$ integrations of

$$\int_{t_0}^{t_f} \sum_{i=1}^{n} x_i^{(m)}(t) f_i(t)\, dt,$$

plus one integration of the $\dot{y}_i$ equations after the initial conditions are found, a total of $2(n-r)+1$ integrations.

In our discussions so far we have stated that (3.6.1) is integrated forward using the r specified initial conditions and the $(n-r)$ trial initial conditions. This is the usual procedure if $r \geqslant n-r$, since fewer trial initial conditions must be assumed. On the other hand, if $r < n-r$, it is better to integrate (3.6.1) backward, as we mentioned in Section 3.3. Under these circumstances the adjoint equations would be integrated forward using the Kronecker delta conditions at the initial time. The fundamental identity would be used to solve for the set of corrections to the trial values of the $y_i(t_f)$ which were not specified by the boundary conditions (3.6.3).

It should be pointed out here that the derivation of the method of adjoints is primarily a formal one depending on the fundamental identity of the method of adjoints and the solution of a set of linear algebraic equations. Nothing so far has been mentioned about convergence properties of the method or error estimates. In Chapter 6, on the Newton-Raphson method, we discuss the underlying structure of the method of adjoints, as described by Roberts and Shipman [3, 4], and state the sufficient conditions for the method to converge. In addition we give estimates for the rate of convergence and bounds on the estimates of the errors.

3.8. IMPLICIT BOUNDARY CONDITIONS

In Section 3.7 we have described the method of adjoints for a very special type of boundary conditions, namely,

$$y_i(t_0) = c_i, \qquad i = 1, 2, \ldots, r, \tag{3.8.1}$$

$$y_{i_m}(t_f) = c_{i_m}, \qquad m = 1, 2, \ldots, n-r. \tag{3.8.2}$$

For convenience we refer to this set of boundary conditions as the standard case.

The method of adjoints is applicable to very general implicit boundary conditions, provided certain modifications are made in the procedure described in Section 3.6. In some cases the implicit boundary conditions reduce to the standard case. We now discuss the solution of the nonlinear ordinary differential equations (3.6.1) for a variety of implicit boundary conditions.

Case A. *Implicit Boundary Conditions, Functions of Both Initial and Terminal Conditions.* The most general implicit boundary conditions are given by the n nonlinear relations which are functions of both the initial and the terminal conditions:

$$q_i(y_1(t_0), y_2(t_0), \ldots, y_n(t_0), y_1(t_f), y_2(t_f), \ldots, y_n(t_f)) = 0, \qquad i = 1, 2, \ldots, n. \tag{3.8.3}$$

If we define the variation in q_i as

$$\delta q_i = q_{i_{\text{true}}} - q_{i_{\text{calc.}}}, \qquad i = 1, 2, \ldots, n, \tag{3.8.4}$$

since $q_{i_{\text{true}}} = 0$ it follows that

$$\delta q_i = -q_{i_{\text{calc.}}}, \qquad i = 1, 2, \ldots, n. \tag{3.8.5}$$

An alternative expression for the variation is

$$\delta q_i = \sum_{j=1}^{n} \frac{\partial q_i}{\partial y_j(t_0)} \delta y_j(t_0) + \sum_{j=1}^{n} \frac{\partial q_i}{\partial y_j(t_f)} \delta y_j(t_f), \qquad i = 1, 2, \ldots, n, \tag{3.8.6}$$

where the $\partial q_i/\partial y_j(t_0)$ and $\partial q_i/\partial y_j(t_f)$, $i, j = 1, 2, \ldots, n$, can be found analytically from (3.8.3). Equations (3.8.6) are a set of n equations in $2n$ variables, $y_i(t_0), y_i(t_f)$, $i = 1, 2, \ldots, n$.

To solve the nonlinear ordinary differential equations (3.6.1) with the boundary conditions (3.8.3), we must first assume a set of trial initial conditions

$$y_i(t_0) = c_i, \qquad i = 1, 2, \ldots, n, \tag{3.8.7}$$

and integrate (3.6.1) from t_0 to t_f. As a consequence, δq_i, $i = 1, 2, \ldots, n$, can be evaluated numerically from the assumed initial conditions and the calculated terminal conditions. In addition the partial derivatives $\partial q_i/\partial y_j(t_0)$, $\partial q_i/\partial y_j(t_f)$, $i, j = 1, 2, \ldots, n$, can be evaluated numerically using their analytical expressions.

For every $y_j(t_f)$ which appears in the set of implicit boundary conditions q_i, $i = 1, 2, \ldots, n$, the adjoint equations must be integrated backward with Kronecker delta terminal conditions:

$$x_i^{(j)}(t_f) = \left.\begin{matrix} 1, & i = j, \\ 0, & i \neq j. \end{matrix}\right\} \tag{3.8.8}$$

By the fundamental identity of the method of adjoints we have the set of n equations:

$$\sum_{i=1}^{n} x_i^{(j)}(t_0)\delta y_i(t_0) = \delta y_j(t_f), \qquad j = 1, 2, \ldots, n, \tag{3.8.9}$$

a set of n equations in $2n$ variables $\delta y_i(t_0)$, $\delta y_j(t_f)$, i, j, $= 1, 2, \ldots, n$. If we substitute (3.8.9) in (3.8.6), we obtain

$$\delta q_i = \sum_{j=1}^{n} \frac{\partial q_i}{\partial y_j(t_0)} \delta y_j(t_0) + \sum_{j=1}^{n} \frac{\partial q_i}{\partial y_j(t_f)} \left\{ \sum_{s=1}^{n} x_s^{(j)}(t_0)\delta y_s(t_0) \right\},$$

$$i = 1, 2, \ldots, n, \tag{3.8.10}$$

a set of n equations in the n unknowns $\delta y_j(t_0)$, $j = 1, 2, \ldots, n$. On rearranging (3.8.10) we have

$$\delta q_i = \sum_{p=1}^{n} \left\{ \frac{\partial q_i}{\partial y_p(t_0)} + \sum_{j=1}^{n} \frac{\partial q_i}{\partial y_j(t_f)} x_p^{(j)}(t_0) \right\} \delta y_p(t_0), \qquad i = 1, 2, \ldots, n. \tag{3.8.11}$$

In matrix form the set of equations (3.8.11) appears as

$$\begin{bmatrix} \dfrac{\partial q_1}{\partial y_1(t_0)} + \sum_{j=1}^{n} \dfrac{\partial q_1}{\partial y_j(t_f)} x_1^{(j)}(t_0), & \dfrac{\partial q_1}{\partial y_2(t_0)} + \sum_{j=1}^{n} \dfrac{\partial q_1}{\partial y_j(t_f)} x_2^{(j)}(t_0), & \ldots, & \dfrac{\partial q_1}{\partial y_n(t_0)} + \sum_{j=1}^{n} \dfrac{\partial q_1}{\partial y_j(t_f)} x_n^{(j)}(t_0) \\ \dfrac{\partial q_2}{\partial y_1(t_0)} + \sum_{j=1}^{n} \dfrac{\partial q_2}{\partial y_j(t_f)} x_1^{(j)}(t_0), & \dfrac{\partial q_2}{\partial y_2(t_0)} + \sum_{j=1}^{n} \dfrac{\partial q_2}{\partial y_j(t_f)} x_2^{(j)}(t_0), & \ldots, & \dfrac{\partial q_2}{\partial y_n(t_0)} + \sum_{j=1}^{n} \dfrac{\partial q_2}{\partial y_j(t_f)} x_n^{(j)}(t_0) \\ \vdots & \vdots & & \vdots \\ \dfrac{\partial q_n}{\partial y_1(t_0)} + \sum_{j=1}^{n} \dfrac{\partial q_n}{\partial y_j(t_0)} x_1^{(j)}(t_0), & \dfrac{\partial q_n}{\partial y_2(t_0)} + \sum_{j=1}^{n} \dfrac{\partial q_n}{\partial y_j(t_f)} x_n^{(j)}(t_0), & \ldots, & \dfrac{\partial q_n}{\partial y_n(t_0)} + \sum_{j=1}^{n} \dfrac{\partial q_n}{\partial y_j(t_f)} x_n^{(j)}(t_0) \end{bmatrix} \begin{bmatrix} \delta y_1(t_0) \\ \delta y_2(t_0) \\ \vdots \\ \delta y_n(t_0) \end{bmatrix} = \begin{bmatrix} \delta q_1 \\ \delta q_2 \\ \vdots \\ \delta q_n \end{bmatrix} \tag{3.8.12}$$

We observe that all the elements of the matrix are known by evaluating the partials using the profiles found by the forward integration of (3.6.1) and by the backward integrations of the adjoint equations. The right-hand side of the equations is known from the assumed initial conditions $y_i(t_0)$, $i = 1, 2, \ldots, n$, and the calculated terminal conditions $y_i(t_f)$, $i = 1, 2, \ldots, n$. Provided the inverse of the matrix exists, the corrections to the trial initial conditions $\delta y_i(t_0)$, $i = 1, 2, \ldots, n$, are found by solving (3.8.12). Once the $\delta y_i(t_0)$, $i = 1, 2, \ldots, n$, are known, new trial values for the initial conditions $y_i(t_0) + \delta y_i(t_0)$, $i = 1, 2, \ldots, n$, are formed.

To recapitulate, the nonlinear two-point boundary value problem with the implicit boundary conditions (3.8.3) is solved by the following procedure:

1. Determine analytically the partial derivatives $\partial g_i/\partial y_j$, $i,j = 1, 2, \ldots, n$, from (3.6.1).

2. Determine analytically the partial derivatives $\partial q_i/\partial y_j(t_0)$, $\partial q_i/\partial y_j(t_f)$, $i,j = 1, 2, \ldots, n$, from (3.8.3).

3. Initialize the counter on the iterative process. Set $k = 0$.

4. For $k = 0$ guess the missing initial conditions

$$y_i^{(0)}(t_0) = c_i^{(0)}, \qquad i = 1, 2, \ldots, n.$$

5. Integrate (3.6.1) with the initial conditions $y_i^{(k)}(t_0) = c_i^{(k)}$, $i = 1, 2, \ldots, n$, and store the profiles. The expression $c_i^{(k)}$ means the kth iterate of the trial initial conditions.

6. Using the trial initial values $y_j^{(k)}(t_0)$, $j = 1, 2, \ldots, \text{n}$, and the calculated terminal values $y_j^{(k)}(t_f)$, $j = 1, 2, \ldots, n$, evaluate δq_i, $i = 1, 2, \ldots, n$ by (3.8.5).

7. If $\max(\delta q_1, \delta q_2, \ldots, \delta q_n)$ is less than a specified tolerance or if the iteration count k exceeds a specified maximum count, terminate the calculation. Otherwise go to item 8.

8. For each $y_j(t_f)$ appearing in the implicit boundary conditions (3.8.4), integrate the adjoint equations backward using the Kronecker delta terminal conditions (3.8.8). The profiles $y_i^{(k)}(t)$ stored in item 5 are used to evaluate the partial derivatives $\partial g_i/\partial y_j$ (determined analytically in item 1) in the adjoint equations. Save $x_i^{(j)}(t_0)$, $i,j = 1, 2, \ldots, n$.

9. On the basis of item 2 and the profiles in item 5 evaluate numerically the partial derivatives $\partial q_i/\partial y_j(t_0)$, $\partial q_i/\partial y_j(t_f)$, $i,j = 1, 2, \ldots, n$, and form the matrix of coefficients in (3.8.12).

10. Solve (3.8.12) for $\delta y_i(t_0)$, $i = 1, 2, \ldots, n$, and call the solution $\delta y_i^{(k)}(t_0)$ for the kth iteration through the process.

11. Form the next set of trial initial conditions:

$$y_i^{(k+1)}(t_0) = y_i^{(k)}(t_0) + \delta y_i^{(k)}(t_0) = c_i^{(k+1)}, \qquad i = 1, 2, \ldots, n.$$

12. Set $k = k+1$. Return to item 5.

Although the boundary conditions in Case A are the most general boundary conditions possible, it is instructive to consider certain subsets of these general boundary conditions which appear commonly in practice.

Case B. *Implicit Boundary Conditions, r Initial Conditions, $n-r$ Terminal Conditions.*

$$q_i(y_1(t_0), y_2(t_0), \ldots, y_n(t_0)) = 0, \qquad i = 1, 2, \ldots, r \tag{3.8.13}$$

$$q_{r+i}(y_1(t_f), y_2(t_f), \ldots, y_n(t_f)) = 0, \qquad i = 1, 2, \ldots, n-r. \tag{3.8.14}$$

The first r implicit boundary conditions are functions of the initial conditions only, while the remaining $(n-r)$ implicit boundary conditions are functions of the terminal conditions only. The procedures for solving this case are essentially the same as for Case A except for some details which we discuss.

Instead of a single group of equations for the variation as in (3.8.6), we now consider two groups of equations, one for the initial and the other for the terminal conditions:

$$\delta q_i = \sum_{j=1}^{n} \frac{\partial q_i}{\partial y_j(t_0)} \delta y_j(t_0), \qquad i = 1, 2, \ldots, r, \tag{3.8.15}$$

$$\delta q_{r+i} = \sum_{j=1}^{n} \frac{\partial q_{r+i}}{\partial y_j(t_f)} \delta y_j(t_f), \qquad i = 1, 2, \ldots, n-r. \tag{3.8.16}$$

For each $y_j(t_f)$ appearing in (3.8.14), the adjoint equations are integrated backward using the Kronecker delta terminal conditions to form

$$\sum_{i=1}^{n} x_i^{(j)}(t_0) \delta y_i(t_0) = \delta y_j(t_f), \qquad j = 1, 2, \ldots, n. \tag{3.8.9}$$

Equation (3.8.9) is substituted in (3.8.16) to give

$$\delta q_{r+i} = \sum_{j=1}^{n} \frac{\partial q_{r+i}}{\partial y_j(t_f)} \sum_{s=1}^{n} x_s^{(j)}(t_0) \delta y_s(t_0), \qquad i = 1, 2, \ldots, n-r \tag{3.8.17}$$

or, on rearranging,

$$\delta q_{r+i} = \sum_{p=1}^{n} \left\{ \sum_{j=1}^{n} \frac{\partial q_{r+i}}{\partial y_j(t_f)} x_p^{(j)}(t_0) \right\} \delta y_p(t_0), \qquad i = 1, 2, \ldots, n-r. \qquad (3.8.18)$$

We may solve the set of n equations consisting of (3.8.18) and (3.8.15) for the n unknowns $\delta y_j(t_0)$, $j = 1, 2, \ldots, n$. These two sets of equations are equivalent to (3.8.12). Except for handling (3.8.13) and (3.8.14) as described in this paragraph, Case B is solved by the same procedure as Case A.

Case C. *Explicit Boundary Conditions at One Boundary and Implicit at the Other.*

$$q_i = y_i(t_0) - c_i = 0, \qquad i = 1, 2, \ldots, r, \qquad (3.8.19)$$

$$q_{r+i}(y_1(t_f), y_2(t_f), \ldots, y_n(t_f)) = 0, \qquad i = 1, 2, \ldots, n-r. \qquad (3.8.20)$$

Since the first r initial conditions are specified in (3.8.19), trial values need be assumed only for $y_{r+i}(t_0)$, $i = 1, 2, \ldots, n-r$. The adjoint equations are integrated backward once for each $y_j(t_f)$ that appears in (3.8.20), and the fundamental identity gives

$$\sum_{i=r+1}^{n} x_i^{(j)}(t_0) \delta y_i(t_0) = \delta y_j(t_f), \qquad j = 1, 2, \ldots, n. \qquad (3.8.21)$$

Since $\delta y_i(t_0) = 0$, $i = 1, 2, \ldots, r$, by virtue of (3.8.19) the summation in (3.8.21) is taken over $i = r+1, \ldots, n$. Substitution of (3.8.21) in the variation of (3.8.20) which is given by (3.8.16) yields

$$\delta q_{r+i} = \sum_{j=1}^{n} \frac{\partial q_{r+i}}{\partial y_j(t_f)} \left\{ \sum_{p=r+1}^{n} x_p^{(j)}(t_0) \delta y_p(t_0) \right\}, \qquad i = 1, 2, \ldots, n-r. \qquad (3.8.22)$$

This is a set of $(n-r)$ equations in $(n-r)$ unknowns $\delta y_{r+i}(t_0)$, $i = 1, 2, \ldots, n-r$.

Case D.

$$q_i = y_i(t_0) - c_i = 0, \qquad i = 1, 2, \ldots, r, \qquad (3.8.23)$$

$$q_{r+i}(y_{i_1}(t_f), y_{i_2}(t_f), \ldots, y_{i_{n-r}}(t_f)) = 0, \qquad i = 1, 2, \ldots, n-r, \qquad (3.8.24)$$

This is a special case of Case C, where the set of $(n-r)$ terminal implicit boundary conditions are a function of $(n-r)$ variables rather than n. Under these circumstances, (3.8.24) can be solved by the Newton-Raphson method to yield

$$y_{i_m}(t_f) = c_{i_m}, \qquad m = 1, 2, \ldots, n-r. \tag{3.8.25}$$

Case D then reduces to the standard case.

Case E. *Linear Differential Equations with Most General Linear Implicit Boundary Conditions.* Case A applies to the most general boundary conditions for nonlinear ordinary differential equations; in Case E we specialize these results to the linear system of equations and linear boundary conditions.

Consider the system of n linear ordinary differential equations with variable coefficients (3.2.1). The most general implicit linear boundary conditions are

$$\sum_{i=1}^{n} \alpha_{p,i} y_i(t_0) + \sum_{i=1}^{n} \beta_{p,i} y_i(t_f) = D_p, \qquad p = 1, 2, \ldots, n, \tag{3.8.26}$$

where $\alpha_{p,i}$, $\beta_{p,i}$, D_p, $i, p = 1, 2, \ldots, n$ are known constants.

Let us integrate the adjoint equations backward once for each $y_i(t_f)$ appearing in (3.8.26), using as the terminal boundary conditions

$$x_i^{(p)}(t_f) = \beta_{p,i}, \qquad i, p = 1, 2, \ldots, n, \tag{3.8.27}$$

where $x_i^{(p)}(t_f)$ = the ith component at t_f for the pth backward integration of the adjoint equation.

Recalling that the fundamental identity for the method of adjoints is

$$\sum_{i=1}^{n} x_i^{(p)}(t_f) y_i(t_f) - \sum_{i=1}^{n} x_i^{(p)}(t_0) y_i(t_0) = \int_{t_0}^{t_f} \sum_{i=1}^{n} x_i^{(p)}(t) f_i(t)\, dt,$$
$$p = 1, 2, \ldots, n, \tag{3.8.28}$$

and substituting (3.8.27) in it, we obtain

$$\sum_{i=1}^{n} \beta_{p,i} y_i(t_f) - \sum_{i=1}^{n} x_i^{(p)}(t_0) y_i(t_0) = \int_{t_0}^{t_f} \sum_{i=1}^{n} x_i^{(p)}(t) f_i(t)\, dt, \qquad p = 1, 2, \ldots, n. \tag{3.8.29}$$

Introducing (3.8.26) into (3.8.29) gives

$$D_p - \sum_{i=1}^{n} \alpha_{p,i} y_i(t_0) - \sum_{i=1}^{n} x_i^{(p)}(t_0) y_i(t_0) = \int_{t_0}^{t_f} \sum_{i=1}^{n} x_i^{(p)}(t) f_i(t)\, dt,$$
$$p = 1, 2, \ldots, n, \tag{3.8.30}$$

or

$$\sum_{i=1}^{n} \{\alpha_{p,i}+x_i^{(p)}(t_0)\}y_i(t_0) = D_p-\int_{t_0}^{t_f} \sum_{i=1}^{n} x_i^{(p)}(t)f_i(t)\,dt, \qquad p = 1,2,\ldots,n. \tag{3.8.31}$$

This is a set of n equations in the n unknowns $y_i(t_0)$, $i = 1,2,\ldots,n$. Thus the linearity of the differential equations, the linearity of the implicit boundary conditions, and the proper choice of the terminal conditions for the adjoint equations permit us to solve this problem with no iterations as for the standard case.

Case F. *Linear Differential Equations with Implicit Linear Boundary Conditions at Each Boundary.* This is a specialization of Case E. Consider once again the system of linear equations (3.2.1) with the implicit boundary conditions at each boundary

$$\sum_{i=1}^{n} \alpha_{p,i}y_i(t_0) = C_p, \qquad p = 1,2,\ldots,r, \tag{3.8.32}$$

$$\sum_{i=1}^{n} \beta_{p,i}y_i(t_f) = K_p, \qquad p = 1,2,\ldots,n-r. \tag{3.8.33}$$

Integrate the adjoint equations backward $(n-r)$ times (once for each specified terminal implicit boundary condition), using the following terminal conditions:

$$x_i^{(p)}(t_f) = \beta_{p,i}, \qquad p = 1,2,\ldots,n-r, \qquad i = 1,2,\ldots,n. \tag{3.8.34}$$

Referring to the first summation in the fundamental identity of the method of adjoints in (3.8.28), we observe that, by (3.8.33) and (3.8.34),

$$\sum_{i=1}^{n} x_i^{(p)}(t_f)y_i(t_f) = \sum_{i=1}^{n} \beta_{p,i}y_i(t_f) = K_p, \qquad p = 1,2,\ldots,n-r. \tag{3.8.35}$$

Equation (3.8.28) may be rearranged as

$$\sum_{i=1}^{n} x_i^{(p)}(t_0)y_i(t_0) = K_p-\int_{t_0}^{t_f} \sum_{i=1}^{n} x_i^{(p)}(t)f_i(t)\,\mathrm{d}t, \qquad p = 1,2,\ldots,n-r. \tag{3.8.36}$$

If we solve the set of equations consisting of the r implicit initial boundary conditions (3.8.32) and the $(n-r)$ equations (3.8.36), we have a set of n equations in the n unknowns $y_i(t_0)$, $i = 1,2,\ldots,n$.

3.9. EXAMPLE: IMPLICIT BOUNDARY CONDITIONS

3.9.1. Nonlinear System Example

Consider the nonlinear differential equation (see Section 5.4, Chapter 5)

$$3y\ddot{y}+\dot{y}^2 = 0$$

with the boundary conditions

$$y(0) = 0, \qquad \dot{y}(1)+y^2(1) = 1.75.$$

Setting

$$y_1 = y, \qquad \dot{y}_1 = y_2,$$

we write the nonlinear differential equation as the set

$$\dot{y}_1-y_2 = 0, \qquad \dot{y}_2 + \frac{y_2^2}{3y_1} = 0,$$

with the boundary conditions

$$y_1(0) = 0, \qquad y_2(1)+y_1^2(1) = 1.75.$$

This is an example of Case C, where

$$q_1 = y_1(0) = 0, \qquad q_2 = y_2(1)+y_1^2(1)-1.75 = 0.$$

To solve the boundary value problem choose a trial value for $y_2(0)$ and integrate the pair of first-order equations over the interval [0, 1]. Evaluate numerically

$$\delta q_2 = 2y_{1\,\text{calc.}}(1)\delta y_1(1)+\delta y_2(1) = -q_{2\,\text{calc.}}$$

Since $y_1(1)$ and $y_2(1)$ both appear in q_2, we integrate the adjoint equations backward twice with the terminal conditions

$$\begin{matrix} x_1^{(1)}(1) = 1, & \text{and} & x_1^{(2)}(1) = 0, \\ x_2^{(1)}(1) = 0, & & x_2^{(2)}(1) = 1, \end{matrix}$$

and form by the fundamental identity of the method of adjoints

$$\delta y_1(1) = \delta y_2(0)x_2^{(1)}(0), \qquad \delta y_2(1) = \delta y_2(0)x_2^{(2)}(0).$$

Substituting these equations in δq_2 gives the correction $\delta y_2(0)$:

$$\delta y_2(0) = \frac{\delta q_{2_{\text{calc.}}}}{2y_{1_{\text{calc.}}}(1)x_2^{(1)}(0)+x_2^{(2)}(0)}.$$

The new trial initial condition is formed as $y_2(0)+\delta y_2(0)$. The process is repeated until $q_{2\text{calc.}}$ is within a small tolerance of zero.

3.9.2. Linear System Example

Consider the linear system of differential equations

$$\begin{pmatrix} \dot{y}_1 \\ \dot{y}_2 \end{pmatrix} = \begin{pmatrix} 0 & 1 \\ 1 & 0 \end{pmatrix} \begin{pmatrix} y_1(t) \\ y_2(t) \end{pmatrix} + \begin{pmatrix} 0 \\ t \end{pmatrix}$$

with the boundary conditions

$$y_1(t_0)+y_2(t_0) = C_1, \qquad y_1(t_f) = K_1.$$

This is an example of Case F, where $\beta_{1,1} = 1$. Using the terminal boundary conditions for the adjoint equations,

$$x_1^{(1)}(t_f) = 1, \qquad x_2^{(1)}(t_f) = 0,$$

the fundamental identity appears as

$$\sum_{i=1}^{2} x_i^{(1)}(t_0)y_i(t_0) = K_1 - \int_{t_0}^{t_f} x_2^{(1)}(t)t\,dt.$$

The simultaneous solution of this equation and the boundary condition

$$y_1(t_0)+y_2(t_0) = C_1$$

yields $y_1(t_0)$ and $y_2(t_0)$.

Now suppose that thc boundary conditions are

$$y_1(t_0) = C_1, \qquad y_1(t_f)+y_2(t_f) = K_1.$$

Using as the terminal boundary conditions for the adjoint equations

$$\beta_{11} = x_1^{(1)}(t_f) = 1, \qquad \beta_{12} = x_2^{(1)}(t_f) = 1,$$

the fundamental identity becomes

$$x_1^{(1)}(t_f)y_1(t_f)+x_2^{(1)}(t_f)y_2(t_f)-x_1^{(1)}(t_0)y_1(t_0)-x_2^{(1)}(t_0)y_2(t_0)$$
$$= \int_{t_0}^{t_f} \sum_{i=1}^{2} x_i^{(1)}(t)f_i(t)\,dt.$$

By the boundary conditions this reduces to

$$K_1 - C_1 x_1^{(1)}(t_0) - x_2^{(1)}(t_0) y_2(t_0) = \int_{t_0}^{t_f} \sum_{i=1}^{2} x_i^{(1)}(t) f_i(t)\, dt.$$

The missing initial condition is found as

$$y_2(t_0) = \frac{1}{x_2^{(1)}(t_0)} \left[K_1 - C_1 x_1^{(1)}(t_0) - \int_{t_0}^{t_f} x_2^{(1)}(t) t\, dt \right].$$

3.10. NUMERICAL EXAMPLES: METHOD OF ADJOINTS

In this section we give several numerical examples of the method of adjoints applied to linear and nonlinear two-point boundary value problems. The first example is a system of three first-order linear ordinary differential equations whose analytical solution is also given. The second example gives the numerical solution of the second-order linear ordinary differential equations discussed in Section 3.5. The third example is a rather complicated third-order nonlinear ordinary differential equation with one boundary condition specified at infinity. In addition the equation is described by two sets of right-hand sides which depend on a parameter. The value of the parameter at which the right side of the equation switches to the alternative right side is itself determined iteratively.

Example 1.

$$\begin{bmatrix} \dot{y}_1 \\ \dot{y}_2 \\ \dot{y}_3 \end{bmatrix} = \begin{bmatrix} 0 & 3 & 1 \\ 1 & 0 & 0 \\ 0 & 2 & 0 \end{bmatrix} \begin{bmatrix} y_1(t) \\ y_2(t) \\ y_3(t) \end{bmatrix} + \begin{bmatrix} e^t \\ 0 \\ t \end{bmatrix}.$$

The initial condition at $t = 0$ is

$$y_3(0) = -0.75.$$

The terminal conditions at $t = 1$ are

$$y_1(1) = 13.7826, \qquad y_3(1) = 5.64783.$$

The tabulated results appear in Table 3.1. As a comparison, the reader might care to evaluate the analytical solution, which is

$$\begin{aligned}
y_1(t) &= \tfrac{1}{2}((b_1 + b_2 t)e^{-t} - 2b_2 e^{-t} + 4b_3 e^{2t} - \tfrac{1}{2}e^t - 1),\\
y_2(t) &= \tfrac{1}{2}(-(b_1 + b_2 t)e^{-t} + b_2 e^{-t} + 2b_3 e^{2t} - \tfrac{1}{2}e^t + \tfrac{3}{2} - t),\\
y_3(t) &= (b_1 + b_2 t)e^{-t} + b_3 e^{2t} - \tfrac{1}{2}e^t + \tfrac{3}{2}t - \tfrac{9}{4}),
\end{aligned}$$

Table 3.1

t	$y_1(t)$	$y_2(t)$	$y_3(t)$
0.00	1.7500201	1.0000039	−0.7500000
0.05	1.9231594	1.0917433	−0.6442348
0.10	2.1189529	1.1926972	−0.5263443
0.15	2.3396354	1.3040532	−0.3953488
0.20	2.5876876	1.4271166	−0.2501437
0.25	2.8658614	1.5633237	−0.0894876
0.30	3.1772083	1.7142554	0.0880114
0.35	3.5251106	1.8816534	0.2839119
0.40	3.9133157	2.0674378	0.4999546
0.45	4.3459745	2.2737255	0.7380825
0.50	4.8276834	2.5028523	1.0004606
0.55	5.3635313	2.7573957	1.2894996
0.60	5.9591507	3.0402011	1.6078812
0.65	6.6207752	3.3544102	1.9585860
0.70	7.3553025	3.7034930	2.3449250
0.75	8.1703638	4.0912820	2.7705741
0.80	9.0744013	4.5220115	3.2396120
0.85	$1.0076753(10^1)$	5.0003598	3.7565627
0.90	$1.1187747(10^1)$	5.5314965	4.3264425
0.95	$1.2418806(10^1)$	6.1211345	4.9548109
1.00	$1.3782563(10^1)$	6.7755870	5.6478286

where the constants b_1, b_2, b_3 are, of course, to be determined from the boundary conditions.

Example 2. The problem

$$\frac{d^2y}{dt^2} = y+t, \qquad y(0) = 0, \qquad y(1) = 1,$$

has been discussed in Chapter 2, Section 2.4, and Chapter 3, Section 3.5. In Table 3.2 are listed the exact answer and the solutions obtained by the method of adjoints for step sizes $h = 0.10$ and 0.01. Good agreement is obtained.

Example 3. In the course of some boundary layer studies, Dukler has developed a tentative model of the fluid velocity profile in the boundary layer which gives rise to an interesting two-point boundary value problem with a third-order nonlinear differential equation [5].

Table 3.2

t	$y(t)$ $h = 0.1$	$y(t)$ $h = 0.01$	$y(t)$ exact
0.0	0.0	0.0	0.0
0.1	7.0467564(10⁻²)	7.0467406(10⁻²)	7.0467410(10⁻²)
0.2	1.4264122(10⁻¹)	1.4264090(10⁻¹)	1.4264090(10⁻¹)
0.3	2.1824415(10⁻¹)	2.1824367(10⁻¹)	2.1824367(10⁻¹)
0.4	2.9903383(10⁻¹)	2.9903320(10⁻¹)	2.9903319(10⁻¹)
0.5	3.8681967(10⁻¹)	3.8681888(10⁻¹)	3.8681887(10⁻¹)
0.6	4.8348108(10⁻¹)	4.8348014(10⁻¹)	4.8348014(10⁻¹)
0.7	5.9098632(10⁻¹)	5.9098524(10⁻¹)	5.9098524(10⁻¹)
0.8	7.1141217(10⁻¹)	7.1141096(10⁻¹)	7.1141095(10⁻¹)
0.9	8.4696472(10⁻¹)	8.4696338(10⁻¹)	8.4696337(10⁻¹)
1.0	1.0000014	1.0000000	1.0000000

The equation and problem are now described:

$$\frac{d}{d\eta}\left[\left(1+\frac{\varepsilon}{\nu}\right)\ddot{f}\right] = -f\ddot{f}+\lambda(1-(\dot{f})^2),$$

where

$$f = f(\eta),$$

$$\dot{f} = \frac{df}{d\eta},$$

$$\ddot{f} = \frac{d^2 f}{d\eta^2}.$$

At $\eta = 0$ the boundary conditions are

$$f(0) = 0, \qquad \dot{f}(0) = 0$$

and, at $\eta = \infty$,

$$\dot{f}(\infty) = 1.0.$$

The ε/ν term is defined for two regions: when $\eta/\eta_\delta \le 0.16$,

$$\frac{\varepsilon}{\nu} = \sqrt{2}\,k^2\,\mathrm{Re}^{1/2}\,\eta^2(1-e^{-\phi\eta})^2\ddot{f},$$

when $\eta/\eta_\delta > 0.16$, ε/ν is constant at the value $(\varepsilon/\nu)_{\eta/\eta_\delta = 0.16}$, where

$$\begin{aligned}
\phi \;&= \frac{(2\,\mathrm{Re})^{1/4}\,(f(0))^{1/2}}{25},\\
\eta_\delta &= \text{value of } \eta \text{ where } \dot{f} = 0.99,\\
k \;&= \text{constant} = 0.4,\\
\mathrm{Re} &= \text{Reynolds number},\\
\lambda \;&= \text{constant}.
\end{aligned}$$

On expanding the left-hand side of the nonlinear differential equation we obtain

$$\dddot{f} + 2A\eta^2(1-e^{-\phi\eta})^2\ddot{f}\dddot{f} + A\{2\eta^2\phi(1-e^{-\phi\eta})e^{-\phi\eta} + 2\eta(1-e^{-\phi\eta})^2\}(\ddot{f})^2$$
$$= -f\ddot{f} - \lambda(1-(f)^2),$$

where $A = \sqrt{2}k^2\,\mathrm{Re}^{1/2}$.

It is convenient to convert the third-order equation into a system of three first-order equations. Let

$$F_1 = f, \qquad F_2 = \dot{F}_1 = f, \qquad F_3 = \dot{F}_2 = \ddot{F}_1 = \ddot{f}.$$

The first-order equations are

$$\begin{aligned}
\dot{F}_1 &= F_2,\\
\dot{F}_2 &= F_3,\\
\dot{F}_3 &= \begin{cases} \dfrac{\alpha(\eta)F_3{}^2 - F_1F_3 + \lambda(1-F_2{}^2)}{1+w(\eta)F_3}, & \dfrac{\eta}{\eta_\delta} \le 0.16,\\[2ex] \dfrac{-F_1F_3 + \lambda(1-F_2{}^2)}{c(\eta)}, & \dfrac{\eta}{\eta_\delta} > 0.16, \end{cases}
\end{aligned}$$

where

$$\begin{aligned}
\alpha(\eta) &= -2A\eta(1-e^{-\phi\eta})((1-e^{-\phi\eta}) + \eta\phi e^{-\phi\eta}),\\
w(\eta) &= 2A\eta^2(1-e^{-\phi\eta})^2,\\
c(\eta) &= \left(1+\frac{\varepsilon}{\nu}\right)_{\eta/\eta_\delta=0.16} = (1 + A\eta^2(1-e^{-\phi\eta})^2F_3)_{\eta/\eta_\delta=0.16}.
\end{aligned}$$

At $\eta = 0$, the boundary conditions are

$$F_1(0) = 0, \qquad F_2(0) = 0,$$

and, at $\eta = \infty$,

$$F_2(\infty) = 1.0.$$

This problem is interesting for several reasons. First of all, the third equation is described by two different right-hand sides, depending on the value of η/η_δ. On the other hand, η_δ is not known until the η is attained at which $\dot{F}_1(\eta) = 0.99$. If the method of adjoints is employed, it is necessary also to evaluate the right-hand side of the adjoint equations for $\eta/\eta_\delta \leq 0.16$ and $\eta/\eta_\delta > 0.16$. Second, since the terminal η is specified at $\eta = \infty$, it is necessary to approximate $\eta = \infty$ by a finite η, designated $\eta(\infty)$. Third, the boundary conditions are stated so that F_3 is unspecified at both the initial and final η.

To solve this problem we choose a value for $\eta(\infty)$ and pick a trial value for η_δ. In conformity with these choices, we solve the problem by the method of adjoints. Then we check if the calculated η_δ, that is, where $\dot{F}_1(\eta) = 0.99$, agrees with the trial η_δ. If it does, the problem is solved. If it does not, the trial η_δ is adjusted, and the two-point boundary value problem is solved again.

Table 3.3A. Fifth Iteration Profiles [a–d]

η	$F_1(\eta)$	$F_2(\eta)$	$F_3(\eta)$
0.0	0.0	0.0	1.2833638
1.0	3.2502398 (10^{-1})	4.4407578 (10^{-1})	9.6928569 (10^{-2})
2.0	8.0484853 (10^{-1})	5.0743059 (10^{-1})	4.4458274 (10^{-2})
3.0	1.3316281	5.4384194 (10^{-1})	3.0445817 (10^{-2})
4.0	1.8894081	5.7061017 (10^{-1})	2.3738287 (10^{-2})
5.0	2.4711379	5.9217732 (10^{-1})	1.9685985 (10^{-2})
6.0	3.0727023	6.1071621 (10^{-1})	1.8388170 (10^{-2})
8.0	4.3321097	6.4923558 (10^{-1})	2.0020003 (10^{-2})
10.0	5.6713635	6.9032948 (10^{-1})	2.0953174 (10^{-2})
12.0	7.0941850	7.3255876 (10^{-1})	2.1153769 (10^{-2})
15.0	9.3860890	7.9486369 (10^{-1})	2.0131258 (10^{-2})
20.0	1.3596412 (10^{1})	8.8553998 (10^{-1})	1.5701291 (10^{-2})
25.0	1.8196472 (10^{1})	9.4963766 (10^{-1})	9.9254488 (10^{-3})
30.0	2.3046651 (10^{1})	9.8634052 (10^{-1})	5.0291746 (10^{-3})
35.0	2.8026553 (10^{1})	1.0030875	1.9839880 (10^{-3})

[a] Re $= 10^6$, $\lambda = 0.3$, trial $\eta_\delta = 35$, $\eta(\infty) = 35$.
[b] Initial vector $F_1(0) = 0$, $F_2(0) = 0$, $F_3(0) = 1.5$.
[c] Integration step size $h = 0.10$.
[d] One-fourth of the calculated correction to the missing initial condition $F_3(0)$ was applied to $F_3(0)$ at each iteration to obtain the next trial value of $F_3(0)$.

Table 3.3B. Fifth Iteration Profiles [a, b]

η	$F_1(\eta)$	$F_2(\eta)$	$F_3(\eta)$
0.0	0.0	0.0	9.4292458(10^{-1})
1.0	2.6655609(10^{-1})	3.7448164(10^{-1})	9.1115909(10^{-2})
2.0	6.7447391(10^{-1})	4.3362852(10^{-1})	4.1294717(10^{-2})
3.0	1.1260847	4.6749361(10^{-1})	2.8414701(10^{-2})
4.0	1.6066296	4.9260339(10^{-1})	2.2400301(10^{-2})
5.0	2.1099430	5.1396968(10^{-1})	2.1788063(10^{-2})
6.0	2.6350956	5.3661467(10^{-1})	2.3462660(10^{-2})
8.0	3.7571514	5.8630292(10^{-1})	2.6045236(10^{-2})
10.0	4.9830034	6.4002702(10^{-1})	2.7477736(10^{-2})
12.0	6.3183553	6.9539291(10^{-1})	2.7682413(10^{-2})
15.0	8.5271932	7.7621047(10^{-1})	2.5781560(10^{-2})
20.0	1.2703768(10^{1})	8.8821276(10^{-1})	1.8417856(10^{-2})
25.0	1.7338584(10^{1})	9.5859116(10^{-1})	9.9360375(10^{-3})
30.0	2.2227427(10^{1})	9.9197377(10^{-1})	3.9751584(10^{-3})
35.0	2.7222307(10^{1})	1.0036177	1.1186431(10^{-3})

[a] Trial $\eta_\delta = 28$.
[b] Notes of Table 3.3A apply.

Table 3.3C. Fifth Iteration Profiles [a, b]

η	$F_1(\eta)$	$F_2(\eta)$	$F_3(\eta)$
0.0	0.0	0.0	8.9817513(10^{-1})
1.0	2.5836679(10^{-1})	3.6468669(10^{-1})	9.0367713(10^{-2})
2.0	6.5619347(10^{-1})	4.2329403(10^{-1})	4.0885519(10^{-2})
3.0	1.0972915	4.5682574(10^{-1})	2.8145236(10^{-2})
4.0	1.5675124	4.8369877(10^{-1})	2.7774836(10^{-2})
5.0	2.0655287	5.1274762(10^{-1})	3.0256650(10^{-2})
6.0	2.5937663	5.4407080(10^{-1})	3.2314876(10^{-2})
8.0	3.7486519	6.1170424(10^{-1})	3.4979191(10^{-2})
10.0	5.0427428	6.8256763(10^{-1})	3.5521820(10^{-2})
12.0	6.4782075	7.5236778(10^{-1})	3.3938600(10^{-2})
15.0	8.8807851	8.4649605(10^{-1})	2.8262989(10^{-2})
20.0	1.3412818(10^{1})	9.5521299(10^{-1})	1.5111138(10^{-2})
25.0	1.8330148(10^{1})	1.0033808	5.1402398(10^{-3})
30.0	2.3388400(10^{1})	1.0163997	8.6294028(10^{-4})
35.0	2.8474823(10^{1})	1.0173168	−1.9883119(10^{-4})

[a] Trial $\eta_\delta = 21$.
[b] Notes of Table 3.3A apply.

In Tables 3.3A, 3B, and 3C we tabulate, for Re $= 10^6$, $\lambda = 0.3$, $\eta(\infty) = 35$, the profiles for the fifth iteration for three cases of $\eta_\delta = 35, 28$, and 21. For engineering purposes the solution at the trial $\eta_\delta = 28$ is satisfactory, even though the calculated value of $\eta_\delta = 29.5$. The correct value of η_δ as interpolated from the following tabulation is 30:

Trial η_δ	Calculated η_δ
35	31.0
28	29.5
21	23.0

REFERENCES

1. T. R. Goodman and G. N. Lance, The Numerical Solution of Two Point Boundary Value Problems, *MTAC*, **10** (1956), 82–86.
2. E. L. Ince, *Ordinary Differential Equations*, reprinted by Dover, New York.
3. S. M. Roberts and J. S. Shipman, The Kantorovich Theorem and Two-Point Boundary-Value Problems, *IBM J. Res. Develop.* (5) **10** (Sept. 1966), 402–406.
4. S. M. Roberts and J. S. Shipman, Some Results in Two-Point Boundary Value Problems, *IBM J. Res. Develop.* (4) **11** (July 1967), 383–388.
5. A. E. Dukler, personal communication, 1967.

Chapter 4

METHOD OF COMPLEMENTARY FUNCTIONS

4.1. INTRODUCTION

The method of complementary functions is another practical shooting method suited to the numerical solution of two-point boundary value problems for linear ordinary differential equations. Like the method of adjoints, it finds the set of missing initial conditions in one pass through the process [1]. Unlike the method of adjoints, it cannot be conveniently applied directly to the solution of nonlinear problems, but can be used to solve the sequence of linear problems in the quasilinearization method (see Chapter 5).

In this chapter we discuss first the solution of linear two-point boundary value problems by forming linear combinations of solutions of n initial value problems for the same system of n differential equations, and then solving a system of n linear algebraic equations. Next we show that, by the proper choice of initial values, the number of initial value problems to be solved and the number of linear algebraic equations can be reduced to $n - r$, where r is the number of initial conditions specified in the original two-point boundary value problem. The method just described is the "method of particular solutions", exploited by Miele and his students [2–5].

We next develop the method of complementary functions and show that it is a still more efficient technique for combining solutions. We show the relationship between the method of adjoints and the method of complementary functions, and present typical numerical experience with the two methods.

Finally, we discuss the problem of numerical dependence of solutions of the $n - r$ initial value problems which must be integrated in the method of complementary functions. This linear dependence leads to poorly conditioned systems of linear algebraic equations for the $n - r$ missing initial conditions, and thus to inaccurate solutions. Orthonormalization procedures are introduced, and two methods of using them to cope with linear dependence are given.

4.2. LINEAR COMBINATIONS OF SOLUTIONS

Consider, as in Chapter 3, Section 3.2, the two-point boundary value problem defined by the set of n linear ordinary differential equations

$$\dot{\mathbf{y}} = \mathbf{A}(t)\mathbf{y}+\mathbf{f}(t), \tag{4.2.1}$$

where

$\mathbf{y}(t) = n\times 1$ vector with components $y_1(t), y_2(t), \ldots, y_n(t)$,

$\mathbf{A}(t) = n\times n$ matrix whose i, j element is $a_{ij}(t)$,

$\mathbf{f}(t) = n\times 1$ vector with components $f_1(t), f_2(t), \ldots, f_n(t)$,

and by the initial conditions

$$y_i(t_0) = c_i, \qquad i = 1,2,\ldots,r, \tag{4.2.2}$$

and the terminal conditions

$$y_{i_m}(t_f) = c_{i_m}, \qquad m = 1,2,\ldots,n-r. \tag{4.2.3}$$

Suppose we construct n linearly independent n-vectors and use them in turn as initial vectors $\mathbf{y}(t_0)$ for n integrations of (4.2.1) from t_0 to t_f to obtain n linearly independent solutions $\mathbf{y}^{(k)}(t)$. Suppose further that these n solutions are stored. (In practice, only the solution at discrete points t_i is of course computed and stored.) In the spirit of the classical variation of parameters let us try to represent the solution of the problem (4.2.1)–(4.2.3) as a linear combination of the $\mathbf{y}^{(k)}(t)$:

$$\mathbf{y}(t) = \sum_{k=1}^{n} b_k\mathbf{y}^{(k)}(t), \qquad t_0 \le t \le t_f, \tag{4.2.4}$$

where

$\mathbf{y}^{(k)}(t) = k$th solution vector with components $y_1{}^{(k)}(t), y_2{}^{(k)}(t), \ldots, y_n{}^{(k)}(t)$,

b_k = constants to be determined by the boundary conditions.

Writing out (4.2.4) and using the boundary conditions (4.2.2) and (4.2.3), we have:

$$
\begin{aligned}
b_1 y_1^{(1)}(t_0) + b_2 y_1^{(2)}(t_0) + \cdots + b_r y_1^{(r)}(t_0) + b_{r+1} y_1^{(r+1)}(t_0) + \cdots + b_n y_1^{(n)}(t_0) &= c_1, \\
b_1 y_2^{(1)}(t_0) + b_2 y_2^{(2)}(t_0) + \cdots + b_r y_2^{(r)}(t_0) + b_{r+1} y_2^{(r+1)}(t_0) + \cdots + b_n y_2^{(n)}(t_0) &= c_2, \\
\vdots \qquad\qquad & \qquad \vdots \\
b_1 y_r^{(1)}(t_0) + b_2 y_r^{(2)}(t_0) + \cdots + b_r y_r^{(r)}(t_0) + b_{r+1} y_r^{(r+1)}(t_0) + \cdots + b_n y_r^{(n)}(t_0) &= c_r, \\
b_1 y_{i_1}^{(1)}(t_f) + b_2 y_{i_1}^{(2)}(t_f) + \cdots + b_r y_{i_1}^{(r)}(t_f) + b_{r+1} y_{i_1}^{(r+1)}(t_f) + \cdots + b_n y_{i_1}^{(n)}(t_f) &= c_{i_1}, \\
\vdots \qquad\qquad & \qquad \vdots \\
b_1 y_{i_{n-r}}^{(1)}(t_f) + b_2 y_{i_{n-r}}^{(2)}(t_f) + \cdots + b_r y_{i_{n-r}}^{(r)}(t_f) + b_{r+1} y_{i_{n-r}}^{(r+1)}(t_f) + \cdots + b_n y_{i_{n-r}}^{(n)}(t_f) &= c_{i_{n-r}},
\end{aligned}
\tag{4.2.5}
$$

which is a system of n linear algebraic equations in the n unknowns $b_1, b_2, \ldots, b_n$. If this system can be solved for the b_i, the solution of the two-point boundary problem will be given by (4.2.4).

For suitably chosen initial conditions, (4.2.5) can be simplified. For example, if for the initial conditions in the ith equation we choose $y_i^{(i)}(t_0) = 1$ and $y_i^{(j)}(t_0) = 0$, $i \neq j$, then $b_i = c_i$, $i = 1, 2, \ldots, r$, and (4.2.5) reduces to

$$\begin{aligned}
\sum_{i=r+1}^{n} b_i y_{i_1}^{(i)}(t_f) &= c_{i_1} - \sum_{i=1}^{r} b_i y_{i_1}^{(i)}(t_f), \\
\sum_{i=r+1}^{n} b_i y_{i_2}^{(i)}(t_f) &= c_{i_2} - \sum_{i=1}^{r} b_i y_{i_2}^{(i)}(t_f), \\
&\vdots \\
\sum_{i=r+1}^{n} b_i y_{i_{n-r}}^{(i)}(t_f) &= c_{i_{n-r}} - \sum_{i=1}^{r} b_i y_{i_{n-r}}^{(i)}(t_f).
\end{aligned} \tag{4.2.6}$$

This is a set of only $n-r$ algebraic equations in the $n-r$ variables b_i, $i = r+1, \ldots, n$. Once these have been found, all the b_i, $i = 1, 2, \ldots, n$, are known, and again the solution to the two-point boundary value problem can be represented over the interval $[t_0, t_f]$ by (4.2.4). Note, however, that n integrations of the initial value problem given by (4.2.1) are still necessary, first to form the terms $y_{i_m}^{(i)}(t_f)$, $m = 1, 2, \ldots, n-r$, $i = 1, 2, \ldots, n$ in (4.2.6), and then to form the representation (4.2.4).

Miele [4, 5] in his "method of particular solutions" has shown how the number of integrations of (4.2.1) may be reduced to $n-r+1$. In this method, Eqs. (4.2.1) are integrated $n-r+1$ times from t_0 to t_f with initial vectors $\mathbf{y}^{(k)}(t_0)$ such that the first r components are

$$y_i^{(k)}(t_0) = c_i, \qquad i = 1, 2, \ldots, r, \qquad k = 1, 2, \ldots, n-r+1 \tag{4.2.7}$$

and such that the vectors are linearly independent. A suitable choice for the remaining components is

$$y_{r+j}^{(k)}(t_0) = \delta_{jk}, \qquad j = 1, 2, \ldots, n-r, \qquad k = 1, 2, \ldots, n-r+1, \tag{4.2.8}$$

where δ_{jk} is the Kronecker delta. The solutions $\mathbf{y}^{(k)}(t)$, $k = 1, 2, \ldots, n-r+1$, are stored.

Then, similarly to (4.2.4), the solution of the two-point boundary value problem is represented by the linear combination

$$\mathbf{y}(t) = \sum_{k=1}^{n-r+1} b_k \mathbf{y}^{(k)}(t). \tag{4.2.9}$$

Substituting this expression and its derivative in (4.2.1) gives

$$\sum_{k=1}^{n-r+1} b_k[\dot{\mathbf{y}}^{(k)} - \mathbf{A}\mathbf{y}^{(k)}] = \mathbf{f} \tag{4.2.10}$$

or, since the $\mathbf{y}^{(k)}(t)$ satisfy (4.2.1),

$$\sum_{k=1}^{n-r+1} b_k \mathbf{f} = \mathbf{f}, \tag{4.2.11}$$

which will be satisfied if

$$\sum_{k=1}^{n-r+1} b_k = 1. \tag{4.2.12}$$

Substituting our representation of $\mathbf{y}(t)$, Eq. (4.2.9), in the initial conditions (4.2.2) gives the same equation. Finally, substituting our representation of $y(t)$ in the final conditions, Eq. (4.2.3), gives the $n-r$ equations

$$\sum_{k=1}^{n-r+1} b_k y_{i_m}^{(k)}(t_f) = c_{i_m}, \qquad m = 1, 2, \ldots, n-r. \tag{4.2.13}$$

With (4.2.12) and (4.2.13), we now have $n-r+1$ linear algebraic equations for the $n-r+1$ coefficients b_k, $k = 1, 2, \ldots, n-r+1$, and, once this set of equations has been solved, the solution $\mathbf{y}(t)$ of the two-point boundary value problem can be found from the linear combination

$$\mathbf{y}(t) = \sum_{k=1}^{n-r+1} b_k \mathbf{y}^{(k)}(t), \qquad t_0 \leq t \leq t_f. \tag{4.2.14}$$

Another way to solve the linear two-point boundary value problem given by (4.2.1), (4.2.2), and (4.2.3), which is closely related to the method of particular solutions, is to form the sum of n linearly independent solutions of the homogeneous equation corresponding to (4.2.1) and one solution of (4.2.1). The n coefficients in the linear combinations are then determined by the requirement that the boundary conditions (4.2.2) and (4.2.3) be satisfied. Let the homogeneous system of equations be written as

$$\dot{\mathbf{u}}^{(j)} = \mathbf{A}(t)\mathbf{u}^{(j)}(t), \tag{4.2.15}$$

where

$\mathbf{u}^{(j)}(t) = n \times 1$ vector, jth solution of the homogeneous equation, with components $u_1^{(j)}(t), u_2^{(j)}(t), \ldots, u_n^{(j)}(t), j = 1, 2, \ldots, n$.

By choosing n sets of linearly independent initial vectors for (4.2.15) and by integrating (4.2.15) forward, once with each set of initial conditions, we obtain n linearly independent solutions $\mathbf{u}^{(j)}(t)$, $j = 1, 2, \ldots, n$; $t_0 \leq t \leq t_f$, which are saved.

The particular solution $\mathbf{v}(t)$ is found by integrating forward one time

$$\dot{\mathbf{v}} = \mathbf{A}(t)\mathbf{v} + \mathbf{f}(t) \tag{4.2.16}$$

with arbitrary initial conditions. The particular solution is saved. The general solution to (4.2.1) is found by a linear combination of the homogeneous solutions plus the particular solution

$$y_i(t) = \sum_{j=1}^{n} b_j u_i^{(j)}(t) + v_i(t), \qquad i = 1,2,\ldots,n, \qquad t_0 \le t \le t_f, \tag{4.2.17}$$

where the constants b_j, $j = 1,2,\ldots,n$, are to be determined from the boundary conditions. On employing the initial and terminal conditions, we write out (4.2.17) as

$$\begin{aligned}
b_1 u_1^{(1)}(t_0) + b_2 u_1^{(2)}(t_0) + \ldots + b_n u_1^{(n)}(t_0) + v_1(t_0) &= c_1, \\
b_1 u_2^{(1)}(t_0) + b_2 u_2^{(2)}(t_0) + \ldots + b_n u_2^{(n)}(t_0) + v_2(t_0) &= c_2, \\
\vdots \qquad\qquad \vdots \qquad\qquad\qquad \vdots \qquad\qquad \vdots \quad & \\
b_1 u_r^{(1)}(t_0) + b_2 u_r^{(2)}(t_0) + \ldots + b_n u_r^{(n)}(t_0) + v_r(t_0) &= c_r, \\
b_1 u_{i_1}^{(1)}(t_f) + b_2 u_{i_1}^{(2)}(t_f) + \ldots + b_n u_{i_1}^{(n)}(t_f) + v_{i_1}(t_f) &= c_{i_1}, \\
\vdots \qquad\qquad \vdots \qquad\qquad\qquad \vdots \qquad\qquad \vdots \quad & \quad \vdots \\
b_1 u_{i_{n-r}}^{(1)}(t_f) + b_2 u_{i_{n-r}}^{(2)}(t_f) + \ldots + b_n u_{i_{n-r}}^{(n)}(t_f) + v_{i_{n-r}}(t_f) &= c_{i_{n-r}}.
\end{aligned} \tag{4.2.18}$$

This is a set of n linear algebraic equations in the n unknowns b_j, $j = 1,2,\ldots,n$. Once the b_j's are obtained, the solution to the two-point boundary value problem is found by using (4.2.17), since $\mathbf{u}^{(k)}(t)$ and $\mathbf{v}(t)$ profiles have been saved.

It might be noted at this point that, if it should be desirable in a particular application to trade computer storage for computing time, only the initial and final values of $\mathbf{y}^{(k)}(t)$ in the method of particular solutions or of $\mathbf{u}^{(k)}(t)$ and $\mathbf{v}(t)$ in the method under discussion need be saved. The b_j can still be solved for, since the linear algebraic equations which determine them have as coefficients only the initial and final values just referred to. Once the b_j are known, (4.2.14) or (4.2.17), can be used to determine the missing initial conditions $y_i(t_0)$, $i = r+1,\ldots,n$, and then an integration of (4.2.1) with the complete set of initial conditions gives the solution of the two-point boundary value problem.

Returning to the method described in (4.2.15) through (4.2.18), it has a potential advantage of speed over the method of particular solutions, since integration of the homogeneous equations (4.2.15) should be faster than integration of (4.2.1). However, in its present formulation the method requires $n+1$ integrations rather than $n-r+1$, as in the method of particular solutions. As might be anticipated, the proper choice of initial conditions $u_i^{(k)}(t_0)$ and $v_i(t_0)$ reduces to $n-r$ the number of integrations of (4.2.15) needed, and thus the total number to $n-r+1$. This choice of initial conditions for (4.2.15) and (4.2.16) leads to the "method of complementary functions", which is discussed in detail in Section 4.3.

4.3. METHOD OF COMPLEMENTARY FUNCTIONS [1]

The method of complementary functions is essentially the last method described in Section 4.2 with the initial conditions chosen for the homogeneous equations and the particular equation in such a way that only $(n-r)$ integrations of the homogeneous system rather than n are required.

We develop in detail the method of complementary functions and then summarize the results. We consider once again the set of n linear ordinary equations with variable coefficients (4.2.1), the initial boundary conditions (4.2.2), and the terminal boundary conditions (4.2.3).

For the homogeneous equations (4.2.15), we select the Kronecker delta initial conditions

$$u_i^{(j)}(t_0) = \left.\begin{matrix} 1, & i=j \\ 0, & i \neq j \end{matrix}\right\}, \qquad i,j = 1,2,\ldots,n, \tag{4.3.1}$$

where

$u_i^{(j)}(t)$ = ith component of the homogeneous solution for the jth set of initial conditions.

Suppose the homogeneous equations are integrated forward n times from t_0 to t_f using the n Kronecker delta initial conditions in (4.3.1), and the homogeneous solutions are saved.

Now suppose the particular solution is found by integrating forward (4.2.16) from t_0 to t_f with the initial conditions

$$\begin{aligned} v_i(t_0) &= y_i(t_0) = c_i, \qquad & i &= 1,2,\ldots,r, \\ v_i(t_0) &= 0, & i &= r+1,\ldots,n. \end{aligned} \tag{4.3.2}$$

We observe that the first set of equations (4.3.2) are the initial boundary conditions on $y_i(t)$, $i = 1,2,\ldots,r$ in (4.2.2). The particular solution is also saved.

The general solution to the system (4.2.1) is found as a linear combination of the n homogeneous solutions plus the particular solution

$$y_i(t) = \sum_{j=1}^{n} b_j u_i^{(j)}(t) + v_i(t), \qquad i = 1,2,\ldots,n, \qquad t_0 \leq t \leq t_f, \tag{4.3.3}$$

where the b_j constants are to be determined by the boundary conditions. In component notation, Eqs. (4.3.3) appear as

$$
\begin{aligned}
y_1(t) &= b_1 u_1^{(1)}(t) + b_2 u_1^{(2)}(t) + \cdots + b_r u_1^{(r)}(t) + b_{r+1} u_1^{(r+1)}(t) + \cdots + b_n u_1^{(n)}(t) + v_1(t),\\
y_2(t) &= b_1 u_2^{(1)}(t) + b_2 u_2^{(2)}(t) + \cdots + b_r u_2^{(r)}(t) + b_{r+1} u_2^{(r+1)}(t) + \cdots + b_n u_2^{(n)}(t) + v_2(t),\\
&\vdots\\
y_r(t) &= b_1 u_r^{(1)}(t) + b_2 u_r^{(2)}(t) + \cdots + b_r u_r^{(r)}(t) + b_{r+1} u_r^{(r+1)}(t) + \cdots + b_n u_r^{(n)}(t) + v_r(t),\\
y_{r+1}(t) &= b_1 u_{r+1}^{(1)}(t) + b_2 u_{r+1}^{(2)}(t) + \cdots + b_r u_{r+1}^{(r)}(t) + b_{r+1} u_{r+1}^{(r+1)}(t) + \cdots + b_n u_{r+1}^{(n)}(t) + v_{r+1}(t),\\
&\vdots\\
y_n(t) &= b_1 u_n^{(1)}(t) + b_2 u_n^{(2)}(t) + \cdots + b_r u_n^{(r)}(t) + b_{r+1} u_n^{(r+1)}(t) + \cdots + b_n u_n^{(n)}(t) + v_n(t).
\end{aligned}
\tag{4.3.4}
$$

Now for the system (4.3.4) at the initial point, all the off-diagonal terms $u_i^{(j)}(t_0) = 0$, $i \neq j$, by (4.3.1). For the first r equations at the initial point we have, by virtue of (4.3.1),

$$y_i(t_0) = b_i u_i^{(i)}(t_0) + v_i(t_0) = b_i + v_i(t_0), \qquad i = 1, 2, \ldots, r. \quad (4.3.5)$$

Since, by (4.3.2), $v_i(t_0) = y_i(t_0) = c_i$, $i = 1, 2, \ldots, r$, we conclude that

$$b_i = 0, \qquad i = 1, 2, \ldots, r. \quad (4.3.6)$$

In addition, for the equations $i = r+1, \ldots, n$ in system (4.3.4) the $v_i(t_0) = 0$, $i = r+1, \ldots n$ by (4.3.2). We conclude therefore that

$$b_{r+i} = y_{r+i}(t_0), \qquad i = 1, 2, \ldots, n-r. \quad (4.3.7)$$

In other words, the constants b_{r+i}, $i = 1, 2, \ldots, n-r$ equal the missing initial conditions.

To determine the constants b_{r+i}, we select from (4.3.4) those $(n-r)$ equations for which terminal values $y_{i_m}(t_f)$ have been specified by the boundary conditions (4.2.3):

$$\begin{aligned}
y_{i_1}(t_f) &= b_{r+1}u_{i_1}^{(r+1)}(t_f) + b_{r+2}u_{i_1}^{(r+2)}(t_f) + \ldots + b_n u_{i_1}^{(n)}(t_f) + v_{i_1}(t_f),\\
y_{i_2}(t_f) &= b_{r+1}u_{i_2}^{(r+1)}(t_f) + b_{r+2}u_{i_2}^{(r+2)}(t_f) + \ldots + b_n u_{i_2}^{(n)}(t_f) + v_{i_2}(t_f),\\
&\vdots\\
y_{i_{n-r}}(t_f) &= b_{r+1}u_{i_{n-r}}^{(r+1)}(t_f) + b_{r+2}u_{i_{n-r}}^{(r+2)}(t_f) + \ldots + b_n u_{i_{n-r}}^{(n)}(t_f) + v_{i_{n-r}}(t_f).
\end{aligned} \quad (4.3.8)$$

In (4.3.8) the $y_{i_m}(t_f)$, $m = 1, 2, \ldots, n-r$, are known from the boundary conditions, the $u_{i_m}^{(r+j)}(t_f)$, $m = 1, 2, \ldots, n-r$, $j = 1, 2, \ldots, n-r$, are known from integrating the homogeneous equations, the $v_{i_m}(t_f)$, $m = 1, 2, \ldots, n-r$, are known from integrating the particular equation. Equations (4.3.8) are, therefore, a system of $(n-r)$ equations in the $(n-r)$ unknowns b_{r+j}, $j = 1, 2, \ldots, n-r$. The solution of (4.3.8) may be expressed as

$$\begin{bmatrix} b_{r+1} \\ b_{r+2} \\ \vdots \\ b_n \end{bmatrix} = \begin{bmatrix} u_{i_1}^{(r+1)}(t_f) & u_{i_1}^{(r+2)}(t_f) & \ldots & u_{i_1}^{(n)}(t_f) \\ u_{i_2}^{(r+1)}(t_f) & u_{i_2}^{(r+2)}(t_f) & \ldots & u_{i_2}^{(n)}(t_f) \\ \vdots & \vdots & & \vdots \\ u_{i_{n-r}}^{(r+1)}(t_f) & u_{i_{n-r}}^{(r+2)}(t_f) & \ldots & u_{i_{n-r}}^{(n)}(t_f) \end{bmatrix}^{-1} \begin{bmatrix} y_{i_1}(t_f) - v_{i_1}(t_f) \\ y_{i_2}(t_f) - v_{i_2}(t_f) \\ \vdots \\ y_{i_{n-r}}(t_f) - v_{i_{n-r}}(t_f) \end{bmatrix} \quad (4.3.9)$$

provided the inverse exists. Once the b_{r+j}, $j = 1, 2, \ldots, n-r$, are found, the missing initial conditions are known by (4.3.7). The general solution to (4.2.1) may be obtained by forming

$$y_i(t) = \sum_{j=1}^{n-r} b_{r+j} u_i^{(r+j)}(t) + v_i(t), \qquad i = 1, 2, \ldots, n, \tag{4.3.10}$$

since the solutions $u_i^{(r+j)}(t)$, $i = 1, 2, \ldots, n$, $j = 1, 2, \ldots, n-r$; $v_i(t)$, $i = 1, 2, \ldots, n$; $t_0 \leq t \leq t_f$; have been saved previously.

If we reexamine the development so far, we observe that in (4.3.10) only the solutions of the homogeneous equations $\mathbf{u}^{(r+j)}(t)$, $j = 1, 2, \ldots n-r$, are required. Instead of solving the homogeneous equations n times with the boundary conditions (4.3.1), we need to solve them only $(n-r)$ times with the initial conditions

$$u_i^{(m)}(t_0) = \left.\begin{matrix} 1, & i = r+m \\ 0, & i \neq r+m \end{matrix}\right\}, \qquad m = 1, 2, \ldots, n-r, \tag{4.3.11}$$

where r is the number of specified initial conditions on $y_i(t_0)$ in (4.2.2).

In view of this, (4.3.8), (4.3.9), and (4.3.10) may be written, respectively, as

$$y_{i_m}(t_f) = \sum_{j=1}^{n-r} \beta_j u_{i_m}^{(j)}(t_f) + v_{i_m}(t_f), \qquad m = 1, 2, \ldots, n-r, \tag{4.3.8a}$$

$$\begin{bmatrix} \beta_1 \\ \beta_2 \\ \vdots \\ \beta_{n-r} \end{bmatrix} = \begin{bmatrix} u_{i_1}^{(1)}(t_f) & u_{i_1}^{(2)}(t_f) & \cdots & u_{i_1}^{(n-r)}(t_f) \\ u_{i_2}^{(1)}(t_f) & u_{i_2}^{(2)}(t_f) & \cdots & u_{i_2}^{(n-r)}(t_f) \\ \vdots & \vdots & & \vdots \\ u_{i_{n-r}}^{(1)}(t_f) & u_{i_{n-r}}^{(2)}(t_f) & \cdots & u_{i_{n-r}}^{(n-r)}(t_f) \end{bmatrix}^{-1} \begin{bmatrix} y_{i_1}(t_f) - v_{i_1}(t_f) \\ y_{i_2}(t_f) - v_{i_2}(t_f) \\ \vdots \\ y_{i_{n-r}}(t_f) - v_{i_{n-r}}(t_f) \end{bmatrix}, \tag{4.3.9a}$$

$$y_i(t) = \sum_{j=1}^{n-r} \beta_j u_i^{(j)}(t) + v_i(t), \qquad i = 1, 2, \ldots, n, \tag{4.3.10a}$$

where

$$\beta_j = b_{r+j}, \qquad j = 1, 2, \ldots, n-r. \tag{4.3.12}$$

This, then, is the method of complementary functions, which is carried out as follows:

1. Integrate the homogeneous equations (4.2.15) $(n-r)$ times with the initial conditions (4.3.11). Save the profiles $u_i^{(m)}(t)$, $i = 1, 2, \ldots, n$, $m = 1, 2, \ldots n-r$, $t_0 \leq t \leq t_f$.

2. Integrate the particular equation (4.2.16) once with the initial conditions (4.3.2). Save the solution $v_i(t)$, $i = 1, 2, \ldots, n$; $t_0 \leq t \leq t_f$.

3. Solve the system of $(n-r)$ linear algebraic equations (4.3.9) or (4.3.9a) for the $(n-r)$ unknowns $\beta_j = b_{r+j}$, $j = 1, 2, \ldots, n-r$.

4. The general solution to the linear two-point boundary value problem is found from (4.3.10) or (4.3.10a).

4.4. EXAMPLE

To illustrate in detail the method of complementary functions we consider the example in Section 3.4, Chapter 3.

We are given a system of five linear ordinary differential equations

$$\dot{\mathbf{y}} = \mathbf{A}(t)\mathbf{y} + \mathbf{f}(t),$$

where

$$\mathbf{y}(t) = 5 \times 1 \text{ vector with components } y_1, y_2, \ldots, y_5,$$
$$\mathbf{f}(t) = 5 \times 1 \text{ vector with components } f_1, f_2, \ldots, f_5,$$
$$\mathbf{A}(t) = 5 \times 5 \text{ matrix with elements } a_{ij}(t),\ i, j = 1, 2, \ldots, 5,$$

with the initial boundary conditions

$$y_1(t_0) = c_1, \qquad y_2(t_0) = c_2, \qquad y_3(t_0) = c_3,$$

and the terminal boundary conditions

$$y_3(t_f) = c_{i_1}, \qquad y_4(t_f) = c_{i_2}.$$

As before, $n = 5$ equations, $r = 3$ initial conditions, $n - r = 2$ terminal conditions.

The initial conditions for the first integration of the homogeneous equations, using the notation of (4.3.11), are

$$u_1^{(1)}(t_0) = 0,$$
$$u_2^{(1)}(t_0) = 0,$$
$$u_3^{(1)}(t_0) = 0,$$
$$u_4^{(1)}(t_0) = 1,$$
$$u_5^{(1)}(t_0) = 0,$$

and the initial conditions for the second integration of the homogeneous conditions are

$$u_1^{(2)}(t_0) = 0,$$

$$u_2^{(2)}(t_0) = 0,$$

$$u_3^{(2)}(t_0) = 0,$$

$$u_4^{(2)}(t_0) = 0,$$

$$u_5^{(2)}(t_0) = 1.$$

The particular solution is found by integrating forward with the initial conditions

$$v_1(t_0) = c_1,$$

$$v_2(t_0) = c_2,$$

$$v_3(t_0) = c_3,$$

$$v_4(t_0) = 0,$$

$$v_5(t_0) = 0.$$

Equations (4.3.9a) appear as

$$\begin{pmatrix} \beta_1 \\ \beta_2 \end{pmatrix} = \begin{pmatrix} u_3^{(1)}(t_f) & u_3^{(2)}(t_f) \\ u_4^{(1)}(t_f) & u_4^{(2)}(t_f) \end{pmatrix}^{-1} \begin{pmatrix} y_3(t_f) - v_3(t_f) \\ y_4(t_f) - v_4(t_f) \end{pmatrix}.$$

On solving for β_1 and β_2, the missing initial conditions are determined since $\beta_1 = y_4(t_0)$ and $\beta_2 = y_5(t_0)$. The general solution can be found from

$$y_i(t) = \sum_{j=1}^{2} \beta_j u_i^{(j)}(t) + v_i(t), \qquad i = 1, 2, \ldots, 5.$$

4.5. DISCUSSION

As described in the theory of linear ordinary differential equations, the method of complementary functions forms the general solution to (4.2.1) by a linear combination of n linearly independent solutions of the homogeneous equations plus one solution to the particular equation. This is the third method given in Section 4.2, as applied to the two-point boundary value problem. The term "method of complementary functions" employed in this book specializes the initial conditions on the homogeneous equations to the Kronecker delta conditions (4.3.11).

As shown in Section 4.3, the n linearly independent Kronecker delta initial conditions on the homogeneous equations (4.3.1) can be reduced to $(n - r)$ Kronecker delta conditions (4.3.11). The homogeneous equations, therefore, are integrated $(n - r)$ times rather than n times, and $(n - r)$ constants are determined from $(n - r)$ linear algebraic equations rather than n constants from n linear algebraic equations. This is an important reduction in work.

The method of complementary functions requires $(n - r)$ integrations of the homogeneous equations and one integration of the particular equation for a total of $(n-r+1)$ integrations. In addition, a system of $(n - r)$ linear algebraic equations must be solved. The method is not iterative. One pass through the process finds the set of missing initial conditions.

To generate the general solution to the two-point boundary value problem by the method of complementary functions, we can use (4.3.10) or (4.3.10a). This requires that we have previously stored the profiles for the homogeneous and particular solutions. An alternative method of solution is to store only the terminal values $u_{i_m}^{(r+j)}(t_f)$, $m = 1, 2, \ldots, n - r$; $j = 1, 2, \ldots, n - r$; and $v_{i_m}(t_f)$, $m = 1, 2, \ldots, n - r$. This gives sufficient information to solve (4.3.9) for $\beta_j = b_{r+j}$, $j = 1, 2, \ldots, n - r$. Since $\beta_j = b_{r+j} = y_{r+j}(t_0)$, $j = 1, 2, \ldots, n - r$, the set of missing initial conditions, we may integrate forward (4.2.1) with the initial conditions $y_i(t_0) = c_i$, $i = 1, 2, \ldots, r$, and $y_{r+j}(t_0) = b_{r+j}$, $j = 1, 2, \ldots, n - r$. In the alternative method we exchange time for space. By relinquishing storage for the profiles of the homogeneous and particular solutions, we must carry out one more integration.

The method of particular solutions treated in Section 4.2 is also noniterative, and also requires $n-r+1$ integrations, but it is the original system of equations, (4.2.1), which is integrated each time. Since, in the general case, Eqs. (4.2.1) are inhomogeneous, $n-r+1$ integrations should take more computing time than the $n - r$ integrations of the corresponding homogeneous equations (4.2.15) and one integration of (4.2.1) or (4.2.16). Furthermore, the constants b_j, $j = r + 1, \ldots, n$ in the method of complementary functions are actually the missing initial conditions $y_{r+j}(t_0)$, while the b_j's in the method of particular solutions do not have this significance. For these reasons, the method of complementary functions would appear to be the more efficient of the two schemes, and accordingly is the one we have used in our numerical work. That the method of particular solutions uses only the original equations (4.2.1), while both the method of complementary functions and the method of adjoints need equations derived from (4.2.1), as Miele [4, 5] has pointed out, does not seem to us to be significant, since the required derivations are almost trivial. Incidentally, Miele [4, 5] exhibits the relation between the method of particular solutions and the method of adjoints.

The method of adjoints has also been used extensively by us and by other investigators in numerical work, and it is therefore of interest to compare it with the method of complementary functions. Both methods are noniterative for linear problems, both require $n-r+1$ integrations of systems of differential equations, and both require the solution of a set of $n-r$ algebraic equations to find the missing initial conditions.

In the following sections we show that both methods may be derived from a common point of departure and that both methods develop the identical set of linear algebraic equations from which the initial conditions are found. Then we compare the two methods with respect to accuracy, assuming that each is to be used to solve a given two-point boundary value problem.

4.6. ALTERNATIVE DERIVATIONS, METHODS OF ADJOINTS AND COMPLEMENTARY FUNCTIONS

In this section we supply first motivation for and an alternative derivation of the method of complementary functions. Next we give an alternative derivation of the method of adjoints. Both alternative derivations have a common point of departure. Each of the methods leads to a system of linear algebraic equations for the missing initial conditions of the two-point boundary value problem, which equations we show to be theoretically identical, although they appear to be different. The two methods are procedurally different, however; this raises the question of whether one or the other is to be preferred in applications. Our numerical experience is that, once a simple modification, which we describe, is made in the programming of the method of adjoints, neither method is preferred to the other as far as accuracy attained is concerned.

Let us consider the system of n linear ordinary differential equations (4.2.1) with the boundary conditions (4.2.2) and (4.2.3). The general solution of (4.2.1) is given by

$$\mathbf{y}(t) = \mathbf{U}(t)\mathbf{y}(t_0)+\int_{t_0}^{t_f} \mathbf{U}(t)\mathbf{U}(s)^{-1}\mathbf{f}(s)\,ds, \tag{4.6.1}$$

where

$\mathbf{y}(t) = n\times 1$ vector with components $y_1(t), y_2(t), \ldots, y_n(t)$,
$\mathbf{f}(t) = n\times 1$ vector with components $f_1(t), f_2(t), \ldots, f_n(t)$,
$\mathbf{U}(t) = n\times n$ matrix solution of the homogeneous matrix-matrix equation (4.6.2),

$$\dot{\mathbf{U}} = \mathbf{A}(t)\mathbf{U}, \quad \mathbf{U}(t_0) = \mathbf{I}, \qquad \text{the identity matrix, } n\times n. \tag{4.6.2}$$

The matrix $\mathbf{U}(t)$ is nonsingular over any interval $[t_0, t_f]$ over which $\|\mathbf{A}(t)\|$ is integrable.

The second term on the right-hand side of (4.6.1) is that particular solution $\mathbf{w}(t)$, $n \times 1$ vector with components $w_i(t)$, $i = 1, 2, \ldots, n$, of the differential equation (4.2.1) with the initial conditions $w_i(t_0) = 0$, $i = 1, 2, \ldots, n$.

Equation (4.6.1) is the common point of departure for the alternative derivations of the method of adjoints and complementary functions.

4.6.1. Method of Complementary Functions

To carry out the solution of the matrix-matrix equation (4.6.2) we integrate the matrix-vector equation

$$\dot{\mathbf{u}} = \mathbf{A}(t)\mathbf{u} \tag{4.6.3}$$

n times with the Kronecker delta initial conditions

$$u_i^{(k)}(t_0) = \left.\begin{matrix} 1, & i = k \\ 0, & i \neq k \end{matrix}\right\}, \qquad i, k = 1, 2, \ldots, n, \tag{4.6.4}$$

where

$\mathbf{u}^{(k)}(t) = n \times 1$ vector for the kth integration of (4.6.3) with the components $u_1^{(k)}(t), u_2^{(k)}(t), \ldots, u_n^{(k)}(t)$. $\mathbf{u}^{(k)}(t)$ is the kth column of $\mathbf{U}(t)$.

Having obtained the n solutions to the homogeneous equation associated with (4.2.1), we obtain numerically a particular solution to (4.2.1) with the initial conditions $\mathbf{y}(t_0) = \mathbf{w}(t_0) = 0$. The general solution of (4.6.1) may be expressed in component form as

$$y_j(t) = \sum_{k=1}^{n} u_j^{(k)}(t) y_k(t_0) + w_j(t), \qquad j = 1, 2, \ldots, n. \tag{4.6.5}$$

From the total of $(n + 1)$ integrations which were required to generate the set (4.6.5), $(n - r)$ linear algebraic equations for the $(n - r)$ missing initial conditions can be found from (4.6.5) by choosing those components for which the final values of $y_{i_m}(t_f)$ are prescribed.

The method of complementary functions exploits the fact that, if the right-hand side of (4.6.5) is rearranged as

$$y_j(t) = \sum_{k=r+1}^{n} u_j^{(k)}(t) y_k(t_0) + \sum_{k=1}^{r} u_j^{(k)}(t) y_k(t_0) + w_j(t), \qquad j = 1, 2, \ldots, n, \tag{4.6.6}$$

then the expression

$$v_j(t) = \sum_{k=1}^{r} u_j^{(k)}(t) y_k(t_0) + w_j(t) \tag{4.6.7}$$

is simply the jth component of the solution $\mathbf{v}(t)$ of the inhomogeneous system (4.2.1) with the initial conditions

$$\begin{aligned} v_i(t_0) &= y_i(t_0), \qquad i = 1, 2, \ldots, r, \\ v_i(t_0) &= 0, \qquad\qquad i = r+1, \ldots, n. \end{aligned} \tag{4.6.8}$$

The general solution (4.6.6) may be written as

$$y_j(t) = \sum_{k=r+1}^{n} u_j^{(k)}(t) y_k(t_0) + v_j(t), \qquad j = 1, 2, \ldots, n, \tag{4.6.9}$$

where Eqs. (4.6.3) are integrated only $(n-r)$ times with the initial conditions

$$u_j^{(k)}(t_0) = \left.\begin{matrix} 1, & j = r+k \\ 0, & j \neq r+k \end{matrix}\right\}, \qquad k = 1, 2, \ldots, n-r, \tag{4.6.10}$$

and the particular solution $\mathbf{v}(t)$ is integrated once with the initial conditions (4.6.8) for a total of $(n-r+1)$ integrations. Thus it is not necessary to integrate (4.6.3) n times in order to determine the missing initial conditions from the set of algebraic equations (4.6.5).

Starting with the general solution to (4.2.1) given by (4.6.1), we have derived the method of complementary functions and have supplied the motivation for the choice of the initial conditions given by Goodman and Lance [1].

4.6.2. Method of Adjoints

Let us now consider the method of adjoints. The equations adjoint to (4.2.1) are

$$\dot{\mathbf{x}} = -\mathbf{A}^T(t)\mathbf{x}(t), \tag{4.6.11}$$

where $\mathbf{x}(t)$ is a $n \times 1$ vector with components $x_i(t)$, $i = 1, 2, \ldots, n$, and $\mathbf{A}^T(t)$ indicates the transpose of the matrix $\mathbf{A}(t)$. In view of (4.6.2), the solution of (4.6.11) is given by

$$\mathbf{x}(t) = \mathbf{U}(t)^{-T}\mathbf{x}(t_0). \tag{4.6.12}$$

If (4.6.1) is premultiplied by $\mathbf{x}(t)^T$, we obtain

$$\mathbf{x}(t)^T\mathbf{y}(t) = \mathbf{x}(t)^T\mathbf{U}(t)\mathbf{y}(t_0)+\mathbf{x}(t)^T\int_{t_0}^{t}\mathbf{U}(t)\mathbf{U}(s)^{-1}\mathbf{f}(s)\,ds. \quad (4.6.13)$$

Substituting the transpose of (4.6.12) in (4.6.13) yields

$$\mathbf{x}(t)^T\mathbf{y}(t) = \mathbf{x}(t_0)^T\mathbf{y}(t_0)+\int_{t_0}^{t}\mathbf{x}(t)^T\mathbf{U}(t)\mathbf{U}(s)^{-1}\mathbf{f}(s)\,ds, \quad (4.6.14)$$

where $\mathbf{x}(t)^T$ has been brought in under the integral sign, since it is not a function of s. Using (4.6.12), a relationship may be found between $\mathbf{x}(s)$ and $\mathbf{x}(t)$;

$$\mathbf{x}(s) = \mathbf{U}(s)^{-T}\,\mathbf{U}(t)^T\mathbf{x}(t). \quad (4.6.15)$$

Substituting the transpose of (4.6.15) in (4.6.14) gives

$$\mathbf{x}(t)^T\mathbf{y}(t) = \mathbf{x}(t_0)^T\mathbf{y}(t_0)+\int_{t_0}^{t}\mathbf{x}(s)^T\mathbf{f}(s)\,ds. \quad (4.6.16)$$

Expressing (4.6.16) in component form with t set equal to t_f, we have

$$\sum_{i=1}^{n} x_i(t_f)y_i(t_f)-\sum_{i=1}^{n} x_i(t_0)y_i(t_0) = \int_{t_0}^{t_f}\sum_{i=1}^{n} x_i(s)f_i(s)\,ds, \quad (4.6.17)$$

which is the basic identity of the method of adjoints. See Section 3.2, Chapter 3.

To solve the two-point boundary value problem by this method, the adjoint equations are integrated backward $(n-r)$ times from t_f to t_0, with the terminal conditions

$$x_i^{(m)}(t_f) = \left.\begin{matrix} 1, & i = i_m \\ 0, & i \neq i_m \end{matrix}\right\}, \qquad m = 1,2,\ldots,n-r, \quad (4.6.18)$$

where i_m are the subscripts on the specified terminal conditions in (4.2.3). The result is a set of $(n-r)$ linear algebraic equations.

$$\sum_{i=r+1}^{n} x_i^{(m)}(t_0)y_i(t_0) = y_{i_m}(t_f)-\sum_{i=1}^{r} x_i^{(m)}(t_0)y_i(t_0)$$

$$-\int_{t_0}^{t_f}\sum_{i=1}^{n} x_i^{(m)}(s)f_i(s)\,ds, \qquad m = 1,2,\ldots,n-r. \quad (4.6.19)$$

Since the right-hand side of (4.6.19) involves only the specified initial and terminal boundary conditions and the known $y_{i_m}(t_f)$, (4.6.19) may be solved for the set of missing initial conditions $y_i(t_0)$, $i = r+1, \ldots, n$. Once the set of missing initial conditions are found, one more integration of (4.2.1) with the complete set of initial conditions gives the desired solution of the two-point boundary value problem. A total of $(n-r+1)$ integrations are required. Thus we have derived the method of adjoints from (4.6.1), which is also the point of departure for the alternative derivation of the method of complementary functions.

4.7. COMPARISON OF THE TWO METHODS

We show first that the methods of adjoints and complementary functions are theoretically identical in the sense that they both lead to the same set of linear algebraic equations for the missing initial conditions.

For convenience we will write (4.6.6) and (4.6.19) in parallel fashion for $t = t_f$ for the set of specified terminal conditions $y_{i_m}(t_f)$, $m = 1, 2, \ldots, n-r$:

$$\sum_{k=r+1}^{n} u_{i_m}^{(k)}(t_f)y_k(t_0) = y_{i_m}(t_f) - \sum_{k=1}^{r} u_{i_m}^{(k)}(t_f)y_k(t_0) - w_{i_m}(t_f),$$

$$m = 1, 2, \ldots, n-r, \tag{4.7.1}$$

$$\sum_{i=r+1}^{n} x_i^{(m)}(t_0)y_i(t_0) = y_{i_m}(t_f) - \sum_{i=1}^{r} x_i^{(m)}(t_0)y_i(t_0)$$

$$- \int_{t_0}^{t_f} \sum_{i=1}^{n} x_i^{(m)}(s)f_i(s)\, ds, \qquad m = 1, 2, \ldots, n-r. \tag{4.7.2}$$

We now demonstrate that the coefficients of $y_i(t_0)$ in both equations are identical. This will imply that

$$w_{i_m}(t_f) = \int_{t_0}^{t_f} \sum_{i=1}^{n} x_i^{(m)}(s)f_i(s)\, ds, \qquad m = 1, 2, \ldots, n-r,$$

which we also show independently.

The matrix-matrix equation (4.6.2) may be written as n matrix-vector equations:

$$\dot{\mathbf{u}}^{(p)} = \mathbf{A}(t)\,\mathbf{u}^{(p)} \tag{4.7.3}$$

with the initial conditions

$$u_i^{(p)}(t_0) = \begin{matrix} 1, & i = p \\ 0, & i \neq p \end{matrix}\Bigg\}, \qquad i, p = 1, 2, \ldots, n. \tag{4.7.4}$$

Corresponding to (4.7.3) for each p is the adjoint system

$$\dot{\mathbf{x}}^{(m)} = -\mathbf{A}(t)^T \mathbf{x}^{(m)} \tag{4.7.5}$$

with the terminal conditions

$$x_i^{(m)}(t_f) = \begin{matrix} 1, & i = i_m \\ 0, & i \neq i_m \end{matrix}\Bigg\}, \qquad m = 1, 2, \ldots, n-r, \qquad i = 1, 2, \ldots, n. \tag{4.7.6}$$

By the fundamental identity of the method of adjoints we have

$$\sum_{i=1}^{n} x_i^{(m)}(t_f) u_i^{(p)}(t_f) = \sum_{i=1}^{n} x_i^{(m)}(t_0) u_i^{(p)}(t_0). \tag{4.7.7}$$

If we substitute in (4.7.7) the initial conditions for $u_i^{(p)}(t_0)$ from (4.7.4) and the terminal conditions for $x_i^{(m)}(t_f)$ from (4.7.6), we find that, for each p,

$$u_{i_m}^{(p)}(t_f) = x_p^{(m)}(t_0), \qquad p = 1, 2, \ldots, n, \qquad m = 1, 2, \ldots, n-r. \tag{4.7.8}$$

This equation demonstrates that the coefficients of $y_i(t_0)$, $i = 1, 2, \ldots, n$, in (4.7.1) and (4.7.3) are identical. Since the solution vector $\mathbf{y}(t_0)$ is identical for (4.7.1) and (4.7.2), Eq. (4.7.8) implies that

$$w_{i_m}(t_f) = \int_{t_0}^{t_f} \sum_{i=1}^{n} x_i^{(m)}(s) f_i(s)\, ds, \qquad m = 1, 2, \ldots, n-r. \tag{4.7.9}$$

This conclusion may be reached in an alternative way as shown below.

Since $\mathbf{w}(t)$ is a solution of (4.2.1) with the initial conditions $\mathbf{w}(t_0) = 0$, we can express (4.6.16) with $\mathbf{y}(t)$ replaced by $\mathbf{w}(t)$:

$$\mathbf{x}(t)^T \mathbf{w}(t) = \int_{t_0}^{t} \mathbf{x}(s)^T \mathbf{f}(s)\, ds, \tag{4.7.10}$$

where $\mathbf{x}(t_0)^T \mathbf{w}(t_0) = 0$ by the boundary conditions. In component form, (4.7.10) appears as

$$\sum_{i=1}^{n} x_i(t) w_i(t) = \int_{t_0}^{t} \sum_{i=1}^{n} x_i(s) f_i(s)\, ds. \tag{4.7.11}$$

In particular, for $t = t_f$ and for the mth integration of (4.7.5), we can write

$$\sum_{i=1}^{n} x_i^{(m)}(t_f)w_i(t_f) = w_{i_m}(t_f) = \int_{t_0}^{t_f} \sum_{i=1}^{n} x_i^{(m)}(s) f_i(s)\, ds. \qquad (4.7.12)$$

By virtue of (4.7.8) and (4.7.12) we have demonstrated that the set of linear algebraic equations generated by the method of complementary functions (4.7.1) and by the method of adjoints (4.7.2) are identical term by term. We conclude, therefore, that the method of adjoints and the method of complementary functions determine the identical set of linear algebraic equations from which the missing initial conditions are found.

Of course, the two methods are not procedurally identical. There is not only the matter of integrating one set of equations forward in one method as against integrating the adjoint set of equations backward in the other. There is also the fact that the method of complementary functions requires an $(n-r+1)$st integration of the original system (4.2.1) with a set of initial conditions which, in general, never occurs in the method of adjoints (that is, the particular solution) before the missing initial conditions can be found. This raises the question of whether one method or the other is to be preferred in practical applications.

4.8. PRACTICAL NUMERICAL EXPERIENCE

Among the criteria used to determine whether one numerical method is to be preferred to another are ease of programming, computation time, accuracy, and stability. With regard to the comparison of the method of adjoints and complementary functions, our belief, supported by that of experienced programmers, is that there is little difference between the ease (or difficulty) of programming the two methods, or their computation time. The computer program we use to solve numerically linear two-point boundary value problems by the method of complementary functions takes longer time than our program for the method of adjoints, since it employs an $(n-r+2)$nd integration of (4.2.1) with the initial conditions $\mathbf{y}(t_0)$ obtained from (4.3.9) to generate the solution $\mathbf{y}(t)$, rather than the summation (4.3.10). The ratio of computer times of the method of complementary functions to the method of adjoints is about $(n-r+2)/(n-r+1)$, the ratio of the number of integrations. This is the result of a pragmatic programming decision that an additional integration of (4.2.1) is preferable to storing the $(n-r)$ solution profiles $\mathbf{u}^{(j)}(t)$, $j = 1, 2, \ldots, n-r$.

Rather than carry out a complete error analysis of the two methods we attempted a comparison of the relative accuracy by means of numerical experiments. The two methods were applied to the same two-point boundary value problems, the criterion of accuracy being the difference between the computed and the specified terminal values. Our first calculations indicated that the method of complementary functions was more accurate in this sense. In the course of studying our method of adjoints program we found that the source of the (relative) inaccuracy was the numerical quadrature of

$$\int_{t_0}^{t_f} \sum_{i=1}^{n} x_i(s) f_i(s)\, ds,$$

which was difficult to perform with the same accuracy as the calculation of the $x_i(t)$ themselves. To overcome this difficulty a new variable $x_{n+1}(t)$ defined by the equation

$$x_{n+1}(t) = \int_{t_0}^{t_f} \sum_{i=1}^{n} x_i(s) f_i(s)\, ds \tag{4.8.1}$$

was introduced. Then an $(n+1)$st differential equation,

$$\dot{x}_{n+1}(t) = \sum_{i=1}^{n} x_i(t) f_i(t), \tag{4.8.2}$$

with the "initial" condition $x_{n+1}(t_f) = 0$, was integrated with the adjoint equations to determine $x_{n+1}(t)$ as the integral on the right-hand side of (4.6.19). This simple programming device improved the accuracy of the method of adjoints to the point where, in our test problems, one method seemed to give better accuracy about as many times as the other method.

With our FORTRAN programs used on the IBM 7094 computer, the elements $u_j^{(k)}(t_f)$, $x_j(t_0)$ and the right-hand sides of (4.7.1) and (4.7.2) agreed to 13 or more decimal digits in double precision calculations (16 digits). The solutions agreed to essentially the same precision. The question then arises: are the observed differences anything but "noise"?

Some further computations were performed to test by the t-test the null hypothesis that the difference in terminal errors between the two methods is zero. For a set of seven equations, linearized versions of the equations of Example 2, Section 7.4, Chapter 7, the simplest two-point boundary value problem was solved: one specified initial condition and one specified terminal condition. For each unspecified initial condition, a series of seven runs was made, each variable in turn being specified at the terminal point. Thus, for each choice of step size (which was varied to remove its influence as a possible source of difference between the two methods), 49 runs were made using a four-point Runge-Kutta integration formula. In all cases the null hypothesis was accepted.

In the problems we have dealt with there seems to be no real advantage in one method over the other. Despite this we believe there are problems for which one method is superior. For example, if a problem is stable on integrating forward but is unstable on integrating backward, then the method of complementary functions is preferred. For the reverse situation, the method of adjoints is preferred.

4.9. NUMERICAL LINEAR INDEPENDENCE

In both of the practical shooting techniques discussed, the method of adjoints and the method of complementary functions (and, it turns out, also quasilinearization, Chapter 5), a set of linear equations must be solved to find either the corrections to the missing initial conditions or the missing initial conditions themselves. For the set of linear equations to have a solution it is necessary that the columns of the matrix of coefficients of the set be linearly independent. If the columns are almost linearly dependent, the set of equations will, in theory, have a solution, but it will be difficult if not impossible in practice to compute this solution with acceptable accuracy. The columns of our matrix of coefficients are the solutions of the adjoint equations or the homogeneous equations corresponding to Kronecker delta initial conditions. Thus the columns of the coefficient matrix are linearly independent at the initial time.

It is known from the theory of linear ordinary differential equations that two solutions which are linearly independent at the initial time will remain linearly independent for all time [6]. But, in the numerical solution of linear ordinary differential equations, two solutions which are linearly independent at the initial time may not remain numerically independent, owing to the numerical process itself. This will be true in particular when the matrix $\mathbf{A}(t)$ of the system has eigenvalues well separated in numerical value. To see this, let us consider as an example the simple linear ordinary differential equation system [7, pp. 314-315]

$$\begin{pmatrix} \dot{y}_1 \\ \dot{y}_2 \end{pmatrix} = \begin{pmatrix} 0 & 1 \\ 10a^2 & 9a \end{pmatrix} \begin{pmatrix} y_1(t) \\ y_2(t) \end{pmatrix}, \qquad a > 0,$$

whose analytical solution is

$$\begin{pmatrix} y_1(t) \\ y_2(t) \end{pmatrix} = \begin{pmatrix} c_1 e^{-at} + c_2 e^{10at} \\ -ac_1 e^{-at} + 10ac_2 e^{10at} \end{pmatrix}.$$

The eigenvalues of the matrix $\mathbf{A}(t)$ are $10a$ and $-a$, so they may be considered to be well separated.

The matrix of solutions corresponding to the Kronecker delta initial conditions at $t = 0$, accordingly, is

$$\begin{pmatrix} y_1^{(1)}(t) & y_1^{(2)}(t) \\ y_2^{(1)}(t) & y_2^{(2)}(t) \end{pmatrix} = 1/11 \left(\begin{array}{c|c} 10e^{-at}+e^{10at} & -(1/a)e^{-at}+(1/a)e^{10at} \\ -10ae^{-at}+10ae^{10at} & e^{-at}+10e^{10at} \end{array} \right).$$

Thus at $t = 0$ the matrix of solutions is the 2×2 identity matrix whose columns are orthogonal, and therefore linearly independent. But, as t increases, the term e^{10at} will dominate, so the matrix of solutions will be nearly

$$(1/11)e^{10at}\begin{pmatrix} 1 & 1/a \\ 10a & 10 \end{pmatrix}.$$

The first column is a times the second, so the two are linearly dependent. Even if the initial conditions were so chosen that $c_2 = 0$ in one of the columns so that the e^{10at} term would theoretically be absent, numerical integration would, because of round-off error, introduce the term e^{10at} after a few integration steps. Thus the columns would still become increasingly dependent.

To ensure the linear independence of the matrix of coefficients of the linear equations developed by the method of adjoints or the method of complementary functions, Godunov [8] suggested a scheme of reorthogonalizing the solutions which has been improved on by Conte [9]. The idea underlying both methods is to integrate in parallel the set of $(n-r)$ homogeneous differential equations using the Kronecker delta initial conditions which are orthogonal. At a certain time the set of solutions is reorthogonalized by the Gram-Schmidt process and the integration continued until the next time specified for orthogonalization. This procedure is carried out sufficiently often that numerical linear dependence among the solutions never occurs.

In Section 4.10 we present the standard Gram-Schmidt process for the convenience of readers unfamiliar with it, and then cast the resulting equations in a form more suitable to our application. In Section 4.11 we give a method similar to Godunov's, mostly to introduce the ideas and to make Conte's method more understandable. Finally Conte's method, which is the recommended one for dealing with the problem of linear dependence in the method of complementary functions, is given in Section 4.12. Conte's method may also be applied to the method of adjoints.

A very different strategy to cope with the growth of solutions in the method of complementary functions (and similar methods) is the projection method of Guderley and Nikolai [10]. Their treatment of the solution of problem (4.2.1)–(4.2.3) is too involved and lengthy to be given here, but it proceeds, (as described by Fox [11] in his review) by transformations which are essentially projection operators, $y_1 = \overline{T}_1 u = \overline{T}_1 \bar{u}_1$, $y_2 = \overline{T}_2 u = \overline{T}_2 \bar{u}_2$, $T = \overline{T}_1 + \overline{T}_2$, $y = Tu$, to obtain differential equations from which u_1 and u_2 can be obtained accurately; u_1 by forward integration from t_0 to t_f, and u_2 by backward integration from t_f to t_0. Here T_1 is an $r \times n$ matrix, T_2 is an $(n-r) \times n$ matrix, u_1 is an $r \times 1$ vector, u_2 is an $(n-r) \times 1$ vector, and the partitioned matrices and vectors are:

$$\overline{T}_1 = [T_1 \vdots 0], \quad \overline{T}_2 = [0 \vdots T_2], \quad T = [T_1 \vdots T_2], \quad \bar{u}_1 = \begin{bmatrix} u_1 \\ \cdots \\ 0 \end{bmatrix}, \quad \bar{u}_2 = \begin{bmatrix} 0 \\ \cdots \\ u_2 \end{bmatrix}.$$

It turns out that the elements of T are obtained from first-order differential equations; therefore the method, although effective, appears to be time-consuming. However, almost any method which attempts to overcome the problems of linear dependence and the growth of solutions will be "time-consuming" in comparison with a straightforward shooting technique such as the method of complementary functions.

4.10. GRAM-SCHMIDT PROCESS [7 p. 347; 12, pp. 34–35]

In the following sections, use is made of the Gram-Schmidt process, which we describe here in terms suitable for our application.

This process generates a set of N orthonormal vectors $\mathbf{z}^{(k)}$, $k = 1, 2, \ldots, N$, from a set of N linearly independent vectors $\mathbf{y}^{(k)}$, $k = 1, 2, \ldots, N$, by forming linear combinations of the $\mathbf{y}^{(k)}$. The orthonormal set $\mathbf{z}^{(k)}$ has the property

$$(\mathbf{z}^{(j)}, \mathbf{z}^{(k)}) = \left.\begin{matrix} 1, & j = k, \\ 0, & j \neq k, \end{matrix}\right\}$$

where $\mathbf{z}^{(k)}$ has the components $(z_1^{(k)}, z_2^{(k)}, \ldots, z_N^{(k)})$ and the parenthesis (,) means the inner product of the vectors $\mathbf{z}^{(j)}$ and $\mathbf{z}^{(k)}$,

$$(\mathbf{z}^{(j)}, \mathbf{z}^{(k)}) = \sum_{i=1}^{N} z_i^{(j)} \cdot z_i^{(k)}.$$

Let $\{\boldsymbol{\eta}^{(k)}\}$ be the set of unnormalized orthogonal vectors which is normalized to $\{\mathbf{z}^{(k)}\}$. To start the Gram-Schmidt orthogonalization process, we set

$$\begin{aligned} \boldsymbol{\eta}^{(1)} &= \mathbf{y}^{(1)}, \\ w_{11} &= (\boldsymbol{\eta}^{(1)}, \boldsymbol{\eta}^{(1)})^{1/2}, \\ \mathbf{z}^{(1)} &= \frac{\boldsymbol{\eta}^{(1)}}{w_{11}}. \end{aligned}$$

The second vector $\boldsymbol{\eta}^{(2)}$, in the set of unnormalized orthogonal vectors is formed by a linear combination of $\mathbf{z}^{(1)}$ and $\mathbf{y}^{(2)}$:

$$\boldsymbol{\eta}^{(2)} = \mathbf{y}^{(2)} - \alpha_{21}\mathbf{z}^{(1)},$$

where α_{21} is a constant to be determined. Forming the inner product with $\mathbf{z}^{(1)}$, we find that

$$(\boldsymbol{\eta}^{(2)}, \mathbf{z}^{(1)}) = (\mathbf{y}^{(2)}, \mathbf{z}^{(1)}) - \alpha_{21}(\mathbf{z}^{(1)}, \mathbf{z}^{(1)}).$$

Since $\boldsymbol{\eta}^{(2)}$ and $\boldsymbol{\eta}^{(1)}$ are required to be orthogonal, it is necessary that

$$(\boldsymbol{\eta}^{(2)}, \mathbf{z}^{(1)}) = 0.$$

By the orthonormal property we have

$$(\mathbf{z}^{(1)}, \mathbf{z}^{(1)}) = 1.$$

Therefore

$$\begin{aligned} \alpha_{21} &= (\mathbf{y}^{(2)}, \mathbf{z}^{(1)}), \\ \boldsymbol{\eta}^{(2)} &= \mathbf{y}^{(2)} - (\mathbf{y}^{(2)}, \mathbf{z}^{(1)})\mathbf{z}^{(1)}, \\ w_{22} &= (\boldsymbol{\eta}^{(2)}, \boldsymbol{\eta}^{(2)})^{1/2}, \\ \mathbf{z}^{(2)} &= \frac{\boldsymbol{\eta}^{(2)}}{w_{22}}. \end{aligned}$$

The third vector, $\boldsymbol{\eta}^{(3)}$ in the set of unnormalized orthogonal vectors is formed as a linear combination of $\mathbf{z}^{(1)}$, $\mathbf{z}^{(2)}$, $\mathbf{y}^{(3)}$ expressed as

$$\boldsymbol{\eta}^{(3)} = \mathbf{y}^{(3)} - \alpha_{31}\mathbf{z}^{(1)} - \alpha_{32}\mathbf{z}^{(2)},$$

where α_{31}, α_{32} are constants to be determined. The constants α_{31} and α_{32} are formed by taking the inner products:

$$\begin{aligned} (\boldsymbol{\eta}^{(3)}, \mathbf{z}^{(1)}) &= (\mathbf{y}^{(3)}, \mathbf{z}^{(1)}) - \alpha_{31}(\mathbf{z}^{(1)}, \mathbf{z}^{(1)}) - \alpha_{32}(\mathbf{z}^{(2)}, \mathbf{z}^{(1)}), \\ (\boldsymbol{\eta}^{(3)}, \mathbf{z}^{(2)}) &= (\mathbf{y}^{(3)}, \mathbf{z}^{(2)}) - \alpha_{31}(\mathbf{z}^{(1)}, \mathbf{z}^{(2)}) - \alpha_{32}(\mathbf{z}^{(2)}, \mathbf{z}^{(2)}). \end{aligned}$$

Since $\boldsymbol{\eta}^{(3)}$ is required to be orthogonal to $\boldsymbol{\eta}^{(1)}$ and $\boldsymbol{\eta}^{(2)}$, we have $(\boldsymbol{\eta}^{(3)}, \mathbf{z}^{(1)})$

$= 0$ and $(\boldsymbol{\eta}^{(3)}, \mathbf{z}^{(2)}) = 0$. By the orthonormal property, $(\mathbf{z}^{(1)}, \mathbf{z}^{(1)}) = 1$, $(\mathbf{z}^{(2)}, \mathbf{z}^{(2)}) = 1$, and $(\mathbf{z}^{(1)}, \mathbf{z}^{(2)}) = 0$. As a consequence the constants are evaluated as

$$\begin{aligned}
\alpha_{31} &= (\mathbf{y}^{(3)}, \mathbf{z}^{(1)}), \\
\alpha_{32} &= (\mathbf{y}^{(3)}, \mathbf{z}^{(2)}), \\
\boldsymbol{\eta}^{(3)} &= \mathbf{y}^{(3)} - (\mathbf{y}^{(3)}, \mathbf{z}^{(1)})\mathbf{z}^{(1)} - (\mathbf{y}^{(3)}, \mathbf{z}^{(2)})\mathbf{z}^{(2)}, \\
w_{33} &= (\boldsymbol{\eta}^{(3)}, \boldsymbol{\eta}^{(3)})^{1/2}, \\
\mathbf{z}^{(3)} &= \frac{\boldsymbol{\eta}^{(3)}}{w_{33}}.
\end{aligned}$$

The general expression for $\boldsymbol{\eta}^{(k)}$ and $\mathbf{z}^{(k)}$ is

$$\begin{aligned}
\boldsymbol{\eta}^{(k)} &= \mathbf{y}^{(k)} - \alpha_{k1}\mathbf{z}^{(1)} - \alpha_{k2}\mathbf{z}^{(2)} - \ldots - \alpha_{k,k-1}\mathbf{z}^{(k-1)}, \\
w_{kk} &= (\boldsymbol{\eta}^{(k)}, \boldsymbol{\eta}^{(k)})^{1/2}, \\
\alpha_{km} &= (\mathbf{y}^{(k)}, \mathbf{z}^{(m)}), \qquad k \geqslant m, \\
\mathbf{z}^{(k)} &= \frac{\boldsymbol{\eta}^{(k)}}{w_{kk}}.
\end{aligned}$$

The transformations of the set of vectors $\mathbf{y}^{(k)}$ into the orthonormal set $\mathbf{z}^{(k)}$ therefore appear in recursive form as

$$\begin{aligned}
\boldsymbol{\eta}^{(1)} &= \mathbf{y}^{(1)}, \\
w_{11} &= (\boldsymbol{\eta}^{(1)}, \boldsymbol{\eta}^{(1)})^{1/2}, \\
\mathbf{z}^{(1)} &= \frac{\boldsymbol{\eta}^{(1)}}{w_{11}}, \\
\boldsymbol{\eta}^{(2)} &= \mathbf{y}^{(2)} - (\mathbf{y}^{(2)}, \mathbf{z}^{(1)})\mathbf{z}^{(1)}, \\
w_{22} &= (\boldsymbol{\eta}^{(2)}, \boldsymbol{\eta}^{(2)})^{1/2}, \\
\mathbf{z}^{(2)} &= \frac{\boldsymbol{\eta}^{(2)}}{w_{22}}, \\
&\ \vdots \\
\boldsymbol{\eta}^{(N)} &= \mathbf{y}^{(N)} - (\mathbf{y}^{(N)}, \mathbf{z}^{(1)})\mathbf{z}^{(1)} - \ldots - (\mathbf{y}^{(N)}, \mathbf{y}^{(N-1)})\mathbf{z}^{(N-1)}; \\
w_{NN} &= (\boldsymbol{\eta}^{(N)}, \boldsymbol{\eta}^{(N)})^{1/2}, \\
\mathbf{z}^{(N)} &= \frac{\boldsymbol{\eta}^{(N)}}{w_{NN}}.
\end{aligned}$$

This set of equations is not in the most useful form for our applications, which will involve not only generating the **z**'s from the **y**'s, but also in effect recovering the **y**'s from the **z**'s at a later stage of the calculation. For our purposes it is desirable to rearrange the equations so that the set of linearly independent variables, the **y**'s, appear on the right-hand side of the equations and the set of orthonormal vectors, the **z**'s, on the left side. By substituting $\mathbf{z}^{(1)}$ in $\mathbf{z}^{(2)}$, and then $\mathbf{z}^{(1)}$, $\mathbf{z}^{(2)}$ into $\mathbf{z}^{(3)}$, etc., in sequence, we rewrite the equations so that the **z**'s are expressed as linear functions of the **y**'s. Specifically the equations appear as

$$\mathbf{z}^{(1)} = \left[\frac{1}{w_{11}}\right]\mathbf{y}^{(1)},$$

$$\mathbf{z}^{(2)} = \left[-\frac{(\mathbf{y}^{(2)}, \mathbf{z}^{(1)})}{w_{22}w_{11}}\right]\mathbf{y}^{(1)} + \left[\frac{1}{w_{22}}\right]\mathbf{y}^{(2)},$$

$$\mathbf{z}^{(3)} = \left[-\frac{(\mathbf{y}^{(3)}, \mathbf{z}^{(1)})}{w_{33}w_{11}} + \frac{(\mathbf{y}^{(3)}, \mathbf{z}^{(2)})}{w_{33}}\frac{(\mathbf{y}^{(2)}, \mathbf{z}^{(1)})}{w_{22}w_{11}}\right]\mathbf{y}^{(1)} + \left[-\frac{(\mathbf{y}^{(3)}, \mathbf{z}^{(2)})}{w_{33}w_{22}}\right]\mathbf{y}^{(2)} + \left[\frac{1}{w_{33}}\right]\mathbf{y}^{(3)},$$

$$\begin{aligned}\mathbf{z}^{(4)} = &\left[-\frac{(\mathbf{y}^{(4)}, \mathbf{z}^{(1)})}{w_{44}w_{11}} + \frac{(\mathbf{y}^{(4)}, \mathbf{z}^{(2)})}{w_{44}}\frac{(\mathbf{y}^{(2)}, \mathbf{z}^{(1)})}{w_{22}w_{11}} + \frac{(\mathbf{y}^{(4)}, \mathbf{z}^{(3)})}{w_{44}}\frac{(\mathbf{y}^{(3)}, \mathbf{z}^{(1)})}{w_{33}w_{11}}\right.\\ &\left.-\frac{(\mathbf{y}^{(4)}, \mathbf{z}^{(3)})(\mathbf{y}^{(3)}, \mathbf{z}^{(2)})(\mathbf{y}^{(2)}, \mathbf{z}^{(1)})}{w_{44}w_{33}w_{22}w_{11}}\right]\mathbf{y}^{(1)}\\ &+\left[-\frac{(\mathbf{y}^{(4)}, \mathbf{z}^{(2)})}{w_{44}w_{22}} + \frac{(\mathbf{y}^{(4)}, \mathbf{z}^{(3)})}{w_{44}}\frac{(\mathbf{y}^{(3)}, \mathbf{z}^{(2)})}{w_{33}w_{22}}\right]\mathbf{y}^{(2)}\\ &+\left[-\frac{(\mathbf{y}^{(4)}, \mathbf{z}^{(3)})}{w_{44}w_{33}}\right]\mathbf{y}^{(3)} + \left[\frac{1}{w_{44}}\right]\mathbf{y}^{(4)},\end{aligned}$$

etc.

The transformation from the **y**'s to the **z**'s may be expressed in partitioned matrix notation as

$$\mathbf{Z} = \mathbf{PY}$$

or in component form as

$$\begin{bmatrix} \mathbf{z}^{(1)} \\ \mathbf{z}^{(2)} \\ \mathbf{z}^{(3)} \\ \cdot \\ \cdot \\ \cdot \\ \mathbf{z}^{(N)} \end{bmatrix} = \begin{bmatrix} p_{11} & & & & \\ p_{21} & p_{22} & & & \\ p_{31} & p_{32} & p_{33} & & \\ \cdot & \cdot & \cdot & \cdot & \\ \cdot & \cdot & \cdot & \cdot & \\ \cdot & \cdot & \cdot & \cdot & \\ p_{N1} & p_{N2} & p_{N3} & \cdots & p_{NN} \end{bmatrix} \begin{bmatrix} \mathbf{y}^{(1)} \\ \mathbf{y}^{(2)} \\ \mathbf{y}^{(3)} \\ \cdot \\ \cdot \\ \cdot \\ \mathbf{y}^{(N)} \end{bmatrix},$$

where

$\mathbf{Z} = N \times 1$ vector, whose elements are the vectors $\mathbf{z}^{(1)}, \mathbf{z}^{(2)}, \ldots, \mathbf{z}^{(N)}$,

$\mathbf{Y} = N \times 1$ vector, whose elements are the vectors $\mathbf{y}^{(1)}, \mathbf{y}^{(2)}, \ldots, \mathbf{y}^{(N)}$,

$\mathbf{P} = N \times N$ matrix of lower triangular form described by

$$p_{jj} = \frac{1}{w_{jj}}, \qquad j = k,$$

$$p_{jk} = \sum_{s=k}^{j-1} -\frac{(\mathbf{y}^{(j)}, \mathbf{z}^{(s)})}{w_{jj}} p_{sk}, \qquad j > k,$$

$$p_{jk} = 0, \qquad j < k,$$

$$w_{jj} = (\boldsymbol{\eta}^{(j)}, \boldsymbol{\eta}^{(j)})^{1/2}.$$

It must be emphasized that, although in the set of equations $\mathbf{Z} = \mathbf{P}Y$ it seems that we have achieved our goal of representing the **z**'s as a transformation of the **y**'s, the form is deceiving. Actually the matrix **P** also involves the **z**'s and accordingly must be developed recursively, column by column, as described in the last set of equations above. However, as we shall see in the following sections, the methods described there call for a succession of orthonormalizations and, thus, the formation of a sequence of matrices $\mathbf{P}^{(q)}$, which will be saved in the course of the computation. Combinations of the matrices $\mathbf{P}^{(q)}$ or their inverses will then be used in the final steps of the methods to generate the solution of the given two-point boundary value problem; hence our form of $\mathbf{Z} = \mathbf{PY}$ is the appropriate one.

4.11. ORTHONORMALIZATION AND DIFFERENTIAL EQUATIONS [8]

We now direct our attention to the orthonormalization procedure as applied to the method of complementary functions. The procedure is applicable also to the method of adjoints. We recall that, for the linear two-point boundary value problem

$$\dot{\mathbf{y}} = \mathbf{A}(t)\mathbf{y}+\mathbf{f}(t) \tag{4.11.1}$$

with the initial conditions

$$y_i(t_0) = c_i, \qquad i = 1,2,\ldots,r, \tag{4.11.2}$$

$$y_{i_m}(t_f) = c_{i_m}, \qquad m = 1,2,\ldots,n-r, \tag{4.11.3}$$

in the method of complementary functions we solve the homogeneous equations $(n-r)$ times:

$$\dot{\mathbf{u}} = \mathbf{A}(t)\mathbf{u} \tag{4.11.4}$$

with the $(n-r)$ sets of initial conditions

$$u_i^{(m)}(t_0) = \left.\begin{matrix} 1, & i = r+m \\ 0, & i \neq r+m \end{matrix}\right\}, \qquad m = 1,2,\ldots,n-r, \tag{4.11.4a}$$

where $u_i^{(m)}(t)$ = the solution for the ith component of $\mathbf{u}$ for the mth integration of the homogeneous equations.

The particular equation

$$\dot{\mathbf{v}} = \mathbf{A}(t)\mathbf{v}+\mathbf{f}(t) \tag{4.11.5}$$

is integrated forward once with the initial conditions

$$v_i(t_0) = y_i(t_0) = c_i, \qquad i = 1,2,\ldots,r, \tag{4.11.6a}$$

$$v_i(t_0) = 0, \qquad i = r+1,r+2,\ldots,n. \tag{4.11.6b}$$

Since the general solution is given by

$$y_i(t) = \sum_{j=1}^{n-r} \beta_j u_i^{(j)}(t)+v_i(t), \qquad i = 1,2,\ldots,n. \tag{4.11.7}$$

The constants β_j are found by solving (4.11.7) for the specified terminal conditions (4.11.3):

$$\begin{bmatrix} \beta_1 \\ \beta_2 \\ \vdots \\ \beta_{n-r} \end{bmatrix} = \begin{bmatrix} u_{i_1}^{(1)}(t_f) & u_{i_1}^{(2)}(t_f) & \dots & u_{i_1}^{(n-r)}(t_f) \\ u_{i_2}^{(1)}(t_f) & u_{i_2}^{(2)}(t_f) & \dots & u_{i_2}^{(n-r)}(t_f) \\ \vdots & \vdots & \vdots & \vdots \\ u_{i_{n-r}}^{(1)}(t_f) & u_{i_{n-r}}^{(2)}(t_f) & \dots & u_{i_{n-r}}^{(n-r)}(t_f) \end{bmatrix}^{-1} \begin{bmatrix} y_{i_1}(t_f) - v_{i_1}(t_f) \\ y_{i_2}(t_f) - v_{i_2}(t_f) \\ \vdots \\ y_{i_{n-r}}(t_f) - v_{i_{n-r}}(t_f) \end{bmatrix} \tag{4.11.8}$$

Equation (4.11.8) is a set of $(n-r)$ linear equations in the $(n-r)$ unknowns $\beta_1, \beta_2, \dots, \beta_{n-r}$ For a solution to exist, it is necessary that the equations be numerically linearly independent. The purpose of the orthonormalization procedure is to guarantee that (4.11.8) will be numerically independent.

Let us define

q = index counter for the number of orthonormalizations $q = 0, 1, 2, \dots, Q$;

t_i = ith discrete point in the interval $[t_0, t_f]$;

$t^{(q)}$ = The discrete point at which the qth orthonormalization occurs; this can be set by the analyst or can be determined automatically by a programming test; the $t^{(q)}$, $q = 0, 1, 2, \dots, Q$, need not be evenly spaced; set $t^{(0)} = t_0$;

$\mathbf{u}^{(m,q)}(t)$ = $(n-r) \times 1$ vector solution of the homogeneous equations, where the m refers to the mth set of initial conditions and q refers to the point $t^{(q)}$ at which the vectors were last orthonormalized, $m = 1, 2, \dots, n-r$; for $q = 0$, $\mathbf{u}^{(m,q)}(t_0) = \mathbf{u}^{(m)}(t_0)$ = Kronecker delta initial conditions in (4.11.4a); the $\mathbf{u}^{(m,q)}(t)$ hold for the interval $t^{(q)} \le t \le t^{(q+1)}$; for each m, $\mathbf{u}^{(m,q)}(t)$ has the components $u_1^{(m,q)}(t), u_2^{(m,q)}(t), \dots, u_n^{(m,q)}(t)$;

$\mathbf{P}^{(q)}$ = $(n-r) \times (n-r)$ lower triangular orthonormalization matrix which is formed at $t^{(q)}$ by the Gram-Schmidt process operating on the $(n-r)$ linearly independent vectors $\mathbf{u}^{(m,q-1)}(t^{(q)})$, $m = 1, 2, \dots, n-r$.

The orthonormalization matrix $\mathbf{P}^{(q)}$ operates on the $(n-r)$ linearly independent vectors $\mathbf{u}^{(m,q-1)}(t^{(q)})$ at the point $t^{(q)}$ to generate the $(n-r)$

orthonormal vectors $\mathbf{u}^{(m,q)}(t^{(q)})$:

$$\begin{bmatrix} \mathbf{u}^{(1,q)}(t^{(q)}) \\ \mathbf{u}^{(2,q)}(t^{(q)}) \\ \vdots \\ \mathbf{u}^{(n-r,q)}(t^{(q)}) \end{bmatrix} = \begin{bmatrix} \\ \mathbf{P}^{(q)} \\ \\ \end{bmatrix} \begin{bmatrix} \mathbf{u}^{(1,q-1)}(t^{(q)}) \\ \mathbf{u}^{(2,q-1)}(t^{(q)}) \\ \vdots \\ \mathbf{u}^{(n-r,q-1)}(t^{(q)}) \end{bmatrix}.$$

The vectors $\mathbf{u}^{(m,q-1)}(t)$ are formed in the interval $t^{(q-1)} \leq t \leq t^{(q)}$ and stored in the interval at discrete points. The vectors $\mathbf{u}^{(m,q-1)}(t^{(q)})$ are not stored at $t^{(q)}$ because they are orthonormalized to yield $\mathbf{u}^{(m,q)}(t^{(q)})$, which are stored.

The matrix $\mathbf{P}^{(q)}$ is called $\mathbf{P}$ in Section 4.10, where it is defined. The linearly independent vectors $\mathbf{y}^{(1)}, \mathbf{y}^{(2)}, \ldots, \mathbf{y}^{(N)}$ in Section 4.10 correspond to $\mathbf{u}^{(1,q-1)}(t^{(q)}), \mathbf{u}^{(2,q-1)}(t^{(q)}), \ldots, \mathbf{u}^{(n-r,q-1)}(t^{(q)})$ here, and the orthonormalized vectors $\mathbf{z}^{(1)}, \mathbf{z}^{(2)}, \ldots, \mathbf{z}^{(N)}$ in Section 4.10 correspond to $\mathbf{u}^{(1,q)}(t^{(q)})$, $\mathbf{u}^{(2,q)}(t^{(q)}), \ldots, \mathbf{u}^{(n-r,q)}(t^{(q)})$ here. In Section 4.10 the number of linearly independent vectors is N. Here N equals $(n-r)$.

The orthonormalization procedure we describe for the method of complementary functions is a two-pass procedure. On the first pass the homogeneous equations are orthonormalized at certain points and the solutions are stored. On the second pass the inverse transformation is applied to the stored solutions. The resulting solutions are the ones we would have obtained if numerical difficulties had not prevented us from directly integrating forward with the Kronecker delta initial conditions.

The reorthogonalization procedure thus consists of the following steps.

First Pass

1. Integrate the particular equation with the initial conditions $v_i t_0) = c_i$, $i = 1,2, \ldots r$; $v_i(t_0) = 0$, $i = r+1, r+2, \ldots, n$, from t_0 to t_f and store the results at discrete points.

2. Set the index $q = 0$. Integrate in parallel the homogeneous equations with the Kronecker delta initial conditions

$$u_i^{(m,0)}(t^{(0)}) = \left.\begin{matrix} 1, & i = r+m \\ 0, & i \neq r+m \end{matrix}\right\}$$

from $t^{(0)}$ to $t^{(1)}$ and store the solutions at discrete points in the interval $[t^{(0)}, t^{(1)}]$. Note that $t^{(0)} = t_0$.

3. Set $q = q+1$. At $t^{(q)}$ form and store from the set of $(n-r)$ linearly independent vectors $\mathbf{u}^{(m,q-1)}(t^{(q)})$, $m = 1, 2, \ldots, n-r$, the set of orthonormal vectors $\mathbf{u}^{(m,q)}(t^{(q)})$, $m = 1, 2, \ldots, n-r$, and the associated matrix $\mathbf{P}^{(q)}$.

4. Integrate in parallel the $(n-r)$ homogeneous equations from $t^{(q)}$ to $t^{(q+1)}$, using as the initial vectors at $t^{(q)}$, $\mathbf{u}^{(m,q)}(t^{(q)})$, $m = 1, 2, \ldots, n-r$. If $q = Q$ and $t^{(Q)} \neq t_f$, integrate the homogeneous equations from $t^{(Q)}$ to t_f.

5. If $q < Q$, go to item 3.

6. If $q = Q$, terminate the first pass of the orthonormalization procedure.

Second Pass

7. To recover the homogeneous profiles that would have been generated from the Kronecker delta initial conditions (4.11.4a), we apply the inverse of the product of the $\mathbf{P}^{(q)}$ matrices to the stored transformed homogeneous equation profiles as follows:

$$\begin{bmatrix} \mathbf{u}^{(1)}(t) \\ \mathbf{u}^{(2)}(t) \\ \vdots \\ \mathbf{u}^{(n-r)}(t) \end{bmatrix} = \begin{bmatrix} \\ \mathbf{P}^{(1)} \\ \\ \end{bmatrix}^{-1} \begin{bmatrix} \mathbf{u}^{(1,1)}(t) \\ \mathbf{u}^{(2,1)}(t) \\ \vdots \\ \mathbf{u}^{(n-r,1)}(t) \end{bmatrix}, \qquad t^{(1)} \leq t < t^{(2)},$$

$$\begin{bmatrix} \mathbf{u}^{(1)}(t) \\ \mathbf{u}^{(2)}(t) \\ \vdots \\ \mathbf{u}^{(n-r)}(t) \end{bmatrix} = \begin{bmatrix} \\ \mathbf{P}^{(2)}\mathbf{P}^{(1)} \\ \\ \end{bmatrix}^{-1} \begin{bmatrix} \mathbf{u}^{(1,2)}(t) \\ \mathbf{u}^{(2,2)}(t) \\ \vdots \\ \mathbf{u}^{(n-r,2)}(t) \end{bmatrix}, \qquad t^{(2)} \leq t < t^{(3)},$$

$$\begin{bmatrix} \mathbf{u}^{(1)}(t) \\ \mathbf{u}^{(2)}(t) \\ \vdots \\ \mathbf{u}^{(n-r)}(t) \end{bmatrix} = \begin{bmatrix} \\ \mathbf{P}^{(Q)}\mathbf{P}^{(Q-1)} \quad \cdots \quad \mathbf{P}^{(1)} \\ \\ \end{bmatrix}^{-1} \begin{bmatrix} \mathbf{u}^{(1,Q)}(t) \\ \mathbf{u}^{(2,Q)}(t) \\ \vdots \\ \mathbf{u}^{(n-r,Q)}(t) \end{bmatrix}, \qquad t^{(Q)} \leq t \leq t_f.$$

The vectors on the left are stored at discrete points over the intervals $[t^{(1)}, t^{(2)}], [t^{(2)}, t^{(3)}], \ldots, [t^{(Q)}, t_f]$ and represent the solutions of the homogeneous equations corresponding to the Kronecker delta initial conditions (4.11.4a).

8. Determine the constants β_j in (4.11.8) by using the stored homogeneous and particular equations solutions at t_f.

9. Construct the general solution at the discrete points from the stored profiles and the β's by applying the general solution formula (4.11.7).

For some problems the particular solution may exhibit numerical linear dependence on the homogeneous solutions. To thwart this behavior we may apply the orthonormalization procedure to both the homogeneous equations and the particular equation. Our procedure can be modified slightly to do this. The homogeneous and the particular equations are integrated in parallel starting with the initial conditions (4.11.4a) for the homogeneous equations and (4.11.6) for the particular equation. At the point $t^{(q)}$, $q = 1, 2, \ldots$, the set of linearly independent vectors $\mathbf{u}^{(m,q-1)}(t^{(q)})$ and the particular solution $\mathbf{v}^{(q-1)}(t^{(q)})$ (where $\mathbf{v}^{(q-1)}(t)$ = the particular solution last orthogonalized at the point $t^{(q-1)}$) are orthonormalized by the Gram-Schmidt process. The $\mathbf{P}^{(q)}$ matrix is now of dimension $(n-r+1) \times (n-r+1)$ since the particular solution is taken to be the $(n-r+1)$st vector. Except for the integration of the particular equation in parallel with the homogeneous equations and the orthonormalization of $(n-r+1)$ vectors, the procedure is identical with that previously described.

The orthonormalization procedure requires more storage than the method of complementary functions because all the piecewise linearly independent solutions must be stored. The orthonormalization procedure requires that at each $t^{(q)}$ the vectors be linearly independent, otherwise the Gram-Schmidt process breaks down. For the investigator to predetermine the $t^{(q)}$, $q = 1, 2, \ldots$, may be difficult, for there is no way to know beforehand whether the vectors will be numerically linearly independent at the specified $t^{(q)}$. Experience has shown that orthonormalizing too few times or too often can result in poor accuracy. Over the interval $[t_0, t_f]$ there is a range of frequency of orthonormalizations which can yield satisfactory results. In fact, where in the interval $[t_0, t_f]$ the orthonormalizations occur is probably more important than the number of orthonormalizations.

Instead of reorthogonalizing at predetermined points $t^{(q)}$, Conte [9] has suggested a method which determines automatically when to reorthogonalize, at the expense of course of some computer time. His procedure is to test the angle between each pair of vectors at each point t_i at which the solution is computed. If the smallest angle is less than a prescribed tolerance α, the solution vectors are orthonormalized, otherwise the integration proceeds to the next step. Conte's test is

$$\min_{i,j} \arccos \left| \frac{(\mathbf{u}^{(i,q-1)}(t), \mathbf{u}^{(j,q-1)}(t))}{((\mathbf{u}^{(i,q-1)}(t), \mathbf{u}^{(i,q-1)}(t))(\mathbf{u}^{(j,q-1)}(t), \mathbf{u}^{(j,q-1)}(t)))^{1/2}} \right| < \alpha.$$

In principle, the tolerance α can be set to any value between 0 and $\pi/2$ radians. The setting 0 will have the effect that reorthogonalization never takes place, while the setting $\pi/2$ will result in reorthogonalization at every integration step t_i. In practice, α is set to some small angle, say less than 0.2 radians. Usually some numerical experimentation is necessary to determine a suitable value of α for a given problem.

It should be noted that the second pass of our reorthogonalizing procedures requires the formation of products of matrices and the taking of inverses of the product matrices. In the same reference mentioned above, Conte [9] has provided a way of reorthogonalizing which does not require the computation of these inverses. We describe his method in Section 4.12.

4.12. CONTE'S METHOD [9]

Conte's method is also a two-pass procedure. On the first pass the homogeneous equations and the particular equation are integrated in parallel and orthonormalized at certain $t^{(q)}$, $q = 1, 2, \ldots$, determined by the criterion at the end of Section 4.11. On the second pass, working backward from the final point t_f, the stored profiles are recombined over each interval $[t^{(q)}, t^{(q-1)}]$ to form the general solution without inverses being taken.

Let us define

$\mathbf{U}^{(q)}(t) = n \times (n-r)$ matrix of solutions of the homogeneous equations $\mathbf{u}^{(m,q)}(t)$, $m = 1, 2, \ldots, n-r$, which were last orthonormalized at $t^{(q)}$; the columns of $\mathbf{U}^{(q)}(t)$ are the vectors $\mathbf{u}^{(1,q)}(t), \mathbf{u}^{(2,q)}(t), \ldots$ $\mathbf{u}^{(n-r,q)}(t)$:

$$\begin{bmatrix} u_1^{(1,q)}(t) & u_1^{(2,q)}(t) & \cdots & u_1^{(n-r,q)}(t) \\ u_2^{(1,q)}(t) & u_2^{(2,q)}(t) & \cdots & u_2^{(n-r,q)}(t) \\ \vdots & \vdots & & \vdots \\ u_n^{(1,q)}(t) & u_n^{(2,q)}(t) & \cdots & u_n^{(n-r,q)}(t) \end{bmatrix}$$

$\mathbf{P}^{(q)T}$ = transpose of $\mathbf{P}^{(q)}$, $(n-r) \times (n-r)$ matrix, which is the orthonormalization matrix of the homogeneous solutions;

$\mathbf{v}^{(q)}(t)$ = the particular solution last orthogonalized at $t^{(q)}$.

The procedure is as follows:

First Pass

1. Set index $q = 0$ and $t^{(0)} = t_0$. Integrate the homogeneous and particular equations in parallel from t_0 to $t^{(1)}$, using the Kronecker delta initial con-

ditions for the homogeneous equations:

$$u_i^{(m,0)}(t^{(0)}) = u_i^{(m)}(t_0) = \left.\begin{matrix} 1, & i = r+m \\ 0, & i \neq r+m \end{matrix}\right\}, \qquad m = 1,2,\ldots,n-r,$$

and for the particular equation:

$$v_i(t_0) = c_i, \quad i = 1,2,\ldots,r; \quad v_i(t_0) = 0, \quad i = r+1, r+2, \ldots, n.$$

Store the solutions at discrete points.

2. Set $q = q+1$. At $t^{(q)}$, form from the set of $(n-r)$ linearly independent vectors $\mathbf{u}^{(m,q-1)}(t^{(q)})$, $m = 1,2,\ldots,n-r$, the set of orthonormal vectors $\mathbf{u}^{(m,q)}(t^{(q)})$, $m = 1,2,\ldots,n-r$, and store. In matrix form this is written

$$\mathbf{U}^{(q)}(t^{(q)}) = \mathbf{U}^{(q-1)}(t^{(q)})\mathbf{P}^{(q)T}. \tag{4.12.1}$$

3. At $t^{(q)}$, form the orthogonal complement of $\mathbf{v}^{(q-1)}(t^{(q)})$ by subtracting out a linear combination of the orthonormal homogeneous vectors $\mathbf{u}^{(m,q)}(t^{(q)})$, $m = 1,2,\ldots,n-r$:

$$\mathbf{v}^{(q)}(t^{(q)}) = \mathbf{v}^{(q-1)}(t^{(q)}) - \mathbf{U}^{(q)}(t^{(q)})\boldsymbol{\omega}^{(q)}, \tag{4.12.2}$$

where

$$\begin{aligned} \boldsymbol{\omega}^{(q)} &= n-r\times 1 \text{ vector with components } \omega_1^{(q)}, \omega_2^{(q)}, \ldots, \omega_{n-r}^{(q)}, \\ \omega_j^{(q)} &= (\mathbf{v}^{(q-1)}(t^{(q)}), \mathbf{u}^{(j,q)}(t^{(q)})). \end{aligned} \tag{4.12.3}$$

For each $t^{(q)}$ the vector $\omega^{(q)}$ is saved. The vectors $\mathbf{v}^{(q)}(t)$ are stored at discrete points over the interval $t^{(q)} \leq t \leq t^{(q+1)}$.

4. Integrate in parallel from $t^{(q)}$ to $t^{(q+1)}$ the $(n-r)$ homogeneous equations starting with the initial vectors at $t^{(q)}$, $\mathbf{u}^{(m,q)}(t^{(q)})$, $m = 1,2,\ldots, n-r$, and the particular vector $\mathbf{v}^{(q)}(t^{(q)})$ and store the solutions at discrete points within the interval $t^{(q)} \leq t \leq t^{(q+1)}$.

5. If $t^{(q)} < t_f$, go to item 2.

6. At $t = t_f$, execute items 2 and 3, where $q+1 = Q$ and $t_f = t^{(Q)}$.

Second Pass

7. Since $t^{(Q)} = t_f$, the general solution at $t^{(Q)}$ is given as the sum of the particular solution plus a linear combination of the homogeneous solutions

$$\mathbf{y}(t^{(Q)}) = \mathbf{v}^{(Q)}(t^{(Q)}) + \mathbf{U}^{(Q)}(t^{(Q)})\boldsymbol{\beta}^{(Q)}, \tag{4.12.4}$$

where

$\boldsymbol{\beta}^{(Q)} = (n-r) \times 1$ vector of constants with components $\beta_1^{(Q)}, \beta_2^{(Q)}, \ldots, \beta_{n-r}^{(Q)}$

The vector $\boldsymbol{\beta}^{(Q)}$ is found by solving (4.12.4) for the specified terminal conditions $y_{i_m}(t_f) = c_{i_m}$, $m = 1, 2, \ldots, n-r$.

8. The general solution is constructed by working backward from $t^{(Q)}$ as follows:

$$\mathbf{y}(t) = \mathbf{v}^{(q-1)}(t) + \mathbf{U}^{(q-1)}(t)\boldsymbol{\beta}^{(q-1)}, \qquad t^{(q-1)} \leqslant t < t^{(q)}, \qquad q = Q, Q-1, \ldots, 1, \tag{4.12.5}$$

$$\boldsymbol{\beta}^{(q-1)} = \mathbf{P}^{(q)T}[\boldsymbol{\beta}^{(q)} - \boldsymbol{\omega}^{(q)}], \tag{4.12.6}$$

where $\mathbf{v}^{(q)}(t)$, $\mathbf{U}^{(q)}(t)$, $\boldsymbol{\omega}^{(q)}$ have been stored on the first pass at discrete points.

We observe that the general solution is obtained piecewise. Except for the interval $[t_0, t^{(1)}]$, the homogeneous and particular solutions which are stored are not the solutions of the original two-point boundary value problem. These stored particular solutions plus the stored homogeneous solutions, when combined linearly by the proper β's, do give the true general solution. Only when $\boldsymbol{\beta}^{(0)}$ is finally produced do we have the true constants corresponding to the original two-point boundary value problem.

To justify the formulation in item 8 we argue as follows. Owing to the continuity of the general solution, we may express at each point of orthogonalization $t^{(q)}$, $q = 1, 2, \ldots, Q$, the general solution in terms of the homogeneous and particular solutions immediately before and after the orthogonalization point. If we start at $t^{(Q)}$, we have the general solution

$$\mathbf{y}(t^{(Q)}) = \mathbf{v}^{(Q)}(t^{(Q)}) + \mathbf{U}^{(Q)}(t^{(Q)})\boldsymbol{\beta}^{(Q)}, \tag{4.12.7}$$

where $\boldsymbol{\beta}^{(Q)}$ is the vector whose components are determined by the specified terminal conditions. This expression gives the general solution in terms of the particular and homogeneous solutions orthogonalized at $t^{(Q)}$. If we substitute in (4.12.7) the expressions (4.12.1) and (4.12.2), we have

$$\mathbf{y}(t^{(Q)}) = \mathbf{v}^{(Q-1)}(t^{(Q)}) - \mathbf{U}^{(Q)}(t^{(Q)})\boldsymbol{\omega}^{(Q)} + \mathbf{U}^{(Q)}(t^{(Q)})\,\boldsymbol{\beta}^{(Q)}, \tag{4.12.8}$$

$$\mathbf{y}(t^{(Q)}) = \mathbf{v}^{(Q-1)}(t^{(Q)}) + \mathbf{U}^{(Q-1)}(t^{(Q)})\mathbf{P}^{(Q)T}[\boldsymbol{\beta}^{(Q)} - \boldsymbol{\omega}^{(Q)})], \tag{4.12.9}$$

or

$$\mathbf{y}(t^{(Q)}) = \mathbf{v}^{(Q-1)}(t^{(Q)}) + \mathbf{U}^{(Q-1)}\boldsymbol{\beta}^{(Q-1)}, \tag{4.12.10}$$

where

$$\boldsymbol{\beta}^{(Q-1)} = \mathbf{P}^{(Q)T}[\boldsymbol{\beta}^{(Q)} - \boldsymbol{\omega}^{(Q)}]. \tag{4.12.11}$$

Equation (4.12.10) determines the general solution in terms of the particular and homogeneous solutions immediately before orthogonalization. Since $\mathbf{v}^{(Q-1)}(t)$ and $\mathbf{U}^{(Q-1)}(t)$ are defined and stored in the interval $t^{(Q-1)} \leq t < t^{(Q)}$, it follows that the vector $\boldsymbol{\beta}^{(Q-1)}$ is constant in the interval $t^{(Q-1)} \leq t < t^{(Q)}$. A similiar argument holds for all $t^{(q)}$, $q = Q, Q-1, \ldots, 1$, which accounts for the formulation in item 8. The continuity of the general solution $\mathbf{y}(t^{(q)})$, $q = Q, Q-1, \ldots, 1$, provides the vehicle by which the constants $\boldsymbol{\beta}^{(q-1)}$ are related to the previous vector $\boldsymbol{\beta}^{(q)}$.

REFERENCES

1. T. R. Goodman and G. N. Lance, The Numerical Solution of Two Point Boundary Value Problems, *MTAC*, **10** (1956), 82–86.
2. J. C. Heideman, Use of the Method of Particular Solutions in Nonlinear Two Point Boundary Value Problems, Part 1, Uncontrolled Systems, Aero-Astronautics Report No. 50, Rice University, 1968.
3. J. C. Heideman, Use of the Method of Particular Solutions in Nonlinear Two Point Boundary Value Problems, Part 2, Controlled Systems, Aero-Astronautics Report No. 51, Rice University, 1968.
4. A. Miele, Method of Particular Solutions for Linear Two Point Boundary Value Problems, Part I, Preliminary Examples, Aero-Astronautics, Report No. 48, Rice University, 1968.
5. A. Miele, Method of Particular Solutions for Linear Two Point Boundary Value Problems, Part II, General Theory, Aero-Astronautics Report No. 49, Rice University, 1968.
6. E. L. Ince, *Ordinary Differential Equations*, reprinted by Dover, New York.
7. J. Todd, *Survey of Numerical Analysis*, McGraw-Hill, New York, 1962.
8. S. Godunov, On The Numerical Solution of Boundary Value Problems for Systems of Linear Ordinary Differential Equations, *Uspekhi Mat. Nauk*, **16** (1961), 171-174.
9. S. D. Conte, The Numerical Solution of Linear Boundary Value Problems, *SIAM Rev.* (3) **8** (July (1966), 309–321.
10. K. G. Guderley and P. J. Nikolai, Reduction of Two Point Boundary Value Problems in a Vector Space to Initial Value Problems by Projection, *Numer. Math.*, **8** (1966), 270–289 (MR 34, No. 2192).
11. L. Fox, Review of Guderley and Nikolai [10], *Math. Rev.*, Revue No. 2192, (2) **34** (Aug. 1967).
12. F. B. Hildebrand, *Methods of Applied Mathematics*, Prentice-Hall, Englewood Cliffs, N. J., 1952.

General References

R. Bellman and R. Kalaba, *Quasilinearization and Nonlinear Boundary Value Problems*, American Elsevier, New York, 1965.

L. Fox, *Numerical Solution of Two-Point Boundary Problems in Ordinary Differential Equations*, Oxford, London, 1957.

Chapter 5

QUASILINEARIZATION

5.1. INTRODUCTION

In this chapter we discuss quasilinearization, a relatively new and important technique for solving nonlinear two-point boundary value problems. Although the technique as originally developed by Bellman and Kalaba [1] was motivated by dynamic programming, it is not necessary to know or to employ dynamic programming in order to use the quasilinear method. In Chapter 6 we show that quasilinearization is one realization of the abstract Newton-Raphson method in Banach space [2]. Included in this chapter is another approach to quasilinearization, namely, the generalized Newton-Raphson operator, as developed by McGill and Kenneth for second-order systems [3, 4].

5.2. QUASILINEARIZATION

Quasilinearization is the process of solving nonlinear two-point boundary value problems by the following steps:

1. Linearizing the nonlinear ordinary differential equations around a nominal solution, which satisfies the specified boundary conditions.

2. Solving a sequence of linear two-point boundary value problems in which the solution of the kth linear two-point boundary value problem satisfies the specified boundary conditions and is taken as the nominal profile for the $(k+1)$st linear two-point boundary value problem.

In the limit the solution of the linear two-point boundary value problems converges to the solution of the original nonlinear two-point boundary value problem.

To spell out in detail the quasilinear process, let us consider the system of n nonlinear ordinary differential equations

$$\dot{y}_i = g_i(y_1, y_2, \ldots, y_n, t), \quad i = 1, 2, \ldots, n, \tag{5.2.1}$$

with the initial conditions

$$y_i(t_0) = c_i, \qquad i = 1, 2, \ldots, r, \tag{5.2.2}$$

and the terminal conditions

$$y_{i_m}(t_f) = c_{i_m}, \qquad m = 1, 2, \ldots, n-r. \tag{5.2.3}$$

Suppose now that we have the kth nominal solution to (5.2.1), $y_1^{(k)}(t), y_2^{(k)}(t), \ldots, y_n^{(k)}(t)$, over the interval $[t_0, t_f]$, which is nominal in the sense that the initial conditions (5.2.2) and the terminal conditions (5.2.3) are satisfied exactly but the profiles $\mathbf{y}^{(k)}(t)$ only satisfy the differential equations (5.2.1) approximately.

Let us now expand the right-hand side of (5.2.1) in a Taylor's series up through first-order terms around the nominal solution $\mathbf{y}^{(k)}(t)$ and let us replace (5.2.1) by

$$\dot{y}_i^{(k+1)} = g_i(y_1^{(k)}(t), y_2^{(k)}(t), \ldots, y_n^{(k)}(t), t) + \sum_{j=1}^{n} \frac{\partial g_i}{\partial y_j} \delta y_j^{(k)}, \quad i = 1, 2, \ldots, n, \tag{5.2.4}$$

where $\partial g_i / \partial y_j$ is evaluated at the nominal $\mathbf{y}^{(k)}(t)$:

$$\delta y_j^{(k)} = y_j^{(k+1)}(t) - y_j^{(k)}(t), \qquad j = 1, 2, \ldots, n. \tag{5.2.5}$$

On rearranging terms, we obtain

$$\dot{y}_i^{(k+1)} = \sum_{j=1}^{n} \frac{\partial g_i}{\partial y_j} y_j^{(k+1)}(t) + g_i(y_1^{(k)}(t), y_2^{(k)}(t), \ldots, y_n^{(k)}(t), t)$$
$$- \sum_{j=1}^{n} \frac{\partial g_i}{\partial y_j} y_j^{(k)}(t), \qquad i = 1, 2, \ldots, n, \qquad k = 0, 1, \ldots. \tag{5.2.6}$$

This is of the form

$$\dot{\mathbf{y}}_i^{(k+1)} = \mathbf{A}(\mathbf{y}^{(k)}(t))\mathbf{y}^{(k+1)}(t) + \mathbf{f}(t), \qquad k = 0, 1, \ldots, \tag{5.2.7}$$

a linear differential equation with variable coefficients, where

$\mathbf{A}(\mathbf{y}^{(k)}(t))$ = $n \times n$ matrix with elements $\partial g_i / \partial y_j$ evaluated at $\mathbf{y}^{(k)}(t)$,

$\mathbf{f}(t)$ = $n \times 1$ vector with elements

$$g_i(y_1^{(k)}(t), y_2^{(k)}(t), \ldots, y_n^{(k)}(t), t) - \sum_{j=1}^{n} \frac{\partial g_i}{\partial y_j} y_j^{(k)}(t),$$

$$i = 1, 2, \ldots, n.$$

Instead of solving the nonlinear two-point boundary value problem (5.2.1) with the boundary conditions (5.2.2) and (5.2.3), we now solve a sequence of linear two-point boundary value problems (5.2.6) for $k = 0, 1, \ldots$. with the boundary conditions (5.2.2′) and (5.2.3′):

$$y_i^{(k)}(t_0) = y_i(t_0) = c_i, \qquad i = 1, 2, \ldots, r, \qquad k = 0, 1, \ldots, \tag{5.2.2′}$$

$$y_{i_m}^{(k)}(t_f) = y_{i_m}(t_f) = c_{i_m}, \qquad m = 1, 2, \ldots, n-r, \qquad k = 0, 1, \ldots. \tag{5.2.3′}$$

The linear two-point boundary value problem may be solved by either the method of adjoints or the method of complementary functions.

Theoretically, for a solution to the nonlinear problem we require

$$\lim_{k \to \infty} y_i^{(k)}(t) = y_i^*(t), \qquad i = 1, 2, \ldots, n, \qquad t_0 \le t \le t_f, \tag{5.2.8}$$

where $y_i^*(t)$, $i = 1, 2, \ldots, n$, is the solution of the nonlinear problem. Numerically we require that

$$|y_i^{(k+1)}(t) - y_i^{(k)}(t)| < \varepsilon, \qquad i = 1, 2, \ldots, n, \qquad t_0 \le t \le t_f, \tag{5.2.9}$$

where ε is a small tolerance prescribed by the investigator.

To recapitulate, the quasilinear process consists of the following steps.

1. Linearize the right-hand side of (5.2.1) to obtain (5.2.6).

2. For the zeroth iteration $k = 0$, provide the nominal profiles, $y_1^{(0)}(t)$, $y_2^{(0)}(t), \ldots, y_n^{(0)}(t)$, $t_0 \le t \le t_f$, which satisfy the boundary conditions.

3. For the kth iteration, using as nominal profiles $y_i^{(k)}(t)$, $i = 1, 2, \ldots, n$, $t_0 \le t \le t_f$, solve the linear two-point boundary value problem (5.2.6) with the boundary conditions (5.2.2′) and (5.2.3′) by either the method of adjoints or the method of complementary functions.

4. Store the profiles.

5. Test whether

$$|y_i^{(k+1)}(t) - y_i^{(k)}(t)| < \varepsilon, \qquad i = 1, 2, \ldots, n; \quad t_0 \le t \le t_f.$$

where ε is small number prescribed by the investigator.

6. If tolerance test (item 5) is not passed, increment the index $k \to k+1$ and go to item 3.

7. If tolerance test is passed, terminate the calculation; the profiles $y_i^{(k+1)}(t)$, $i = 1, 2, \ldots, n$; $t_0 \le t \le t_f$, are the numerical solution to the nonlinear two-point boundary value problem.

As a practical matter, of course, we always set an upper limit on the iteration count k.

5.3. DISCUSSION

In Chapter 6 we show that the method of adjoints and quasilinearization are both realizations of the Newton-Raphson method in Banach space. Each method, however, has its own particular Banach space. For the method of adjoints the Banach space is the Cartesian product of the reals; that is, each point in the Banach space is an $(n - r)$-tuple of missing initial conditions. For the quasilinear process the Banach space is the set of all continuous functions which have continuous first derivatives and satisfy the prescribed boundary conditions*. In other words, each "point" in this Banach space is a continuous curve which has continuous first derivatives and which satisfies the prescribed boundary conditions. In the method of adjoints we iterate on the set of $(n - r)$ missing initial conditions. On the other hand, in quasilinearization we iterate on the entire profile $y_i^{(k)}(t)$, $i = 1, 2, \ldots, n$, $t_0 \leq t \leq t_f$.

Once we realize that quasilinearization is a Newton-Raphson type process, we should recognize that the initial point (that is, the initial profiles in quasilinearization) is crucial to the success of the method. There are several ways to obtain the nominal profiles, a few of which we mention here. The simplest way is merely to guess them. Sometimes just straight line guesses are adequate. Such straight line profiles may be attempted if the same variables are specified at the initial and terminal points. In trajectory problems for example, the initial position and terminal position are often prescribed, and straight line initial profiles have been used successfully.

A second method for finding the initial profiles requires the investigator to be familiar enough with the problem to know the general shape of the profiles.

A third technique for generating nominal profiles is to assume values for the missing initial conditions for the nonlinear differential equations (5.2.1) and integrate forward. Although the profiles in general will not satisfy the terminal conditions, the profiles can be deformed by the investigator to satisfy the terminal conditions. A fourth method is described in Chapter 7 and is an application of continuation.

In order of increasing complexity of problem preparation, problem analysis, and computer implementation, two-point boundary value problems can be solved by shooting methods, quasilinearization, and finite difference

* Strictly speaking, the boundary conditions must be zero for a Banach space. See page 144.

methods. For the numerical solution of sensitive (unstable) problems, shooting methods may give unsatisfactory answers due to round-off errors. In view of this, many investigators have turned to finite difference methods in order to obtain a satisfactory solution, even though the problem preparation may be arduous and even though large sets of nonlinear algebraic equations are generated which subsequently are linearized and then solved. Quasilinearization is in a sense a compromise between shooting methods and finite difference methods. Quasilinearization bypasses the nonlinear differential equations, which may be difficult to solve, in favor of a sequence of linear differential equations, which are easier to solve. As in the finite difference method, the quasilinear process "clamps" the trial solutions at the specified initial and terminal boundary conditions so the boundary conditions are always satisfied. As in the methods of adjoints and complementary functions, quasilinearization requires the determination of the partial derivatives $\partial g_i/\partial y_j$, $i,j = 1,2,\ldots,n$, and a "standard" integration code to solve the linear boundary value problems. (See Chapter 8 for finite difference methods.)

Since the linearized version of the nonlinear differential equations may be less sensitive numerically than the original nonlinear differential equations, quasilinearization may be employed rather than finite difference approximations to the original nonlinear differential equations. It should be mentioned, however, that the linearization of numerically sensitive nonlinear differential equations does not guarantee the existence of numerically stable differential equations. We discuss in Section 7.9, Chapter 7, two numerically sensitive linear differential equations that are derived from numerically sensitive nonlinear differential equations.

It should also be mentioned that quasilinearization is not limited to the boundary conditions of the type (5.2.2) and (5.2.3) of Section 5.2. See Bellman and Kalaba for a discussion of more complicated boundary conditions including the multipoint boundary conditions [1].

As another practical computing matter, we may find that the kth quasilinear iteration produces profiles for the system which do not satisfy the terminal conditions exactly. If the error at the terminal conditions is small, the $(k+1)$st profiles are adjusted to satisfy the specified terminal conditions and the process is continued. We have found that, for problems in which the quasilinear process fails, the failure is characterized by two types of events. First, the set of missing initial conditions found by quasilinearization gives rise to profiles which miss the specified terminal conditions by larger and larger errors in the sequence of problems. This occurs even when the terminal conditions are adjusted to meet the specified terminal conditions. Second, overflow occurs in the machine solution.

In our discussions we have assumed that we have been dealing with systems of first-order nonlinear differential equations. Higher-order systems can be reduced to first order, as we showed in Section 2.3, Chapter 2. Quasilinearization can also be applied directly to higher-order systems by writing the equations so that the highest derivative is the only term on the left-hand side of the equation and then expanding the right-hand side around a nominal solution. Thus, if the nonlinear differential equation of the mth order can be written as

$$\frac{d^m y}{dt^m} = f\left(y, \frac{dy}{dt}, \frac{d^2 y}{dt^2}, \ldots, \frac{d^{m-1} y}{dt^{m-1}}\right),$$

the quasilinear process equations appear as

$$(y^m)^{(k+1)} = f(y, (y^1)^{(k)}, (y^2)^{(k)}, \ldots, (y^m)^{(k)}) + \frac{\partial f}{\partial y}(\delta y)^{(k)} + \\ + \frac{\partial f}{\partial y^1}(\delta y^1)^{(k)} + \ldots + \frac{\partial f}{\partial y^m}(\delta y^m)^{(k)},$$

where

$$y^m = \frac{d^m y}{dt^m},$$

$(y^m)^{(k)}$ = mth derivative of y at kth iteration,

$$(\delta y^m)^{(k)} = (y^m)^{(k+1)} - (y^m)^{(k)}.$$

The two approaches are illustrated in the example of Section 5.4.

To summarize: Quasilinearization possesses several advantages.

1. It avoids solving directly nonlinear ordinary differential equation boundary value problems, which may be difficult, by solving instead a sequence of linear two-point boundary problems which are easier to solve.
2. The quasilinear process equations may be formulated without an elaborate analysis of the differential equations.
3. When the process converges, it does so quadratically. (See Section 6.2, Chapter 6.)
4. With the exception of the subroutine into which the right-hand side of the quasilinear equations are entered, a standard linear two-point boundary value computer program may be used, such as a code for the method of adjoints or the method of complementary functions.

5. For numerically sensitive two-point boundary value problems, the quasilinear equations may be more stable than the original nonlinear differential equations. Since the quasilinear process "clamps" the trial solutions at the specified initial and terminal conditions, the quasilinear process may succeed where shooting methods fail.

Quasilinearization also possesses some disadvantages.

1. Selection of the initial nominal profiles which lead to convergence may be a difficult task.
2. The computer program must store profiles for both $\mathbf{y}^{(k)}(t)$ and $\mathbf{y}^{(k-1)}(t)$ at discrete points within the interval $t_0 \leq t \leq t_f$ and must test the error $|y_i^{(k+1)}(t) - y_i^{(k)}(t)|$ against ε, $i = 1, 2, \ldots, n$; $t_0 \leq t \leq t_f$.
3. For each problem the linearized equations must be developed. (The partial derivatives $\partial g_i / \partial y_j$, $i, j = 1, 2, \ldots, n$, however, must be developed also for techniques such as the method of adjoints and Newton-Raphson method applied to nonlinear finite difference equations.)

5.4. EXAMPLE

In this section we show in detail the quasilinear process equations for a particular second-order nonlinear differential equation by the two approaches mentioned in Section 5.3: first, by transforming the second-order system into a first-order system and then applying quasilinearization; second, by applying quasilinearization directly to the second-order equation. Numerical results illustrate the first approach.

Consider the second-order nonlinear ordinary differential equation

$$3y\ddot{y} + \dot{y}^2 = 0 \tag{5.4.1}$$

with the boundary conditions

$$y(0) = 0, \qquad y(1) = 1. \tag{5.4.2}$$

This problem, which arose in optimization theory, is easily solved analytically; the solution is $y(t) = t^{3/4}$ which we will use for purposes of comparison.

To transform (5.4.1) into a system of first-order equations, let

$$y_1 = y, \qquad y_2 = \dot{y}_1, \tag{5.4.3}$$

and replace (5.4.1) with the set

$$\dot{y}_1 = y_2 = g_1(y_2), \qquad \dot{y}_2 = -\frac{{y_2}^2}{3y_1} = g_2(y_1, y_2), \tag{5.4.4}$$

with the boundary conditions

$$y_1(0) = 0, \qquad y_1(1) = 1. \tag{5.4.5}$$

On expanding the right-hand side of (5.4.4) in a Taylor's series around nominal solutions $y_1^{(k)}(t)$, $y_2^{(k)}(t)$, we have

$$g_i(y_1, y_2) = g_i({y_1}^{(k)}(t), {y_2}^{(k)}(t)) + \left(\frac{\partial g_i}{\partial y_1}\right)\delta {y_1}^{(k)} + \left(\frac{\partial g_i}{\partial y_2}\right)\delta {y_2}^{(k)}, \qquad i = 1, 2, \tag{5.4.6}$$

where the partial derivatives are evaluated using $y_1^{(k)}(t)$, $y_2^{(k)}(t)$. From (5.4.4) it follows that

$$\begin{aligned}
\frac{\partial g_1}{\partial y_1} &= 0, & \frac{\partial g_1}{\partial y_2} &= 1, \\
\frac{\partial g_2}{\partial y_1} &= \frac{1}{3}\left(\frac{{y_2}^{(k)}(t)}{{y_1}^{(k)}(t)}\right)^2, & \frac{\partial g_2}{\partial y_2} &= -\frac{2}{3}\left(\frac{{y_2}^{(k)}(t)}{{y_1}^{(k)}(t)}\right),
\end{aligned} \tag{5.4.7}$$

$$\delta {y_1}^{(k)} = {y_1}^{(k+1)}(t) - {y_1}^{(k)}(t), \qquad \delta {y_2}^{(k)} = {y_2}^{(k+1)}(t) - {y_2}^{(k)}(t).$$

The set of nonlinear differential equations (5.4.4) is then replaced by the system of linear equations, using (5.4.6) and (5.4.7):

$$\begin{aligned}
{\dot{y}_1}^{(k+1)} &= {y_2}^{(k)}(t) + ({y_2}^{(k+1)}(t) - {y_2}^{(k)}(t)), \\
{\dot{y}_2}^{(k+1)} &= -\frac{{y_2}^{(k)}(t)^2}{3{y_1}^{(k)}(t)} + \frac{1}{3}\left(\frac{{y_2}^{(k)}(t)}{{y_1}^{(k)}(t)}\right)^2 (y_1^{(k+1)}(t) - y_1^{(k)}(t)) \\
&\quad - \frac{2}{3}\left(\frac{{y_2}^{(k)}(t)}{{y_1}^{(k)}(t)}\right)({y_2}^{(k+1)}(t) - {y_2}^{(k)}(t)).
\end{aligned} \tag{5.4.8}$$

This is a system linear in $y_1^{(k+1)}(t)$ and $y_2^{(k+1)}(t)$. Equations (5.4.8) can be written more compactly as

$$\begin{bmatrix} {\dot{y}_1}^{(k+1)} \\ \\ {\dot{y}_2}^{(k+1)} \end{bmatrix} = \begin{bmatrix} 0 & 1 \\ \frac{1}{3}\left(\frac{{y_2}^{(k)}(t)}{{y_1}^{(k)}(t)}\right)^2 & -\frac{2}{3}\left(\frac{{y_2}^{(k)}(t)}{{y_1}^{(k)}(t)}\right) \end{bmatrix} \begin{bmatrix} {y_1}^{(k+1)}(t) \\ \\ {y_2}^{(k+1)}(t) \end{bmatrix} \tag{5.4.9}$$

with the boundary conditions

$$y_1^{(k)}(0) = 0, \qquad y_1^{(k)}(1) = 1, \qquad k = 0, 1, \ldots. \tag{5.4.10}$$

Equations (5.4.9) with the boundary conditions (5.4.10) are the equations of the quasilinear process. Once satisfactory profiles for $y_1^{(0)}(t), y_2^{(0)}(t)$, $0 \leq t \leq 1$, are assumed in this problem, the sequence of linear two-point boundary value problems will converge to the solution of the original nonlinear problem, (5.4.1) and (5.4.2).

The alternative derivation of the quasilinear process will now be given in terms of second-order linear ordinary differential equations. We solve (5.4.1) for the highest-order derivative:

$$\ddot{y} = -\frac{\dot{y}^2}{3y} = f(y, \dot{y}). \tag{5.4.11}$$

The Taylor's series expansion of the right-hand side of (5.4.11) through first-order terms around the nominal solution $y^{(k)}(t)$, $\dot{y}^{(k)}(t)$ gives

$$f(y, \dot{y}) = f(y^{(k)}(t), \dot{y}^{(k)}(t)) + \frac{\partial f}{\partial y}\,\delta y^{(k)} + \frac{\partial f}{\partial \dot{y}}\,\delta \dot{y}^{(k)}, \tag{5.4.12}$$

where

$$\frac{\partial f}{\partial y} = \frac{1}{3}\left(\frac{\dot{y}^{(k)}(t)}{y^{(k)}(t)}\right)^2, \qquad \frac{\partial f}{\partial \dot{y}} = -\frac{2}{3}\left(\frac{\dot{y}^{(k)}(t)}{y^{(k)}(t)}\right),$$

$$\delta y^{(k)} = y^{(k+1)}(t) - y^{(k)}(t), \qquad \delta \dot{y}^{(k)} = \dot{y}^{(k+1)}(t) - \dot{y}^{(k)}(t). \tag{5.4.13}$$

Equation (5.4.11) is replaced by the linear ordinary differential equation

$$\begin{aligned}\ddot{y}^{(k+1)} = &-\frac{\dot{y}^{(k)}(t)^2}{3y^{(k)}(t)} + \frac{1}{3}\left(\frac{\dot{y}^{(k)}(t)}{y^{(k)}(t)}\right)^2 (y^{(k+1)}(t) - y^{(k)}(t)) \\ &-\frac{2}{3}\left(\frac{\dot{y}^{(k)}(t)}{y^{(k)}(t)}\right)(\dot{y}^{(k+1)}(t) - \dot{y}^{(k)}(t)),\end{aligned} \tag{5.4.14}$$

which reduces to

$$\ddot{y}^{(k+1)} = \frac{1}{3}\left(\frac{\dot{y}^{(k)}(t)}{y^{(k)}(t)}\right)^2 y^{(k+1)}(t) - \frac{2}{3}\left(\frac{\dot{y}^{(k)}(t)}{y^{(k)}(t)}\right)\dot{y}^{(k+1)}(t). \tag{5.4.15}$$

The equation is linear in $y^{(k+1)}(t)$ with the boundary conditions

$$y^{(k)}(0) = 0, \qquad y^{(k)}(1) = 1, \qquad k = 0, 1, \ldots. \tag{5.4.16}$$

Once satisfactory initial profiles are obtained for $y^{(0)}(t), \dot{y}^{(0)}(t)$, $0 \le t \le 1$, the quasilinear process will again converge to the solution of the nonlinear two-point boundary value problem, (5.4.1) and (5.4.2). As a check on the two quasilinear representations, we notice that (5.4.9) and (5.4.15) are consistent if we use the defining equations (5.4.3).

It can be expected that the nonlinear two-point boundary value problem, (5.4.1), (5.4.2), will present numerical difficulties near the origin, since both $\dot{y}(0)$ and $\ddot{y}(0)$ are infinite there. Even if we did not know that the exact solution is $t^{3/4}$, the presence of $y_1(t)$ in the denominator of the second of equations (5.4.9) must cause numerical problems in the neighborhood of $t = 0$ in view of the initial condition $y_1(0) = 0$. In the three cases presented in Table 5.1, a simple device was used to handle this problem, namely,

Table 5.1.

t	$y(t)$ exact	$y^{(3)}(t)$ (case 1)	$y^{(3)}(t)$ (case 2)	$y^{(3)}(t)$ (case 3)
0.00	0.0	$1.0000000(10^{-8})$	$1.0000000(10^{-4})$	$1.0000000(10^{-2})$
0.05	$1.0573712(10^{-1})$	$1.0966162(10^{-1})$	$1.1019137(10^{-1})$	$1.6364506(10^{-1})$
0.10	$1.7782794(10^{-1})$	$1.8087364(10^{-1})$	$1.8129937(10^{-1})$	$2.2307465(10^{-1})$
0.15	$2.4102852(10^{-1})$	$2.4360867(10^{-1})$	$2.4397770(10^{-1})$	$2.7940928(10^{-1})$
0.20	$2.9906975(10^{-1})$	$3.0131881(10^{-1})$	$3.0164820(10^{-1})$	$3.3258867(10^{-1})$
0.25	$3.5355339(10^{-1})$	$3.5553890(10^{-1})$	$3.5583736(10^{-1})$	$3.8322186(10^{-1})$
0.30	$4.0536004(10^{-1})$	$4.0712280(10^{-1})$	$4.0739558(10^{-1})$	$4.3178725(10^{-1})$
0.35	$4.5504152(10^{-1})$	$4.5660888(10^{-1})$	$4.5685949(10^{-1})$	$4.7863470(10^{-1})$
0.40	$5.0297337(10^{-1})$	$5.0436500(10^{-1})$	$5.0459593(10^{-1})$	$5.2402425(10^{-1})$
0.45	$5.4942622(10^{-1})$	$5.5065696(10^{-1})$	$5.5087010(10^{-1})$	$5.6815439(10^{-1})$
0.50	$5.9460355(10^{-1})$	$5.9568502(10^{-1})$	$5.9588183(10^{-1})$	$6.1118053(10^{-1})$
0.55	$6.3866338(10^{-1})$	$6.3960494(10^{-1})$	$6.3978658(10^{-1})$	$6.5322705(10^{-1})$
0.60	$6.8173161(10^{-1})$	$6.8254099(10^{-1})$	$6.8270843(10^{-1})$	$6.9439541(10^{91})$
0.65	$7.2391073(10^{-1})$	$7.2459440(10^{-1})$	$7.2474843(10^{-1})$	$7.3476976(10^{-1})$
0.70	$7.6528557(10^{-1})$	$7.6584906(10^{-1})$	$7.6599036(10^{-1})$	$7.7442082(10^{-1})$
0.75	$8.0592744(10^{-1})$	$8.0637552(10^{-1})$	$8.0650469(10^{-1})$	$8.1340881(10^{-1})$
0.80	$8.4589701(10^{-1})$	$8.4623386(10^{-1})$	$8.4635138(10^{-1})$	$8.5178551(10^{-1})$
0.85	$8.8524645(10^{-1})$	$8.8547575(10^{-1})$	$8.8558208(10^{-1})$	$8.8959588(10^{-1})$
0.90	$9.2402108(10^{-1})$	$9.2414612(10^{-1})$	$9.2424166(10^{-1})$	$9.2687926(10^{-1})$
0.95	$9.6226060(10^{-1})$	$9.6228430(10^{-1})$	$9.6236940(10^{-1})$	$9.6367036(10^{-1})$
1.00	1.0000000	$9.9992503(10^{-1})$	1.0000000	1.0000000

the initial condition $y_1(0) = 0$ was replaced respectively by the condition $y_1(0) = 10^{-8}$, 10^{-4} and 10^{-2}. In Table 5.1 the numerical quasilinear profiles $y^{(k)}(t)$, obtained by a four-point Runge-Kutta integration formula with a step size of 0.05, are tabulated for $k = 3$. The initial profiles in each case were based on a straight line between the initial and final values, and the method of adjoints was used to solve the sequence of linear two-point boundary value problems. As would be expected, the quasilinear profiles are least accurate near $t = 0$, but the agreement with the exact solution seems to be quite satisfactory for $t > 0.15$, particularly in the case $y_1(0) = 10^{-8}$ (third column). In practice, where the exact solution would not be available, the reasonably close agreement between the solution corresponding to $y_1(0) = 10^{-4}$ and $y_1(0) = 10^{-8}$. would be taken as evidence that the problem of the zero initial condition had been circumvented.

5.5. CONVERGENCE OF THE QUASILINEARIZATION METHOD

5.5.1. Introduction

So far we have described the quasilinearization method of solving two-point boundary value problems, and we have illustrated the method with an example, but we have not established the conditions under which the method converges, nor have we given estimates of the rate of convergence. There are several methods for proving the convergence of quasilinearization. Bellman and Kalaba [1] make use of convexity and monotonicity arguments. In Chapter 6 we show that quasilinearization is a particular realization of the abstract Newton-Raphson-Kantorovich method, which allows us to infer convergence of quasilinearization from the proof of the Newton-Raphson-Kantorovich method. We give below yet another proof of convergence due to McGill and Kenneth [3] which may be of some interest in itself. The discussion makes use of some concepts from the theory of metric spaces. Brief definitions of these concepts have been gathered in the Appendix for the convenience of the reader.

5.5.2. Preliminaries

McGill and Kenneth have provided a proof of the convergence of the quasilinear process defined on a metric space for the second-order two-point boundary value problem

$$\frac{d^2u}{dt^2} = f(u, t), \tag{5.5.1}$$

$$u(a) = a_1, \qquad u(b) = b_1. \tag{5.5.2}$$

The function $f(u,t)$ is continuous in u and t and satisfies the Lipschitz condition

$$|f(u_1,t)-f(u_2,t)| \leq K|u_2-u_1| \tag{5.5.3}$$

in the domain $-\infty \leq u(t) \leq \infty$; $a \leq t \leq b$, $K > 0$.

While their proof in many ways runs parallel to the classical proof of the existence, uniqueness, and convergence properties of this problem via Green's function and an integral equation, it does exhibit certain novel aspects for quasilinearization. McGill and Kenneth use a new type of implicit operator defined on a metric space and do not require monotone operators or convexity of the space, as some quasilinear process proofs demand.

First we sketch briefly the classical strategy for proving the existence and uniqueness of the solution to (5.5.1) and (5.5.2) [5, 6]. Then the problem will be considered in a metric space and solved by quasilinearization.

The homogeneous two-point boundary value problem

$$\frac{d^2v}{dt^2} = f(v,t), \tag{5.5.4}$$

$$v(a) = 0, \qquad v(b) = 0, \tag{5.5.5}$$

with $f(v,t)$ continuous in v and t and satisfying the Lipschitz condition

$$|f(v_1,t)-f_2(v_2,t)| \leq K|v_1-v_2|, \tag{5.5.6}$$

may be cast into an integral equation form:

$$v(t) = \int_a^b G(t,s)\,f(v,s)\,ds, \tag{5.5.7}$$

The $G(t,s)$ is the Green's function associated with $-d^2v/dt^2$ and the homogenous boundary conditions, and is defined by

$$G(t,s) = \left.\begin{matrix} \dfrac{(b-t)(s-a)}{b-a}, & a \leq s \leq t, \\[2ex] \dfrac{(t-a)(b-s)}{b-a}, & t \leq s \leq b. \end{matrix}\right\} \tag{5.5.8}$$

$G(t, s)$ is symmetric, that is, $G(t, s) = G(s, t)$; continuous in s and t; non-negative, $G(t, s) \geq 0$; and for fixed t, takes its maximum value at $s = t$. $G(s, s)$ assumes its maximum value of $(b-a)/4$ at $s = (b+a)/2$.

The boundary value problem (5.5.1) and (5.5.2) can be reduced to that of (5.5.4) and (5.5.5) by setting

$$v(t) = u(t) - u_{ab}(t), \tag{5.5.9}$$

where

$$u_{ab}(t) = a_1 + \left(\frac{b_1 - a_1}{b - a}\right)(t - a). \tag{5.5.10}$$

The $u_{ab}(t)$ is a linear function joining the boundary conditions $u(a)$ and $u(b)$. The Lipschitz constant in (5.5.6) is the same as that in (5.5.3).

To solve (5.5.7) we may employ an iteration procedure, since v appears on both sides of the equation:

$$v^{(k)}(t) = \int_a^b G(t, s)\, f(v^{(k-1)}, s)\, ds. \tag{5.5.11}$$

If we define the space X as the set of all continuous functions on the closed interval $[a, b]$ with the norm

$$\|w\| = \max_{a \le t \le b} |w(t)|, \tag{5.5.12}$$

then X is a Banach space, that is, a complete normed linear space. In the sequence of functions $v^{(k)}(t)$ in (5.5.11), all the members belong to X.

We may express (5.5.11) in operator notation as

$$v^{(k)}(t) = Pv^{(k-1)}(t), \tag{5.3.13}$$

where P is the operation of integrating the right-hand side of (5.5.11). Given a $v^{(k-1)}(t)$, a member of X, we see that the element $Pv^{(k-1)}(t)$ is also a member of X. Under the appropriate conditions, P is a contraction mapping.

To see this let v_1 and v_2 be any members of X; then

$$Pv_1(t) - Pv_2(t) = \int_a^b G(t, s)\, \{f(v_1, s) - f(v_2, s)\}\, ds. \tag{5.5.14}$$

Taking the norm of both sides, we have

$$\|Pv_1 - Pv_2\| \le \int_a^b \|G(t, s)\|\, \|f(v_1, s) - f(v_2, s)\|\, \|ds\|, \tag{5.5.15}$$

$$\|Pv_1 - Pv_2\| \le \frac{K(b-a)^2}{4}\, \|v_1 - v_2\|, \tag{5.5.16}$$

where we have used the Lipschitz condition and the maximum value of $G(t,s)$. If

$$\alpha = \frac{K(b-a)^2}{4} < 1, \tag{5.5.17}$$

then

$$\|Pv_1 - Pv_2\| \leq \alpha\|v_1 - v_2\| \tag{5.5.18}$$

and P is a contraction mapping. The existence and uniqueness of the solution of (5.5.4) and (5.5.5) follow from the properties of the contraction mapping. From the contraction mapping we can obtain the estimate of the rate of convergence to the solution v^* by

$$\|v^{(k)} - v^*\| \leq \frac{\alpha^k}{1-\alpha}\|v^{(1)} - v^{(0)}\|, \qquad k = 1, 2, \ldots. \tag{5.5.19}$$

The results may be summarized in the following theorem.

THEOREM 5.1. *If, in* (5.5.1) *and* (5.5.2), $f(u,t)$ *is continuous over the domain* $-\infty \leq u \leq \infty$ *and* $a \leq t \leq b$, *and* $f(u,t)$ *is Lipschitzian with nonnegative constant* K, *then a solution exists and is unique provided* $K(b-a)^2/4 < 1$. *In addition, the iterative solution technique of* (5.5.11) *converges to the solution* u^* *at the rate*

$$\|u^{(k)} - u^*\| \leq \frac{\alpha^k}{1-\alpha}\|u^{(1)} - u^{(0)}\|, \qquad k = 1, 2, \ldots. \tag{5.5.20}$$

To recapitulate, the pattern of development has been:

1. To cast the second-order differential equation into an integral equation using the appropriate Green's function.
2. To define a solution space X for $v^{(k)}(t)$ on the closed interval $[a,b]$ and an appropriate norm and to define an operator P on the space.
3. To demonstrate that, if $v^{(k-1)}(t)$ is a member of X, then the element $Pv^{(k-1)} = v^{(k)}$ is also a member of X; this shows that the members of the sequence $\{v^{(k)}(t)\}$ belong to X. P, therefore, is a mapping from X to X.
4. To show that P is a contraction mapping.

In Section 5.5.3 we carry out similar steps for the proof of quasilinearization in a metric space.

5.5.3. Theorem of McGill and Kenneth [3]

We now state and prove, following the outline described above, the McGill and Kenneth theorem for second-order boundary value problems in a complete metric space.

THEOREM 5.2. *Consider the nonlinear second-order two-point boundary value problem*

$$\frac{d^2u}{dt^2} = f(u,t), \tag{5.5.1}$$

$$u(a) = a_1, \qquad u(b) = b_1, \tag{5.5.2}$$

(i) *where* $u(t)$ *is a member of the complete metric space* S, *defined as* $S = \{u(t)|u(t)$ *is continuous on* $[a,b]$, $u(a) = a_1$, $u(b) = b_1$, $\rho(u(t), u_{ab}(t)) \le C\}$.

(ii) *the metric* ρ *is defined as*

$$\rho(u_1,u_2) = \max_{a\le t\le b} |u_1(t)-u_2(t)|, \qquad \forall u_1,u_2 \in S; \tag{5.5.21}$$

(iii)

$$u_{ab}(t) = a_1 + \left(\frac{b_1-a_1}{b-a}\right)(t-a); \tag{5.5.10}$$

(iv) C *is a nonnegative constant.*
Assume:

(1) $f(u,t)$ *is continuous in* t,
(2) $\partial f(u,t)/\partial u = f_u$ *exists and is continuous in* u,
(3) $\partial f/\partial u$ *is Lipschitzian in* u *with the Lipschitz constant* $M_1, M_1 > 0$; *that is,*

$$|f_u(u_1,t)-f_u(u_2,t)| < M_1|u_1-u_2|, \qquad \forall u_1, u_2 \in S.$$

Associated with (5.5.1) *and* (5.5.2) *is the sequence of linear two-point boundary value problems*

$$\frac{d^2u^{(k+1)}}{dt^2} = f_u(u^{(k)},t)\,[u^{(k+1)}(t)-u^{(k)}(t)]+f(u^{(k)},t), \qquad k = 0,1,\ldots, \tag{5.5.22}$$

$$u^{(k)}(a) = a_1, \qquad u^{(k)}(b) = b_1, \tag{5.5.23}$$

where $u^{(0)}(t)$ *is a member of* S.

As a consequence of the preceding conditions for a sufficiently small interval $[a, b]$, *the following results hold.*

1. *The unique solution of* (5.5.1) *and* (5.5.2) *exists.*
2. *The sequence* $\{u^{(k)}(t)\}$ *generated by* (5.5.22) *and* (5.5.23) *converges uniformly in t to the unique solution of* (5.5.1) *and* (5.5.2).
3. *The convergence of* $\{u^{(k)}(t)\}$ *is at least quadratic, that is,*

$$\rho(u^{(k+1)}, u^*) \leq \frac{\alpha}{C+1} \rho^2(u^{(k)}, u^*), \tag{5.5.24}$$

where u^* *is the unique solution to* (5.5.1) *and* (5.5.2), α, *a constant* < 1, *is given in* (5.5.34), *and C is the constant in the definition of the space S.*

4. *A bound on the error is given by*

$$\max_{a\leq t\leq b} |u^{(k+1)}(t) - u^*(t)| \leq \frac{\alpha}{(1-\alpha)^2 (C+1)} [\max_{a\leq t\leq b} |u^{(k+1)}(t) - u^{(k)}(t)|]^2 \tag{5.5.25}$$

PROOF. *Part* 1. Since $f(u,t)$ and $\partial f(u,t)/\partial u$ are defined on the closed and bounded domain $a \leq t \leq b$, $u_{ab}(t) - C \leq u \leq u_{ab}(t) + C$, then $f(u,t)$ and $f_u(u,t)$ are bounded; that is, $|f(u,t)| \leq M_2$, $|f_u(u,t)| \leq M_3$, $M_2, M_3 > 0$.

By the continuity of $f(u,t)$ and existence of $\partial f/\partial u$,

$$|f(u_2, t) - f(u_1, t)| \leq \frac{\partial f}{\partial u} |u_2 - u_1| \leq M_3 |u_2 - u_1|.$$

Let $m = \max(M_1, M_2, M_3)$.

The solution of the linear second-order two-point boundary value problem (5.5.22) and (5.5.23) may be expressed in an integral equation form with Green's function, as was done in (5.5.1) through (5.5.11). We now define an implicit operator A:

$$Au(t) = u_{ab}(t) - \int_a^b G(t,s) \{f_u(u,s)[Au(s) - u(s)] + f(u,s)\} \, ds, \tag{5.5.26}$$

whose form is patterned on the integral equation representation of (5.5.22) and (5.5.23). The Green's function $G(t,s)$ is defined in (5.5.8).

We show first that, if $u(t) \in S$, then $Au(t) \in S$. That is, A is a mapping of S into S. Transposing $u_{ab}(t)$ to the left-hand side and taking norms yields

$$\max_{a\leq t\leq b} |Au(t) - u_{ab}(t)| \leqslant \max_{a\leq t\leq b} \int_a^b |G(t,s)| \{|f_u| \, |Au(s) - u(s)| + |f(u,s)|\} \, |ds|. \tag{5.5.27}$$

On using the bounds above, we have

$$\max_{a\le t\le b} |Au(t)-u_{ab}(t)| \le \frac{m}{4}(b-a)^2 \max_{a\le t\le b}\{|Au(t)-u_{ab}(t)|+\rho(u,u_{ab})+1\} \tag{5.5.28}$$

or

$$\max_{a\le t\le b} |Au(t)-u_{ab}(t)| \le \frac{(m/4)(b-a)^2(C+1)}{1-(m/4)(b-a)^2}. \tag{5.5.29}$$

If $(b-a)$ is sufficiently small, the right-hand side of (5.5.29) is less than C. Under these circumstances,

$$\max_{a\le t\le b} |Au(t)-u_{ab}(t)| \le C. \tag{5.5.30}$$

Since it is easy to show that $Au(t)$ is continuous, $Au(a) = a_1$, and $Au(b) = b_1$, we have just proved that, if $u \in S$, then $Au \in S$, so A is a mapping of S into S.

Since by definition the initial solution of (5.5.22) and (5.5.23) namely, $u^{(0)}(t)$ is a member of S, it follows that operator A, applied to (5.5.22) generates a sequence of solutions $\{u^{(k)}(t)\}$ whose members also belong to S.

Part 2. We next show that A is a contraction mapping. For any u_1 and $u_2 \in S$, we may write

$$Au_1 - Au_2 = \int_a^b G(t,s)\{f_u(u_2,s)[Au_2(s)-u_2(s)]-f_u(u_1,s)[Au_1(s)-u_1(s)]- \\ -[f(u_1,s)-f(u_2,s)]\}\,ds. \tag{5.5.31}$$

On adding and subtracting $f_u(u_1,s)Au_2(s)$ and $f_u(u_2,s)u_1$ to, and from, the expression in the parentheses on the right-hand side of (5.5.31), we write (5.5.31) as

$$Au_1 - Au_2 = \int_a^b G(t,s)\{f_u(u_1,s)[Au_2(s)-Au_1(s)]+f_u(u_2,s)[u_1(s)-u_2(s)]+ \\ +[f_u(u_1,s)-f_u(u_2,s)][u_1(s)-Au_2(s)]-[f(u_1,s)-f(u_2,s)]\}\,ds. \tag{5.5.32}$$

Taking norms in (5.5.32) leads to

$$\rho(Au_1,Au_2) \le \frac{m}{4}(b-a)^2[\rho(Au_1,Au_2)+\rho(u_1,u_2)+ \\ +\rho(u_1,u_2)\rho(u_1,Au_2)+\rho(u_1,u_2)]$$

or

$$\rho(Au_1, Au_2) \le \frac{(m/2)(b-a)^2(C+1)}{1-(m/4)(b-a)^2}\,\rho(u_1, u_2). \tag{5.5.33}$$

If $(b-a)$ is sufficiently small so that

$$\alpha = \frac{(m/2)(b-a)^2(C+1)}{1-(m/4)(b-a)^2} < 1, \tag{5.5.34}$$

then

$$\rho(Au_1, Au_2) \le \alpha\rho(u_1, u_2), \qquad \alpha < 1; \tag{5.5.35}$$

this proves that A is a contraction mapping.

By the contraction mapping principle every contraction mapping defined on a complete metric space S has one and only one fixed point. In other words, $u = Au$ has one and only one solution, u^*. Since A is a contraction mapping, the sequence $\{u^{(k)}(t)\}$, $k = 0, 1, 2, \ldots$, $t \in [a, b]$, generated by $u^{(k)}(t) = Au^{(k-1)}(t)$ (where A is the operation of solving the linear two-point boundary value problems in (5.5.22) and (5.5.23)) is a Cauchy sequence. In a complete metric space the limit of a Cauchy sequence exists and is a member of the space; that is,

$$\lim_{k\to\infty} u^{(k)}(t) = u^*(t), \qquad \forall t \in [a, b].$$

If we substitute $u^*(t) = Au^*(t)$ in the left-hand side and the right-hand side of (5.5.26), it reduces to

$$u^*(t) = Au^*(t) = u_{ab}(t) - \int_a^b G(t, s)\, f(u^*, s)\, ds, \tag{5.5.36}$$

which is the integral equation representation of the solution of (5.5.1) and (5.5.2), the original nonlinear two-point boundary value problem.

We have proved that A is a contraction mapping and that the sequence $\{u^{(k)}(t)\}$ converges uniformly to the fixed point $u^*(t)$, which is the solution of the original two-point boundary value problem.

Part 3. We now exhibit the quadratic convergence of the $\{u^{(k)}(t)\}$. On forming the difference $|Au^{(k)} - Au^*|$, we have

$$|u^{(k+1)}(t) - u^*(t)| = \left| \int_a^b G(t, s)\{f_u(u^{(k)}, s)[u^{(k+1)}(s) - u^{(k)}(s)] + \right.$$
$$\left. + [f(u^{(k)}, s) - f(u^*, s)]\}\, ds \right| . \tag{5.5.37}$$

By the mean value theorem, the last term in (5.5.37) may be expressed as

$$f(u^{(k)},s)-f(u^*,s) = f_u(\bar{u},s)\,[u^{(k)}(s)-u^*(s)], \tag{5.5.38}$$

where $\bar{u}(s)$ is such that $\rho(\bar{u},u^*) \le \rho(u^{(k)},u^*)$.

Adding and subtracting $f_u(\bar{u},s)u^{(k+1)}(s)$ to, and from, the expression in the parentheses on the right-hand side of (5.5.37) and introducing (5.5.38) into (5.5.37) give

$$|u^{(k+1)}(t)-u^*(t)| = \left|\int_a^b G(t,s)\{[f_u(u^{(k)},s)-f_u(\bar{u},s)]\,[u^{(k+1)}(s)-u^{(k)}(s)]+ \right.$$

$$\left. +f_u(\bar{u},s)\,[u^{(k+1)}(s)-u^*(s)]\}\,ds\right|. \tag{5.5.39}$$

Taking norms on both sides and introducing the bounds yield

$$\rho(u^{(k+1)},u^*) \le \frac{m}{4}(b-a)^2\,[\rho(u^{(k)},u^*)\,\rho(u^{(k+1)},u^{(k)})+\rho(u^{(k+1)},u^*)]. \tag{5.5.40}$$

Since by the contraction mapping $\rho(u^{(k+1)},u^*) < \rho(u^{(k)},u^*)$, it follows by the triangle inequality

$$\rho(u^{(k+1)},u^{(k)}) \le \rho(u^{(k+1)},u^*)+\rho(u^{(k)},u^*) \le 2\rho(u^{(k)},u^*). \tag{5.5.41}$$

Introducing the inequality (5.5.41) into (5.5.40) gives

$$\rho(u^{(k+1)},u^*) \le \frac{(m/2)(b-a)^2}{1-(m/4)(b-a)^2}\,\rho^2(u^{(k)},u^*) \le \frac{\alpha}{C+1}\,\rho^2(u^{(k)},u^*), \tag{5.5.42}$$

where α is defined in (5.5.34). Thus we have exhibited the quadratic convergence of the method.

Part 4. The bound on the error is estimated next. From the contraction mapping principle we have

$$\rho(u^{(k)},u^{(k+1)}) \le \alpha^k \rho(u^{(0)},u^{(1)}), \tag{5.5.43}$$

$$\rho(u^{(k)},u^{(k+p)}) \le \frac{\alpha^k-\alpha^{k+p}}{1-\alpha}\,\rho(u^{(0)},u^{(1)}). \tag{5.5.44}$$

As $p \to \infty$,

$$\rho(u^{(k)},u^*) \le \frac{\alpha^k}{1-\alpha}\,\rho(u^{(0)},u^{(1)}). \tag{5.5.45}$$

If, instead of expressing the inequalities in (5.5.43)–(5.5.45) in terms of the metric $\rho(u^{(0)}, u^{(1)})$, we use the metric $\rho(u^{(k)}, u^{(k+1)})$, we can show

$$\rho(u^{(k)}, u^*) \leq \frac{1}{1-\alpha} \rho(u^{(k)}, u^{(k+1)}). \tag{5.5.46}$$

Substituting (5.5.46) in (5.5.42) yields

$$\rho(u^{(k+1)}, u^*) = \frac{\alpha}{(C+1)(1-\alpha)^2} \rho^2(u^{(k)}, u^{(k+1)}). \tag{5.5.47}$$

This establishes the estimate of the error bound.

5.5.4. Extension to Systems of Second-Order Differential Equations

For systems of second-order differential equations, the McGill and Kenneth proof is readily extended with a proof similiar to the single-equation proof given in Theorem 5.2.

THEOREM 5.3. *Consider the system of N nonlinear second order differential equations and the two-point boundary value conditions:*

$$\frac{d^2\mathbf{U}}{dt^2} = \mathbf{F}(\mathbf{U}, t), \tag{5.5.48}$$

$$\mathbf{U}(a) = \mathbf{a}_1, \qquad \mathbf{U}(b) = \mathbf{b}_1, \tag{5.5.49}$$

where

(i) $\mathbf{U}$ = *$N \times 1$ vector with elements $u_1(t), u_2(t), \ldots, u_N(t)$,*
$\mathbf{F}$ = *$N \times 1$ vector with elements $f_1(\mathbf{U}, t), f_2(\mathbf{U}, t), \ldots, f_N(\mathbf{U}, t)$,*
$\mathbf{a}_1$ = *$N \times 1$ vector with elements $a_{11}, a_{12}, \ldots, a_{1N}$,*
$\mathbf{b}_1$ = *$N \times 1$ vector with elements $b_{11}, b_{12}, \ldots, b_{1N}$,*
$\mathbf{U}_{a,b}$ = *$N \times 1$ vector with elements $u_{ab1}(t), u_{ab2}(t), \ldots, u_{abN}(t)$, and $u_{abi}(t)$ is defined similarly to $u_{ab}(t)$ in (5.5.10) for the ith component;*

(ii) $\mathbf{U}(t)$ *is a member of the complete metric space S defined as*
$S = \{\mathbf{U}(t) | \mathbf{U}(t)$ *is continuous on* $[a, b]$, $\mathbf{U}(a) = \mathbf{a}_1, \mathbf{U}(b) = \mathbf{b}_1, \rho(\mathbf{U}(t), \mathbf{U}_{ab}(t)) \leq C_1\}$;

(iii) *the metric ρ is defined as*

$$\rho(\mathbf{U}_1, \mathbf{U}_2) = \sum_{i=1}^{N} \max_{a \leq t \leq b} |u_{1,i}(t) - u_{2,i}(t)|, \qquad \forall \mathbf{U}_1, \mathbf{U}_2 \in S, \tag{5.5.50}$$

and vector $\mathbf{U}_p$ has components $u_{p,i}$, $i = 1, 2, \ldots, N$;

(iv) *C_1 is a nonnegative constant.*

Assume:

1. $f_i(u_1, u_2, \ldots, u_N, t)$ *are continuous,* $i = 1, 2, \ldots, N$.

2. $\dfrac{\partial f_i}{\partial u_j}(u_1, u_2, \ldots, u_N, t)$ *exist and are continuous,* $i, j = 1, 2, \ldots, N$.

3. $$\left|\frac{\partial f_i}{\partial u_j}(u_{1,1}, u_{1,2}, \ldots, u_{1,N}, t) - \frac{\partial f_i}{\partial u_j}(u_{2,1}, u_{2,2}, \ldots, u_{2,N}, t)\right|$$

$$\leq M_{ij} \sum_{i=1}^{N} |u_{1,i} - u_{2,i}|, \qquad \forall \mathbf{U}_1, \mathbf{U}_2 \in S; \qquad M_{ij} > 0.$$

Associated with (5.5.48) *and* (5.5.49) *is the sequence of linear two-point boundary value problems*

$$\frac{d^2\mathbf{U}^{(k+1)}}{dt^2} = J(\mathbf{U}^{(k)}, t)[\mathbf{U}^{(k+1)}(t) - \mathbf{U}^{(k)}(t)] + \mathbf{F}(\mathbf{U}^{(k)}, t), \qquad k = 0, 1, \ldots, \tag{5.5.51}$$

$$\mathbf{U}^{(k)}(a) = \mathbf{a}_1, \qquad \mathbf{U}^{(k)}(b) = \mathbf{b}_1, \tag{5.5.52}$$

where

$\mathbf{U}^{(k)} = N \times 1$ *vector with elements* $u_1^{(k)}, u_2^{(k)}, \ldots, u_N^{(k)}$,
$\mathbf{F}^{(k)} = N \times 1$ *vector with elements* $f_1(\mathbf{U}^{(k)}, t), f_2(\mathbf{U}^{(k)}, t), \ldots, f_N(\mathbf{U}^{(k)}, t)$,
$\mathbf{J}(\mathbf{U}^{(k)}, t) = N \times N$ *Jacobian matrix, which appears as*

$$\begin{bmatrix} \dfrac{\partial f_1}{\partial u_1} & \dfrac{\partial f_1}{\partial u_2} & \cdots & \dfrac{\partial f_1}{\partial u_N} \\ \vdots & \vdots & & \vdots \\ \dfrac{\partial f_N}{\partial u_1} & \dfrac{\partial f_N}{\partial u_2} & \cdots & \dfrac{\partial f_N}{\partial u_N} \end{bmatrix},$$

and $\mathbf{U}^{(0)}(t)$ *is a member of* S.

As a consequence of the conditions above, for a sufficiently small interval $[a, b]$, *the following results hold:*

1. *The unique solution of* (5.5.48) *and* (5.5.49) *exists.*
2. *The sequence* $\{\mathbf{U}^{(k)}(t)\}$ *generated by* (5.5.51) *and* (5.5.52) *converges uniformly in* t *to the unique solution of* (5.5.48) *and* (5.5.49).

3. *The convergence of* $\{\mathbf{U}^{(k)}(t)\}$ *is at least quadratic:*

$$\rho(\mathbf{U}^{(k+1)}, \mathbf{U}^*) < \left(\frac{\alpha}{C_1+1}\right) \rho^2(\mathbf{U}^{(k)}, \mathbf{U}^*),$$

where $\mathbf{U}^*$ *is the unique solution of* (5.5.48) *and* (5.5.49); α, *the contraction mapping constant, equals*

$$\frac{(Nm/2)(b-a)^2(NC_1+1)}{1-(Nm/4)(b-a)^2}$$

and is less than 1; C_1 *is the constant defined in the definition of the space* S; m *is the maximum of the bounds on* f_i, $\partial f_i/\partial u_j$, *and the constants* M_{ij}, $i,j = 1, 2, \ldots, N$.

4. *A bound on the error is given by*

$$\rho(\mathbf{U}^{(k+1)}, \mathbf{U}^*) < \frac{\alpha}{(1-\alpha)^2(NC_1+1)} \rho^2(\mathbf{U}^{(k+1)}, \mathbf{U}^{(k)}).$$

The proof of these statements parallels exactly that given for the single second-order equation. For further details see McGill and Kenneth [3.]

5.5.5. Extension to Systems of First-Order Differential Equations

The proofs given above are valid only for systems of second-order equations in which half of the boundary conditions are specified at the initial point and the remaining half for the same variables at the final point. It is possible to extend in a straightforward manner the method of proof to the systems generally considered in this book; that is, first-order systems of differential equations with an arbitrary division of boundary conditions between the initial and final points. In this case the matrix Green's function turns out to be piecewise constant, but other than this the proofs are essentially identical.

REFERENCES

1. R. Bellman and R. Kalaba, *Quasilinearization and Nonlinear Boundary Value Problems*, American Elsevier, New York, 1965.
2. L. V. Kantorovich and G. P. Akilov, *Functional Analysis in Normed Spaces*, English translation, Pergamon Press, 1964.

3. R. McGill and P. Kenneth, A Convergence Theorem on the Iterative Solution of Non-Linear Two-Point Boundary-Value Systems, *Proc. XIV Internat. Astronautical Cong., Paris, IV* **12** (1963), 173–186.
4. R. McGill and P. Kenneth, Solution of Variational Problems by Means of a Generalized Newton-Raphson Operator, *AIAA J.*, (10) **2** (1964), 1176–1176.
5. P. B. Bailey, L. F. Shampine, and P. E. Waltman, *Nonlinear Two Point Boundary Value Problems*, Academic, New York, 1968.
6. R. Courant and D. Hilbert, *Methods of Mathematical Physics*, Vol. **1**, Interscience, New York, 1953.

General References

R. Bellman, H. Kagiwada, and R. Kalaba, Quasilinearization and the Calculation of Eigenvalues, *Comm. ACM* (7) **9** (July 1966), 522–523.

J. M. Lewallen, A Modified Quasi-linearization Method for Solving Trajectory Optimization Problems, *AIAA J.*, (5) **5** (May 1957), 962–965.

S. M. Roberts, J. S. Shipman, and C. V. Roth, Continuation in Quasilinearization, *J. Optimization Theory and Appl.*, (3) **2** (May 1968), 164–178.

R. Sylvester and F. Meyer, Two-Point Boundary Value Problems by Quasilinearization, *SIAM J. Appl. Math.*, (2) **13** (June 1965), 586–602.

Chapter 6

NEWTON'S METHOD AND TWO-POINT BOUNDARY VALUE PROBLEMS

6.1. INTRODUCTION [1-21]

The main purpose of this chapter is to place shooting methods, particularly the method of adjoints, on a rigorous basis. In earlier chapters our treatment was purely formal and, as a matter of fact, up to a few years ago the acceptance of the method of adjoints as a valid tool for solving nonlinear two-point boundary value problems was based essentially on the plausibility of the scheme and sufficient numerical experience to warrant its use. In this chapter we give sufficient conditions for the convergence of the method of adjoints as applied to nonlinear problems, and we give estimates of the rate of convergence and of the error. We do this by showing first that the method of adjoints is a particular realization of a very general form of the familiar Newton-Raphson method. Then we apply a theorem of Kantorovich on the convergence of the generalized Newton's method.

In order to motivate the discussion, we first review the Newton-Raphson method as it is applied to a finite set of algebraic or transcendental equations. We then give Kantorovich's theorem in this setting, and use it to establish the convergence of the method of adjoints. Next we show how Newton's method can be extended to the more general setting of nonlinear operator equations on Banach spaces, and give the version of Kantorovich's theorem appropriate for this context. We then show that quasilinearization is also a realization of the abstract Newton-Raphson method. It follows that Kantorovich's theorem will give sufficient conditions for convergence and error estimates for quasilinearization as well.

The deeper insight into the method of adjoints which is gained by viewing it as a form of the Newton-Raphson method enables us to modify the method to treat problems which it formerly failed to handle. In particular, we can develop for numerically sensitive problems the continuation method and the perturbation method, which are discussed in Chapter 7.

In our discussion of the abstract Newton-Raphson method we must perforce use some concepts from functional analysis. Thus we talk about operators, Banach spaces, contraction mappings, etc. [4, 7, 8, 10, 16, 22–26].

For the reader unfamiliar with the terms we have included an Appendix where the most frequently used ideas from functional analysis are defined and briefly discussed. The material on the method of adjoints can still be comprehended if such a reader will replace "operator" by "matrix", and if, in the section on quasilinearization, he can think of "operator" as meaning "the process of solving the linear differential equations".

6.2. NEWTON-RAPHSON METHOD, FINITE SET OF NONLINEAR EQUATIONS [2,3,5,9]

As commonly used, the Newton-Raphson method is a scheme to solve a finite set of nonlinear equations by solving a sequence of linear equations. The solution to the set of nonlinear equations is (under favorable conditions) the limit of the sequence of the solutions of the linear equations. When the equations have more than one solution (one root), the Newton-Raphson method can be applied to each of the roots. The method is essentially a local exploration which finds the root which is sufficiently close to the initial trial guess. For a multiroot problem, we must assume different initial trial values to find each of the roots.

6.2.1. Single Nonlinear Equation

Suppose we want to solve the scalar equation

$$\varphi(x) = 0. \tag{6.2.1}$$

We could expand $\varphi(x)$ in a Taylor's series around a nominal trial solution $x^{(0)}$:

$$\varphi(x) = \varphi(x^{(0)}) + \varphi'(x^{(0)})\,(x - x^{(0)}) + \text{higher-order terms}, \tag{6.2.2}$$

and as a first-order approximation neglect terms higher than first order. If we set $\varphi(x) = 0$ and then write the linear part of (6.2.2) as

$$x = x^{(0)} - \frac{\varphi(x^{(0)})}{\varphi'(x^{(0)})}, \tag{6.2.3}$$

we can compute x, since in this equation all the terms on the right-hand side, $x^{(0)}$, $\varphi(x^{(0)})$, and $\varphi'(x^{(0)})$, are known.

The x in (6.2.3) is an approximation to the true but unknown root x^* of the equation $\varphi(x) = 0$ which should be a better approximation than $x^{(0)}$ if the higher-order terms in (6.2.2) can be neglected. Using the x just found, which we denote $x^{(1)}$, we can repeat the process to find a still better approximation, $x^{(2)}$, to x^*. The process of generating the sequence of approximations $x^{(k)}$ by the relations

$$x^{(k+1)} = x^{(k)} - \frac{\varphi(x^{(k)})}{\varphi'(x^{(k)})}, \qquad k = 0, 1, \ldots. \tag{6.2.4}$$

is the *Newton-Raphson method.*

Under appropriate conditions first adequately prescribed by Kantorovich and summarized in the theorems of Section 6.3, the solution x^* of the nonlinear equation $\varphi(x) = 0$ is given by

$$\lim_{k \to \infty} x^{(k)} = x^*. \tag{6.2.5}$$

6.2.2. Sets of Nonlinear Equations

Now suppose we want to find the solution of the two nonlinear equations

$$\varphi_1(x_1, x_2) = 0, \qquad \varphi_2(x_1, x_2) = 0. \tag{6.2.6}$$

As before, we could write a Taylor's series expansion for each equation around a trial solution $(x_1^{(0)}, x_2^{(0)})$:

$$\varphi_i(x_1, x_2) = \varphi_i(x_1^{(0)}, x_2^{(0)}) + \left(\frac{\partial \varphi_i}{\partial x_1}\right)(x_1 - x_1^{(0)}) + \left(\frac{\partial \varphi_i}{\partial x_2}\right)(x_2 - x_2^{(0)}) \tag{6.2.7}$$

$$+ \text{higher-order terms}; \qquad i = 1, 2$$

and, neglecting the terms higher than first order and assuming $\varphi_1(x_1, x_2) = 0$ and $\varphi_2(x_1, x_2) = 0$, we can express the linear approximation to (6.2.7) as

$$\begin{bmatrix} x_1 \\ x_2 \end{bmatrix} = \begin{bmatrix} x_1^{(0)} \\ x_2^{(0)} \end{bmatrix} - \begin{bmatrix} \dfrac{\partial \varphi_1}{\partial x_1} & \dfrac{\partial \varphi_1}{\partial x_2} \\ \dfrac{\partial \varphi_2}{\partial x_1} & \dfrac{\partial \varphi_2}{\partial x_2} \end{bmatrix}^{-1} \begin{bmatrix} \varphi_1(x_1^{(0)}, x_2^{(0)}) \\ \varphi_2(x_1^{(0)}, x_2^{(0)}) \end{bmatrix} \tag{6.2.8}$$

provided the inverse exists. As for the scalar equation, the solution (x_1^*, x_2^*) of (6.2.8) is found under the appropriate conditions as

$$\lim_{k \to \infty} \begin{pmatrix} x_1^{(k)} \\ x_2^{(k)} \end{pmatrix} = \begin{pmatrix} x_1^* \\ x_2^* \end{pmatrix}. \tag{6.2.9}$$

Similarly, to solve a system of n nonlinear equations

$$\boldsymbol{\varphi}(\mathbf{x}) = \mathbf{0}, \tag{6.2.10}$$

where

$\boldsymbol{\varphi}$ = system (or vector) of n nonlinear equations with equations (components) $\varphi_1(\mathbf{x}), \varphi_2(\mathbf{x}), \ldots, \varphi_n(\mathbf{x})$,
$\mathbf{x}$ = $n \times 1$ vector with components $(x_1, x_2, \ldots, x_n)$,

by Newton's method we have

$$\mathbf{x}^{(k+1)} = \mathbf{x}^{(k)} - \mathbf{A}^{-1}\boldsymbol{\varphi}(\mathbf{x}^{(k)}), \qquad k = 0, 1, \ldots. \tag{6.2.11}$$

provided the inverse exists, where

$\mathbf{x}^{(k)}$ = kth iteration vector, $n \times 1$ vector with components $(x_1^{(k)}, x_2^{(k)}, \ldots, x_n^{(k)})$,
$\mathbf{A}$ = $n \times n$ matrix whose i,j element is $\partial\varphi_i/\partial x_j$ evaluated at $\mathbf{x}^{(k)}$.

As before, the true solution $\mathbf{x}^*$ is found under the appropriate conditions as

$$\lim_{k \to \infty} \mathbf{x}^{(k)} = \mathbf{x}^*. \tag{6.2.12}$$

If we examine (6.2.11) we may look at the vector $\mathbf{A}^{-1}\boldsymbol{\varphi}(\mathbf{x}^{(k)})$ as a correction term to the kth approximation $\mathbf{x}^{(k)}$. The existence of the correction term depends of course on the existence of the inverse $\mathbf{A}^{-1}$ and $\boldsymbol{\varphi}(\mathbf{x}^k)$. In practice we never take (or in fact *can* take numerically) the limit in (6.2.12), but we continue the iteration process until the correction term is so small that $\mathbf{x}^{(k+1)} \approx \mathbf{x}^{(k)}$.

Inherent in the Newton-Raphson method are certain situations which may cause the process to fail. If at least one of the elements of the function $\boldsymbol{\varphi}(\mathbf{x})$ possesses a maximum or a minimum at or near the trial value of $\mathbf{x}$, then the partials $\partial\varphi_i/\partial x_j \approx 0$, $j = 1, 2, \ldots, n$, so the inverse of the matrix $\mathbf{A}$ becomes unbounded. If one or more of the elements of the function $\boldsymbol{\varphi}(\mathbf{x})$ possess a point of inflection in an interval including the trial value of $\mathbf{x}$, then the sequence of $\mathbf{x}$'s may not converge or may not converge to the root nearest the trial value of $\mathbf{x}$. In some pathological cases it is possible for the correction term $\mathbf{A}^{-1}\boldsymbol{\varphi}(\mathbf{x})$ to be of constant magnitude and alternating sign so that with each iteration the trial solution "ping-pongs" between two values and the problem never converges. See Collatz [4, p. 284].

The success of the method is crucially dependent on the initial guess vector. In general, there is no way to ascertain whether the first guess vector is satisfactory without iterating by (6.2.11) or without evaluating the norms in the Kantorovich theorem (see Section 6.3), which is not a practical thing to do. If the Newton-Raphson process fails to converge for some initial trial root, we simply pick another trial root and repeat the process. Because of the local nature of the Newton-Raphson method, finding an initial point that ultimately leads to convergence may be a formidable task. Even if the initial guess does not satisfy the conditions of Kantorovich's theorem, we may still use the iteration process of (6.2.11) as a device to generate new trial initial values.

Newton's method is important because, when it converges, it does so quadratically. That is,

$$\|\mathbf{x}^*-\mathbf{x}^{(k+1)}\| \leq C\|\mathbf{x}^*-\mathbf{x}^{(k)}\|^2, \tag{6.2.13}$$

where x^* is the root, $\| \ \|$ is some norm, and C is a constant. The error in the $(k+1)$st iterate is proportional to the square of the error in the kth iterate.

To show that (6.2.13) is true let us write (6.2.11) as

$$\mathbf{x}^*-\mathbf{x}^{(k+1)} = \mathbf{x}^*-\mathbf{x}^{(k)}-\mathbf{A}^{-1}[\boldsymbol{\varphi}(\mathbf{x}^*)-\boldsymbol{\varphi}(\mathbf{x}^{(k)})]. \tag{6.2.14}$$

By a Taylor's series expansion around the root $\mathbf{x}^*$ we have

$$\boldsymbol{\varphi}(\mathbf{x}^*)-\boldsymbol{\varphi}(\mathbf{x}^{(k)}) = \boldsymbol{\varphi}'(\mathbf{x}^{(k)})\,(\mathbf{x}^*-\mathbf{x}^{(k)})+\tfrac{1}{2}\boldsymbol{\varphi}''(\boldsymbol{\eta})\,(\mathbf{x}^*-\mathbf{x}^{(k)})^2, \tag{6.2.15}$$

where $\boldsymbol{\eta}=\mathbf{x}^*+\theta(\mathbf{x}^*-\mathbf{x}^{(k)}), 0<\theta<1, \boldsymbol{\varphi}'=\mathbf{A}$, the matrix $\boldsymbol{\varphi}''$, which is the array of the second-order partials, is continuous in the interval $(\mathbf{x}^*,\mathbf{x}^{(k)})$. See Chapter 9 for the description of $\boldsymbol{\varphi}''$. Substitution of (6.2.15) in (6.2.14) gives

$$\mathbf{x}^*-\mathbf{x}^{(k+1)} = -\tfrac{1}{2}[\boldsymbol{\varphi}'(\mathbf{x}^{(k)})]^{-1}\boldsymbol{\varphi}''(\boldsymbol{\eta})(\mathbf{x}^*-\mathbf{x}^{(k)})^2. \tag{6.2.16}$$

Taking the norm of both sides yields

$$\|\mathbf{x}^*-\mathbf{x}^{(k+1)}\| \leq C\|\mathbf{x}^*-\mathbf{x}^{(k)}\|^2, \tag{6.2.17}$$

where

$$C = \|-\tfrac{1}{2}[\boldsymbol{\varphi}'(\mathbf{x}^{(k)})]^{-1}\boldsymbol{\varphi}''(\boldsymbol{\eta})\|. \tag{6.2.18}$$

6.3. KANTOROVICH THEOREM [7,8]

It may surprise the reader to learn that, although the Newton-Raphson method was first published in 1685, and has been used and developed by generations of mathematicians since then, a really satisfactory treatment of its convergence properties was not available until in 1937, when the Russian mathematician L. V. Kantorovich published the first of a series of papers on the subject.

It may also surprise the reader to find out that it turns out to be more practical to study the Newton-Raphson method in its greatest generality, that is, as a method for solving operator equations on Banach spaces, for then the various applications are all special cases of the same general result. For example, when the Banach space is the n-dimensional Cartesian product of the reals and the operator is the n-vector function of n variables, Kantorovich's proof applies to the solution of finite sets of nonlinear equations. As another example, quasilinearization can be shown to be a realization of the Newton-Raphson method in which the Banach space is all continuously differentiable functions on a given interval satisfying prescribed boundary conditions. Because of Kantorovich's fundamental contributions to the theory, Newton's method is often referred to as the Newton-Raphson-Kantorovich method in the modern literature, and we will sometimes do this also.

Kantorovich considers the problem of finding the solution x^* of the operator equation

$$P(x) = 0, \tag{6.3.1}$$

where P is a nonlinear operator mapping the open set Ω of the Banach space X into X:

$$P: \Omega \to X, \Omega \subset X,$$

and 0 is the null element of the range space (in this case X itself). If we assume that P has a continuous Fréchet derivative in Ω, $P'(x)$, and if $x^{(0)}$ is any arbitrary element of X, we may write

$$P(x^{(0)}) = P(x^{(0)}) - P(x^*). \tag{6.3.2}$$

The right hand side of (6.3.2) can be approximated by $P'(x^{(0)})(x^{(0)} - x^*)$, so that the equation can be expressed as

$$P(x^{(0)}) = P'(x^{(0)})(x^{(0)} - x^*). \tag{6.3.3}$$

If $x^{(1)}$ is an approximation to x^*, an approximation to (6.3.2) can be written

$$P(x^{(0)}) = P'(x^{(0)})(x^{(0)} - x^{(1)}). \tag{6.3.4}$$

Equation (6.3.4) is a linear operator equation, since the Fréchet derivative $P'(x^{(0)})$ is a linear operator, so that, if the inverse operator $P'(x^{(0)})^{-1}$ exists, (6.3.4) can be solved:

$$x^{(1)} = x^{(0)} - P'(x^{(0)})^{-1} P(x^{(0)}). \tag{6.3.5}$$

Continued repetition of the process

$$x^{(k+1)} = x^{(k)} - P'(x^{(k)})^{-1} P(x^{(k)}), \qquad k = 0, 1, \ldots \tag{6.3.6}$$

(which is possible if each $P'(x^{(k)})^{-1}$ exists) is the Newton-Raphson-Kantorovich method. The method furnishes a sequence $\{x^{(k)}\}$ which we hope will converge to the solution x^* of $P(x) = 0$. Sufficient conditions to ensure this convergence will be given shortly.

If $P(x) = 0$ is the single nonlinear algebraic equation $\varphi(x) = 0$, then (6.3.6) reduces to (6.2.4), the familiar form of the Newton-Raphson method in which $P(x) = \varphi(x)$, $P'(x) = \varphi'(x)$, $P'(x)^{-1} = 1/\varphi'(x)$, and the null element equals the scalar 0.

Similarly, if $P(x) = 0$ is a set of nonlinear equations $\boldsymbol{\varphi}(\mathbf{x}) = \mathbf{0}$, then (6.3.6) becomes (6.2.11), where $\varphi(x) = P(x)$, $P'(x^{(k)})$ is the matrix $\mathbf{A}$, $P'(x^{(k)})^{-1} = \mathbf{A}^{-1}$, and the null element equals the vector of zeros, $\mathbf{0}$.

In addition to the Newton-Raphson method of (6.3.6), Kantorovich also considered the modified Newton-Raphson method

$$x^{(k+1)} = x^{(k)} - P'(x^{(0)})^{-1} P(x^{(k)}), \qquad k = 0, 1, \ldots \tag{6.3.7}$$

where the inverse of the derivative $P'(x^{(0)})^{-1}$ is evaluated once and only once, at the initial point $x^{(0)}$. It will be seen that the modified Newton-Raphson method requires less computation per iteration than the Newton-Raphson method, but may require more iterations because its rate of convergence is slower.

A convenient form of the conditions sufficient for convergence and estimates of the rate of convergence of the Newton-Raphson-Kantorovich method are given in the following.

THEOREM 6.1 (Kantorovich [8]). *Let the operation P be defined as the mapping of the open set $\Omega \subset X$, a Banach space, into X, and let P have a continuous second derivative in Ω_0, the closed set $\|x - x^{(0)}\| \leqq r$, where r is described below. Moreover, let:*

(i) *the linear operation $\Gamma_0 = [P'(x^{(0)})]^{-1}$ exist;*

(ii) $\|\Gamma_0(P(x))\| \leqq \eta$;

(iii) $\|\Gamma_0 P''(x)\| \leqq K, \qquad x \in \Omega_0.$

Now, if

$$h = K\eta \leqq \tfrac{1}{2} \tag{6.3.8}$$

and

$$r \geqq r_0 = \frac{1 - \sqrt{1 - 2h}}{h}\eta, \tag{6.3.9}$$

then

$$P(x) = 0 \tag{6.3.10}$$

will have a solution x^* to which the Newton-Raphson method (original or modified) is convergent. Here

$$\|x^* - x^{(0)}\| \leqq r_0. \tag{6.3.11}$$

Furthermore, if, for $h < \frac{1}{2}$,

$$r < r_1 = \frac{1+\sqrt{1-2h}}{h}\eta \tag{6.3.12}$$

or, for $h = \frac{1}{2}$,

$$r \leqq r_1 \tag{6.3.13}$$

the solution x^* will be unique in the sphere Ω_0.

The speed of convergence of the original method is characterized by the inequality

$$\|x^* - x^{(n)}\| \leqq \frac{1}{2^n}(2h)^{2^n}\frac{\eta}{h}, \qquad n = 0, 1, 2, \ldots. \tag{6.3.14}$$

and that of the modified method for $h < \frac{1}{2}$ by

$$\|x^* - x^{(n)}\| \leqq \frac{\eta}{h}(1-\sqrt{1-2h})^{n+1}, \qquad n = 0, 1, 2, \ldots. \tag{6.3.15}$$

Kantorovich has supplied various other versions of his sufficiency theorem. One form, which we find convenient in discussing a theorem of Friedrichs in Section 6.10, is the following.

THEOREM 6.2 (Kantorovich [8]). *Let the operator P be defined as the mapping of the open set $\Omega \subset X$, a Banach space, into X and let P have a continuous second derivative in Ω_0, the closed set $\|x - x^{(0)}\| \leqq r$, where r is described below. Moreover, let*

(i) $\|\Gamma_0\| \leqq B'$, where $\Gamma_0 = P'(x^{(0)})^{-1}$;

(ii) $\|P(x^{(0)})\| \leqq \eta'$;

(iii) $\|P''(x)\| \leqq K'$, $x \in \Omega_0$.

Now, if

$$h = K'B'^2\eta' \leqq \tfrac{1}{2} \tag{6.3.16}$$

and

$$r \geq r_0 = \frac{1-\sqrt{1-2h}}{h}B'\eta', \tag{6.3.17}$$

then

$$P(x) = 0 \tag{6.3.18}$$

will have a solution x^* to which the Newton method (original or modified) is convergent. Here

$$\|x^* - x^{(0)}\| \leqq r_0. \tag{6.3.19}$$

Furthermore, if, for $h < \frac{1}{2}$,

$$r < r_1 = \frac{1+\sqrt{1-2h}}{h} B'\eta' \tag{6.3.20}$$

or, for $h = \frac{1}{2}$,

$$r \leqq r_1, \tag{6.3.21}$$

the solution x^* will be unique in the sphere Ω_0.

The speed of convergence of the original method is characterized by the inequality

$$\|x^* - x^{(n)}\| \leqq \frac{1}{2^n}(2h)^{2^n}\frac{\eta'}{h}, \qquad n = 0, 1, 2, \ldots, \tag{6.3.22}$$

and that of the modified method for $h < \frac{1}{2}$ by

$$\|x^* - x^{(n)}\| \leqq \frac{\eta'}{h}(1-\sqrt{1-2h})^{n+1}, \qquad n = 0, 1, 2, \ldots. \tag{6.3.23}$$

If we examine (6.3.3) in the light of Theorem 6.2, we find that conditions (i) and (ii) require that the norms of $P'(x^{(0)})^{-1}$ and $P(x^{(0)})$ be bounded. We call these norms "point norms" because they are specified only at the initial "point" $x^{(0)}$ (which may actually be, we remind the reader, a vector or a function). The third condition of the theorem puts a bound on the operator $P''(x)$ for x in the sphere $\Omega_0 = \{x| \; \|x - x^{(0)}\| \leqq r\}$. We call this an "interval norm" since it must hold for a set of points x. It is, by the way, plausible that a condition on the second derivative should be required, since the Newton-Raphson method is a first-order approximation, and the only way a first-order approximation can be justified is to ensure that higher-order terms are negligible. It is one of the noteworthy features of Kantorovich's proofs, as Henrici [27] has emphasized, that bounds on the interval norm $\|P''(x)\|$ and the point norms $\|P(x^{(0)})\|$ and $\|P'(x^{(0)})^{-1}\|$ imply existence of all $P(x^{(k)})$ and $P'(x^{(k)})^{-1}$ as well as estimates of their respective norms.

The actual evaluation of the bound on $\|P''(x)\|$, K', is a rather involved process because it requires determining the radius r_1 of the sphere of uniqueness Ω_0 which in turn depends on the constants B', η', and K' itself through h in (6.3.16) and (6.3.20). Despite this circle of dependency, K' may be determined by an iterative process. First $\|P''(x)\|$ is evaluated for some x to find a trial value of K'; call it K_0'. Assuming that $x^{(0)}$ satisfies the theorem, then, h is found by (6.3.16), using the trial K'. If $h > \frac{1}{2}$, the choice of x is incorrect and a new trial value of x must be picked to evaluate $\|P''(x)\|$. If $h \leqq \frac{1}{2}$, r_1 is evaluated by (6.3.20). Since x must lie in Ω_0, by (6.3.20) we test if x lies in $\|x^* - x^{(0)}\| \leqq r_1$. If not, the choice of x is incorrect and a new trial value of x must be picked, a new trial $\|P''(x)\|$ determined again, and a new r_1 evaluated. If x does lie in $\|x^* - x^{(0)}\| \leqq r_1$, we still do not know that the trial value of $\|P''(x)\|$ is correct so we must explore systematically $\|x^* - x^{(0)}\| \leqq r_1$ for $\|P''(x)\|$ to find its largest value. In the course of choosing various x's to evaluate $\|P''(x)\|$ it is quite likely that r_1 will change with each trial value of K'.

Probably a better route to determine K' is not to presume that the trial $x^{(0)}$ satisfies the theorem but rather to show that $x^{(0)}$ satisfies the theorem. For a trial value of $x^{(0)}$, trial values for B' and η' are computed. The trial K' is computed for $\|P''(x)\|$ evaluated at $x^{(0)}$. If $h > \frac{1}{2}$, then the trial $x^{(0)}$ is not satisfactory and a new trial $x^{(0)}$ must be chosen. If $h \leqq \frac{1}{2}$, r_1 is found by (6.3.20). The sphere $\|x^* - x^{(0)}\| \leq r_1$ is determined. At this time, various values of x in $\|x^* - x^{(0)}\| \leq r_1$ are chosen systematically to evaluate $\|P''(x)\|$ to determine the sphere of uniqueness and K'.

An interesting sidelight to the Kantorovich theorem is a comparison of the error bound estimates of the Newton-Raphson method and the modified Newton-Raphson method, (6.3.14) and (6.3.15). The development of the modified Newton-Raphson method error bound estimate as given by Kantorovich is such that the error bound is true for both methods. On the other hand, the development of the error bound estimate for the Newton-Raphson method takes advantage of properties which are distinctive to that method alone. Therefore we should expect the error bounds developed for the Newton-Raphson method to be sharper than those of the modified Newton-Raphson method applied to the Newton-Raphson method. Although this is true in general, it is not true for $n = 0$ and 1 for a wide range of h.

Let us observe that, if we operate on the same initial data for both the Newton-Raphson method and the modified Newton-Raphson method, $x^{(0)}$ and $x^{(1)}$ are identical for both cases (that is, $n = 0$ and 1). It is therefore fair to compare both error bounds (since h and η are the same for both methods) by rearranging (6.3.14) and (6.3.15) and computing the right-hand sides of the Newton-Raphson method bound,

$$\frac{\|x^* - x^{(n)}\|}{\eta} \leq \frac{1}{2^n h}(2h)^{2^n}, \tag{6.3.24}$$

Table 6.1. Error Bound Estimates for Newton-Raphson and Modified Newton-Raphson Methods

n	h	*Newton-Raphson Method Error Bound (right-hand side* (6.3.24))	*Modified Newton-Raphson Method Error Bound (right-hand side* (6.3.25))	*Modified Newton-Raphson Method Error Bound/Newton-Raphson Method Error Bound*
0	10^{-9}	2.	7.45058060	3.72529039
1		2.(10^{-9})	5.55111510(10^{-8})	2.77555776(10^{1})
2		4.(10^{-27})	4.13590306(10^{-16})	1.03397594(10^{11})
0	10^{-8}	2.	1.49011612	7.45058060(10^{-1})
1		2.(10^{-8})	2.22044602(10^{-8})	1.11022304
2		4.(10^{-24})	3.30872244(10^{-16})	8.27180648(10^{7})
0	10^{-7}	2.	1.04308128	5.21540624(10^{-1})
1		2.(10^{-7})	1.08801855(10^{-7})	5.44009256(10^{-1})
2		4.(10^{-21})	1.13489179(10^{-14})	2.83722925(10^{6})
0	10^{-6}	2.	1.00582840	5.02914220(10^{-1})
1		2.(10^{-6})	1.01169074(10^{-6})	5.05845433(10^{-1})
2		4.(10^{-18})	1.01758726(10^{-12})	2.54396880(10^{5})
0	10^{-5}	2.	1.00061299	5.00306493(10^{-1})
1		2.(10^{-5})	1.00122634(10^{-5})	5.00613195(10^{-1})
2		4.(10^{-15})	1.00184005(10^{-10})	2.50460047(10^{4})
0	10^{-4}	2.	1.00009145	5.00045764(10^{-1})
1		2.(10^{-4})	1.00018288(10^{-4})	5.00091559(10^{-1})
2		4.(10^{-12})	1.00027432(10^{-8})	2.50068700(10^{3})
0	10^{-3}	2.	1.00050868	5.00254339(10^{-1})
1		2.(10^{-3})	1.00101760(10^{-3})	5.00508815(10^{-1})
2		4.(10^{-9})	1.00152676(10^{-6})	2.50381714(10^{2})
0	10^{-2}	2.	1.00505131	5.02525657(10^{-1})
1		2.(10^{-2})	1.01012811(10^{-2})	5.05064070(10^{-1})
2		4.(10^{-6})	1.01523055(10^{-4})	2.53807661(10^{1})
0	10^{-1}	2.	1.05572815	5.27864087(10^{-1})
1		2.(10^{-1})	1.11456190(10^{-1})	5.57280993(10^{-1})
2		4.(10^{-3})	1.17667434(10^{-2})	2.94168642
0	5.(10^{-1})	2.	1.99961403	9.99807012(10^{-1})
1		1.	1.99922808	1.99922816
2		5.(10^{-1})	1.99884218	3.99768490

and the modified Newton-Raphson method bound,

$$\frac{\|x^*-x^{(n)}\|}{\eta} \leq (1-\sqrt{1-2h}\,)^{n+1}/h. \qquad (6.3.25)$$

In Table 6.1 are tabulated the right-hand sides of (6.3.24) and (6.3.25) and the ratio of the right-hand side of (6.3.25) to (6.3.24) for $n = 0, 1, 2$, and for $10^{-9} \leq h \leq 0.5$. For $10^{-7} \leq h \leq 0.4$ and for $n = 0, 1$ the error bound estimate by the modified Newton-Raphson method is smaller than by the Newton-Raphson method. For $n \geq 2$ the estimate of the error bound by the Newton-Raphson method is smaller. We may conclude therefore that the error bound estimate for the Newton-Raphson method when applied to that method is not uniformly the smallest bound.

6.4. KANTOROVICH THEOREM FOR A FINITE SET OF NONLINEAR EQUATIONS

Because of its importance in the study of the method of adjoints as well as finite difference methods, we give explicitly the translation of the theorem of Kantorovich, stated in Section 6.3, into terms applicable to a finite set of nonlinear equations. Thus we wish to consider the set of n nonlinear algebraic or transcendental equations

$$\varphi_i(x_1, x_2, \ldots, x_n) = 0, \qquad i = 1, 2, \ldots n. \qquad (6.4.1)$$

The solution to this set of nonlinear equations by the Newton-Raphson method leads to the successive approximation

$$\mathbf{x}^{(k+1)} = \mathbf{x}^{(k)} - [\mathbf{A}(x^{(k)})]^{-1}\boldsymbol{\varphi}^{(k)}, \qquad (6.4.2)$$

where

$\mathbf{x}^{(k)}$ = the kth iteration vector, $n \times 1$ vector with components $x_1^{(k)}, x_2^{(k)}, \ldots, x_n^{(k)}$,

$\mathbf{A}^{(k)}$ = $\mathbf{A}(\mathbf{x}^{(k)}) = n \times n$ matrix, where the component in the ith row, jth column is $\partial\varphi_i/\partial x_j$, evaluated at $\mathbf{x}^{(k)}$; $\mathbf{A}^{(k)}$ is assumed nonsingular,

$\boldsymbol{\varphi}^{(k)}$ = $\boldsymbol{\varphi}(\mathbf{x}^{(k)})$, $n \times 1$ vector with components $\varphi_i(\mathbf{x}^{(k)})$, $i = 1, 2, \ldots, n$.

The following norms are employed. For a vector $\mathbf{v}$ whose elements are $v_1, v_2, \ldots, v_n$, define a norm

$$\|\mathbf{v}\| = \max_{1 \leqq i \leqq n} |v_i|. \qquad (6.4.3)$$

For an $n \times n$ matrix $\mathbf{A}$ with elements a_{ij}, define the norm

$$\|\mathbf{A}\| = \max_{1 \leqq i \leqq n} \sum_{j=1}^{n} |a_{ij}|. \tag{6.4.4}$$

The vector and matrix norms are consistent. (See Collatz [4, pp. 169–171].)

The Kantorovich theorem may be stated in a form given by Henrici [27, pp. 366–371]:

THEOREM 6.3. (i) *For the initial approximation* $\mathbf{x}^{(0)}$ *to the solution of* (6.4.1),

$$\mathbf{A}^{(0)} = \mathbf{A}(\mathbf{x}^{(0)}) \text{ has an inverse } \mathbf{\Gamma}_0 \text{ such that } \|\mathbf{\Gamma}_0\| \leq B_0; \tag{6.4.5}$$

(ii) $\mathbf{x}^{(0)}$ *satisfies* (6.4.1) *approximately in the sense that*

$$\|\mathbf{\Gamma}_0 \boldsymbol{\varphi}(\mathbf{x}^{(0)})\| \leq \eta_0; \tag{6.4.6}$$

(iii) *in the region defined by the inequality* (6.4.9), *the components of the vector* $\boldsymbol{\varphi}(\mathbf{x})$ *are twice continuously differentiable with respect to the components of* $\mathbf{x}$ *and satisfy*

$$\sum_{j,s=1}^{n} \left| \frac{\partial^2 \varphi_i}{\partial x_j \partial x_s} \right| \leqq K \qquad \text{for each } i; \tag{6.4.7}$$

(iv) *the constants* B_0, η_0, K *satisfy*

$$h_0 = B_0 \eta_0 K \leqq \tfrac{1}{2}. \tag{6.4.8}$$

When hypotheses (i)–(iv) *are satisfied, the system of equations* (6.4.1) *has a solution* $\mathbf{x}^*$ *which is located in the cube*

$$\|\mathbf{x} - \mathbf{x}^{(0)}\| \leqq \frac{1 - \sqrt{1 - 2h_0}}{h_0} \eta_0. \tag{6.4.9}$$

Moreover, the successive approximations $\mathbf{x}^{(k)}$ *defined by* (6.4.2) *exist and converge to* $\mathbf{x}^*$, *and the speed of convergence may be estimated by the inequality*

$$\|\mathbf{x}^{(k)} - \mathbf{x}^*\| \leqq \frac{1}{2^{k-1}} (2h_0)^{2^k - 1} \eta_0. \tag{6.4.10}$$

In this form of the Kantorovich theorem we make the following identification: the operator equation $P(x) = 0$ is the set of equations (6.4.1), the $P'(x)$ is the matrix $\mathbf{A}$ with elements $\partial \varphi_i / \partial x_j$, $i, j = 1, 2, \ldots, n$; $[P'(x)]^{-1}$ is the inverse of $\mathbf{A}$, the Banach space X is the collection of points called the Cartesian product space of reals $\{x_1, x_2, \ldots, x_n\}$; the r_0 and the sphere of existence are given by (6.4.9).

6.5. NEWTON-RAPHSON METHOD AS CONTRACTION MAPPING

In Sections 6.3 and 6.4 we stated several versions of Kantorovich's theorem on the convergence of Newton-Raphson method, but did not give proofs because they can be found in accessible sources. The gist of Kantorovich's method is to show that the sequence of Newton iterates $\{x^{(n)}\}$ is bounded in norm (majorized) by a scalar function, and to show that the scalar function converges.

We give here a proof of the convergence of the modified Newton-Raphson method by showing that it is a contraction mapping, and then applying the contraction mapping theorem.

Recall that a mapping $P(x)$ of a complete metric space X onto itself is *contracting* if there exists a constant $0 < \alpha < 1$ such that

$$d(P(x), P(x')) \leqq \alpha d(x, x'), \qquad x, x' \in X,$$

where $d(x, y)$ is the metric on X. The contraction mapping theorem, which was also used in Section 5, Chapter 5, in McGill and Kenneth's proof of the convergence of quasilinearization, states that, if P is a contraction mapping, then the equation $P(x) = x$ has a unique solution, namely, x^*, which is the limit of the sequence $\{x^{(n)}\}$, where $x^{(n+1)} = P(x^{(n)})$ and $x^{(0)}$ is any point in X. Since every Banach space is also a metric space, the theorem may have applicability in the Banach space context as well, as our proof illustrates.

The contents of this section and the following section are not essential to an understanding of the principal results of this chapter: that the method of adjoints is a realization of the Newton-Raphson method. Therefore the reader interested only in that result and its significance in applications may skip to Section 6.7.

We proceed to prove the following:

THEOREM 6.4. *Let P be an operator which maps the closed sphere $\gamma_0 = \{x \mid \|x - x_0\| \leq \rho\}$, where ρ is defined in* (6.5.10), *into itself. If*

(i) *P has a derivative $P'(x)$ at every point $x \in \gamma_0$;*

(ii) *$[P'(x_0)]^{-1}$ exists so that the operator*

$$S = I - [P'(x_0)]^{-1} P \tag{6.5.1}$$

exists (*where I is the identity operator*);

(iii) *and*

$$\alpha = \sup_{0<\nu<1} \|I - [P'(x^{(0)})]^{-1} P'(x^{(0)} + \nu(x - x^{(0)}))\| < 1 \tag{6.5.2}$$

for $x, x^{(0)} \in \gamma_0$; *then the equation*

$$P(x) = 0 \tag{6.5.3}$$

has a solution $x^* \in \gamma_0$ *to which the iterations* $x^{(i)}$ *of the modified Newton-Raphson method,*

$$x^{(i+1)} = x^{(i)} - [P'(x_0)]^{-1} P(x^{(i)}), \tag{6.5.4}$$

starting with the initial value $x^{(0)}$, *will converge.*

PROOF. By (6.5.1) and (6.5.4) we can write the modified Newton-Raphson method as

$$x^{(j+1)} = S(x^{(j)}). \tag{6.5.5}$$

Now, for any two points $x, x' \in \gamma_0$,

$$\|S(x) - S(x')\| \leqq \sup_{0<\nu<1} \|S'(x+\nu(x'-x))\| \, \|x-x'\|, \tag{6.5.6}$$

by the mean value theorem ("formula of finite increments," Kantorovich and Akilov [8, p. 660]), providing of course that $S(x)$, $S(x')$, and $S'(x+\nu(x'-x))$ are defined. This, however, is assured by assumptions (i) and (ii). Furthermore, by assumption (iii), Eq. (6.5.6) can be written

$$\|S(x) - S(x')\| \leqq \alpha \|x - x'\|, \qquad 0 < \alpha < 1. \tag{6.5.7}$$

Therefore the operator S is a contraction mapping, and hence by the contraction mapping theorem (Kantorovich and Akilov [8, p. 627]) the iterates $x^{(i)}$ given by (6.5.4) or its equivalent (6.5.5) converge to a point $x^* \in \gamma_0$. (Recall that a Banach space X is also a complete metric space with the metric $d(x, x') = \|x - x'\|$.) Thus, by (6.5.4),

$$x^* = x^* - [P'(x_0)]^{-1} P(x^*) \tag{6.5.8}$$

or

$$[P'(x_0)]^{-1} P(x^*) = 0. \tag{6.5.8'}$$

Applying the linear operator $P'(x_0)$ to both sides of (6.5.8), it follows that

$$P(x^*) = 0;$$

that is, x^* is a solution of (6.5.3) since, for any linear operator U, $U(0) = 0$.

Since S is a contraction mapping, the rate of convergence of the modified Newton-Raphson method is given by

$$\|x^* - x^{(n)}\| \leq \frac{\alpha^n}{1-\alpha} \|x^{(1)} - x^{(0)}\|, \qquad n = 0, 1, 2, \ldots, \tag{6.5.9}$$

and the domain of the solution is bounded by

$$\|x^*-x^{(0)}\| \leqq \frac{1}{1-\alpha} \|x^{(1)}-x^{(0)}\| = \rho. \tag{6.5.10}$$

6.6. COMPARISON OF KANTOROVICH'S THEOREM AND CONTRACTION MAPPING

In Section 6.3 we stated a theorem due to Kantorovich on the convergence of the modified Newton-Raphson method, and in Section 6.5 we stated and proved a theorem on its convergence which makes use of the contraction mapping principle. In this section we compare estimates of the rate of convergence and sphere of existence of the solution given by the two theorems, and we find that Kantorovich's theorem not only gives sharper estimates, but also allows a larger value of h. This state of affairs suggests that, while the contraction mapping principle is simple and useful, certain other methods, such as Kantorovich's, are more powerful.

The rate of convergence of the modified Newton-Raphson method as estimated by our application of the contraction mapping theorem in Section 6.5 is

$$\|x^*-x^{(n)}\| \leq \frac{\alpha^n}{1-\alpha} \|x^{(1)}-x^{(0)}\|, \tag{6.6.1}$$

and the estimate given by Kantorovich's theorem (Theorem 6.1) is

$$\|x^*-x^{(n)}\| \leq \frac{\eta}{h}(1-\sqrt{1-2h})^{n+1}. \tag{6.6.2}$$

Therefore, to compare the two estimates, we have to express α and $\|x^{(1)}-x^{(0)}\|$ in terms of η and K, or in terms of $h = K\eta$. Assume, then, that

$$\|\Gamma_0 P(x^{(0)})\| \leq \eta, \qquad \|\Gamma_0 P''(x)\| \leq K, \tag{6.6.3}$$

for $x \in \gamma_0$, where γ_0 is the sphere of existence and uniqueness given by the contraction mapping theorem. Since by (6.5.4), Section 6.5, with $i = 0$,

$$x^{(1)} = x^{(0)}-[P'(x^{(0)})]^{-1}P(x^{(0)}) = x^{(0)}-\Gamma_0 P(x^{(0)}), \tag{6.6.4}$$

it follows that

$$\|x^{(1)}-x^{(0)}\| \leq \|\Gamma_0 P(x^{(0)})\| \leq \eta. \tag{6.6.5}$$

Furthermore, since

$$S = I - \Gamma_0 P \tag{6.6.6}$$

by (6.5.1), Section 6.5, it follows that the first and second derivatives of S at $x \in \gamma_0$ are given, respectively, by the linear and bilinear operators

$$S'(x) = I - \Gamma_0 P'(x), \tag{6.6.7}$$

$$S''(x) = -\Gamma_0 P''(x). \tag{6.6.8}$$

By the mean value theorem,

$$\|S'(x) - S'(x^{(0)})\| \le \sup_{\theta} \|S''(\bar{x})\| \, \|x - x^{(0)}\| \le K\|x - x^{(0)}\|, \tag{6.6.9}$$

where $\bar{x} = x^{(0)} + \theta(x - x^{(0)})$, $0 < \theta < 1$.

But, since

$$S'(x^{(0)}) = I - \Gamma_0 P'(x^{(0)}) = I - I = 0, \tag{6.6.10}$$

Eq. (6.6.9) can be written

$$\|S'(x)\| \le K\|x - x^{(0)}\|. \tag{6.6.9'}$$

Recalling the definition of α given by (6.5.2), Section 6.5, we have

$$\begin{aligned} \alpha &= \sup_{0<\nu<1} \|I - \Gamma_0 P'(x^{(0)} + \nu(x - x^{(0)}))\| \\ &= \sup_{\nu} \|S'(x^{(0)} + \nu(x - x^{(0)}))\| \\ &\le K\|x^{(0)} + \nu(x - x^{(0)}) - x^{(0)}\| \le K\rho. \end{aligned} \tag{6.6.11}$$

Thus

$$\alpha \le K\rho = \frac{K}{1-\alpha}\|x^{(1)} - x^{(0)}\| \le \frac{K\eta}{1-\alpha} = \frac{h}{1-\alpha} \tag{6.6.12}$$

or

$$\alpha^2 - \alpha + h \ge 0. \tag{6.6.12'}$$

The first inequality, (6.6.12), is satisfied if $0 < \alpha < 1$; thus it is essentially vacuous; but the second, (6.6.12'), implies that the smallest value α can have is

$$\alpha = \frac{1 - \sqrt{1 - 4h}}{2}. \tag{6.6.13}$$

This is the desired relation between α and h.

The first conclusion to be drawn from (6.6.13) is that the conditions for the applicability of the contraction mapping theorem are more restrictive than for Kantorovich's theorem, since the former obviously requires that $h \le \frac{1}{4}$, while the latter needs only $h \le \frac{1}{2}$. Next, to compare the estimates of the rate of convergence we would form the ratio

$$\frac{\text{Bound on estimated rate of convergence, contraction mapping theorem}}{\text{Bound on estimated rate of convergence, Kantorovich theorem}}$$

$$= \frac{\dfrac{\alpha^n}{1-\alpha}\|x^{(1)}-x^{(0)}\|}{\dfrac{\eta}{h}(1-\sqrt{1-2h})^{n+1}}$$

$$= \left[\frac{1-\sqrt{1-4h}}{2(1-\sqrt{1-2h})}\right]^n \frac{2h}{(1+\sqrt{1-4h})(1-\sqrt{1-2h})}. \tag{6.6.14}$$

Note that, for $n = 0$, (6.6.14) also gives the ratio of the radii of the spheres in which the solution is guaranteed to lie for the contraction mapping and the Kantorovich theorems. A numerical study shows that the ratio (6.6.14) is always equal to or greater than 1. Therefore the bounds given by Kantorovich's theorem are sharper than the bounds for the contraction mapping theorem. The accompanying tabulation illustrates this point for $n = 0$.

h	*Ratio* (6.6.14) *for* $n = 0$
0.000	1.000
0.050	1.056
0.100	1.072
0.125	1.092
0.200	1.230
0.250	1.707

To sum up, Kantorovich's theorem gives sharper estimates for the bounds on the sphere in which the solution lies and for the rate of convergence, and also is less restrictive with regard to permissible values of h.

6.7. METHOD OF ADJOINTS AND NEWTON-RAPHSON METHOD [17]

Recall that in Section 3.6, Chapter 3, we discussed in a purely formal way the solution of the nonlinear two-point boundary value problem

$$\dot{y}_i = g_i(y_1, y_2, \ldots, y_n, t), \qquad i = 1, 2, \ldots, n \tag{6.7.1}$$

$$y_i(t_0) = c_i, \qquad i = 1, 2, \ldots, r, \tag{6.7.2}$$

$$y_{i_m}(t_f) = c_{i_m}, \qquad m = 1, 2, \ldots, n-r \tag{6.7.3}$$

by the method of adjoints. An estimate of the missing initial conditions is to be made, and corrections to the trial values of the missing initial conditions $\delta_{r+1}^{(k)}(t_0), \ldots, \delta y_n^{(k)}(t_0)$ are to be found by solving

$$\begin{bmatrix} \delta y_{i_1}^{(k)}(t_f) \\ \vdots \\ \delta y_{i_{n-r}}^{(k)}(t_f) \end{bmatrix} = \begin{bmatrix} x_{r+1}^{(1)}(t_0) & x_{r+2}^{(1)}(t_0) & \cdots & x_n^{(1)}(t_0) \\ \vdots & \vdots & & \vdots \\ x_{r+1}^{(n-r)}(t_0) & x_{r+2}^{(n-r)}(t_0) & \cdots & x_n^{(n-r)}(t_0) \end{bmatrix} \begin{bmatrix} \delta y_{r+1}^{(k)}(t_0) \\ \vdots \\ \delta y_n^{(k)}(t_0) \end{bmatrix}, \tag{6.7.4}$$

where

$\delta y_{r+i}^{(k)}(t_0)$ = the correction to the trial value $y_{r+i}^{(k)}(t_0)$, for the kth iteration, $i = 1, 2, \ldots, n-r$,

$$\delta y_{i_m}^{(k)}(t_f) = y_{i_m}(t_f) - y_{i_m}^{(k)}(t_f) \tag{6.7.5}$$

= the miss distance between the specified terminal condition and the calculated terminal condition, for the kth iteration, $m = 1, 2, \ldots, n-r$, and

$x_{r+i}^{(m)}(t_0)$ = the components of the initial vector found by integrating backward the adjoint equations for the mth set of Kronecker delta terminal conditions, $i, m = 1, 2, \ldots, n-r$.

Further, we have discussed in Section 6.2. the Newton-Raphson method for the solution of a finite set of nonlinear algebraic or transcendental equations, and then we described in Section 6.3 several versions of Kantorovich's theorem on the convergence of the method. The point of this section is that the Newton-Raphson method and hence the Kantorovich theorem can be applied to the solution of the two-point boundary value problems (6.7.1)–(6.7.3) by the method of adjoints, if the problem is thought of as the search for those missing initial conditions $y_r(t_0), \ldots, y_n(t_0)$ which drive to zero the final miss distances, φ_m:

$$\varphi_m = Y_m - w_m^{(k)}, \qquad m = 1, 2, \ldots, n-r, \tag{6.7.6}$$

where

$Y_m = y_{i_m}(t_f)$ = specified terminal conditions, $m = 1, 2, \ldots, n-r$ in (6.7.3),

$w_m^{(k)} = y_{i_m}^{(k)}(t_f)$ = the calculated terminal conditions for $m = 1, 2, \ldots, n-r$ found by integrating (6.7.1) with the initial conditions $y_1(t_0), y_2(t_0), \ldots, y_r(t_0), y_{r+1}^{(k)}(t_0), \ldots, y_n^{(k)}(t_0)$.

Since the final values of the solution $y_1(t), y_2(t), \ldots, y_n(t)$ of (6.7.1) are continuous functions of the initial values, φ_m may be considered to be functions of $\mathbf{y}(t_0)$ and, solving (6.7.1), may be considered equivalent to finding the solution to the system

$$\varphi_m(\mathbf{y}(t_0)) = 0, \qquad m = 1, 2, \ldots, n-r. \tag{6.7.7}$$

Assume that the kth approximation to the vector of $(n-r)$ missing initial conditions $\mathbf{z}^{(k)}$, where, by definition,

$$z_j^{(k)} = y_{r+j}^{(k)}(t_0), \qquad j = 1, 2, \ldots, n-r, \tag{6.7.8}$$

has been found. The Newton-Raphson method gives the $(k+1)$st approximation

$$\mathbf{z}^{(k+1)} = \mathbf{z}^{(k)} - [\mathbf{A}^{(k)}]^{-1}\boldsymbol{\varphi}(\mathbf{z}^{(k)}), \tag{6.7.9}$$

where

$\boldsymbol{\varphi}(\mathbf{z}^{(k)})$ = column vector with components $\varphi_m^{(k)} = \varphi_m[\mathbf{y}^{(k)}(t_0)]$, $m = 1, 2, \ldots, n-r$,

$\mathbf{A}^{(k)}$ = $(n-r)\times(n-r)$ matrix whose element $A_{mj}^{(k)}$ in the mth row, jth column is $(\partial\varphi_m/\partial z_j)^{(k)}$ evaluated at $\mathbf{y}^{(k)}(t_0)$.

The Kantorovich theorem is thus applicable to the two-point boundary value problem.

We now show that the method of adjoints is entirely equivalent to the Newton-Raphson method as applied to the two-point nonlinear ordinary differential equation boundary value problem in the formulation above.

Since $w_m^{(k)}$, the calculated terminal value $y_{i_m}^{(k)}(t_f)$, is a function of $y_1(t_0)$, $y_2(t_0), \ldots, y_n(t_0)$ the total variation in $w_m^{(k)}$ can be expressed as

$$\delta w_m^{(k)} = \sum_{j=1}^{n-r} \left(\frac{\partial w_m}{\partial z_j}\right)^{(k)} \delta z_j^{(k)}, \qquad m = 1, 2, \ldots, n-r. \tag{6.7.10}$$

Forming the partial derivative of (6.7.6) and substituting in (6.7.10):

$$\delta w_m^{(k)} = -\sum_{j=1}^{n-r} \left(\frac{\partial \varphi_m}{\partial z_j}\right)^{(k)} \delta z_j^{(k)}, \qquad m = 1, 2, \ldots, n-r \tag{6.7.11}$$

or, in vector form,

$$\delta \mathbf{w}^{(k)} = -\mathbf{A}^{(k)} \delta \mathbf{z}^{(k)}. \tag{6.7.12}$$

If we note that, by the definition for z and w, (6.7.4) can be written

$$\delta w_m^{(k)} = \sum_{j=1}^{n-r} x_{r+j}^{(m)}(t_0) \delta z_j^{(k)}, \qquad m = 1, 2, \ldots, n-r. \tag{6.7.13}$$

If we equate the mth equation of (6.7.11) with the mth equation of (6.7.13),

$$x_{r+j}^{(m)}(t_0) = -\left(\frac{\partial \varphi_m}{\partial z_j}\right)^{(k)} = -A_{mj}^{(k)}, \qquad m, j = 1, 2, \ldots, n-r, \tag{6.7.14}$$

Eq. (6.7.4) may be expressed as

$$\delta \mathbf{w}^{(k)} = -\mathbf{A}^{(k)} \delta \mathbf{z}(k) \tag{6.7.15}$$

with the solution

$$\delta \mathbf{z}^{(k)} = -[\mathbf{A}^{(k)}]^{-1} \delta \mathbf{w}^{(k)}. \tag{6.7.16}$$

On comparing the definition of the miss distance in (6.7.5) and (6.7.6) we observe that $\delta y_{im}^{(k)}(t_f) = \delta w_m^{(k)}$ and therefore

$$\delta w_m^{(k)} = \varphi_m^{(k)} = \varphi_m[\mathbf{y}^{(k)}(t_0)]. \tag{6.7.17}$$

Substituting (6.7.17) in (6.7.16) yields

$$\delta \mathbf{z}^{(k)} = -[\mathbf{A}^{(k)}]^{-1} \boldsymbol{\varphi}^{(k)} \tag{6.7.18}$$

as the correction vector in the method of adjoints.

Referring to (6.7.9) we see that this is the same correction vector which applies in the Newton-Raphson method. Thus we have shown that the method of adjoints for solving nonlinear two-boundary value problems is identically the Newton-Raphson method. The Kantorovich sufficiency theorem can then be applied to the method of adjoints to give a sound theoretical basis for the method of adjoints. In particular, we can determine by the sufficiency theorem if the initial point (in this case the set $y_{r+i}(t_0)$, $i = 1, 2, \ldots, n-r$) will lead to convergence, and then determine the sphere of existence of the solution, the sphere of uniqueness, and the rate of convergence.

Instead of calculating by numerical differentiation the partial derivatives in the **A** matrix required by the Newton-Raphson method, we obtain by the method of adjoints these partials as a consequence of integrations. The initial value of the adjoint vector, which is a consequence of integrating backward the adjoint equations, is the negative of the partial derivatives in (6.7.14). From a numerical point of view this is a tremendous advantage in accuracy over forming the partials by numerical differentiation. In addition, forming the partials numerically may be expensive in computer time. See Chapter 9 for more discussion of the partial derivatives.

The form of the Kantorovich theorem described in Section 6.4 for non-linear algebraic equations can be used here for the two-point boundary value problem. Alternatively, we may invoke Theorem 6.1 of Kantorovich in Section 6.3. Under these circumstances the underlying Banach space is the $(n-r)$-dimensional real space of missing initial conditions $\mathbf{z}$. The operator equation is $P(z) = 0$, where $P(z) = \boldsymbol{\varphi}(\mathbf{z})$; the derivative operator $P'(z) = \mathbf{A}^{(k)}$, the matrix with elements $(\partial\varphi_m/\partial z_j)^{(k)}$; and the null element is the vector of zeros.

In practice, assuming that an initial estimate of the missing initial values $y_{r+i}(t_0)$, $i = 1, 2, \ldots, n-r$, is available, the matrix $\boldsymbol{\Gamma}_0 = P'(\mathbf{y}^{(0)})^{-1}$ can be calculated after one integration of (6.7.1) and $(n-r)$ integrations of the adjoint equations. Then the norm $\|\boldsymbol{\Gamma}_0\| \leq B_0$ can be determined. Since $\varphi_i(\mathbf{y}^{(0)})$ is also known at this point, the vector $\boldsymbol{\Gamma}_0\boldsymbol{\varphi}(\mathbf{y}^{(0)})$ can be formed and also its norm $\|\boldsymbol{\Gamma}_0\boldsymbol{\varphi}(\mathbf{y}^{(0)})\| \leq \eta_0$. However, there are practical problems in computing the norm of the second derivative K, as discussed in Section 6.3. A short-cut compromise estimate of K is described in Section 6.8.

6.8. NUMERICAL RESULTS [17,19]

It should be emphasized that the Kantorovich theorems are sufficiency theorems. That is, if the conditions of the theorems are met, then convergence can be guaranteed and estimates of the rate of convergence made. If, however, the conditions are not met, no conclusions can be drawn. It is not unusual in practice for the initial guess to be such that the conditions of the Kantorovich theorem are not met, and yet for the initial guess to lead ultimately to a satisfactory solution to the problem. Under these circumstances the "initial" guess for the Newton-Raphson iteration recipe is not really the "initial" point for the Kantorovich theorem. Several iterations of the Newton-Raphson recipe may be necessary to find that point, which for the purposes of the theorem can be called the "initial" point. Thus the results of Section 6.7 are somewhat more important theoretically than practically in that they establish the validity of the method of adjoints. However, there

Table 6.2. Kantorovich Norms and Calculated Initial Velocity Vector and Final Position Vector, Two-Point Two-Body Boundary Value Problem

Goodman-Lance Method Iteration No.	*Kantorovich Norms*			
	h_i	β_i	η_i	K_i
0	6.763579 (10^{2})	3.659561	1.069325	1.728374 (10^{2})
1	6.536509 (10^{1})	3.444090	1.663449	1.140937 (10^{1})
2	5.233003 (10^{-1})	7.149846 (10^{-1})	2.365853 (10^{-1})	3.093615
3	8.472580 (10^{-1})	1.124802	3.827557 (10^{-2})	1.967965 (10^{1})
4	3.273678 (10^{-2})	1.017268	2.250358 (10^{-3})	1.430042 (10^{1})
5	4.037952 (10^{-5})	1.013542	2.822407 (10^{-6})	1.411560 (10^{1})
6	4.537566 (10^{-8})	1.013534	3.171723 (10^{-9})	1.411526 (10^{1})

	Calculated Missing Initial Conditions		
	$\dot{x}(0)$	$\dot{y}(0)$	$\dot{z}(0)$
0	−5.379999 (10^{-1})	2.879999 (10^{-1})	4.988299 (10^{-1})
1	−1.376313	5.242722 (10^{-1})	9.080641 (10^{-1})
2	3.236898 (10^{-1})	5.779965 (10^{-1})	1.001111
3	8.720452 (10^{-2})	4.489073 (10^{-1})	7.775289 (10^{-1})
4	1.017630 (10^{-1})	4.709831 (10^{-1})	8.157667 (10^{-1})
5	1.016559 (10^{-1})	4.722822 (10^{-1})	8.180167 (10^{-1})
6	1.016588 (10^{-1})	4.722833 (10^{-1})	8.180185 (10^{-1})

	Calculated Terminal Conditions		
	$x(2)$	$y(2)$	$z(2)$
0	6.086336 (10^{-1})	−3.733722 (10^{-1})	−6.466988 (10^{-1})
1	−2.218209	−1.837729 (10^{-1})	−3.183034 (10^{-1})
2	7.165604 (10^{-1})	9.196269 (10^{-1})	1.592828
3	−8.591061 (10^{-2})	5.013211 (10^{-1})	8.683122 (10^{-1})
4	−2.289911 (10^{-3})	5.726597 (10^{-1})	9.918756 (10^{-1})
5	−9.427341 (10^{-6})	5.759951 (10^{-1})	9.976526 (10^{-1})
6	−7.212622 (10^{-9})	5.759999 (10^{-1})	9.976609 (10^{-1})

are problems for which the Kantorovich theorem can be used to estimate the accuracy of the computed solution, as is shown below.

To test the Kantorovich theorem numerically we solved a two-body two-point boundary value problem by the Goodman-Lance method of adjoints [6] and evaluated the various norms. Since it is not practical to seek K over an interval, we evaluated instead a point norm K_i, that is, we computed the expression in (6.4.7) at the trial initial vector $\mathbf{y}^{(i)}(t_0)$. In other words, B_i, η_i, K_i, and h_i are all evaluated at each iteration at the vector $\mathbf{y}^{(i)}(t_0)$. Although in all probability K_0 will not be the upper bound needed for applying the theorem, the computations of K_i are useful. For a process in which the sequence $\{\mathbf{y}^{(i)}(t_0)\}$ converges, the sequence $\{K_i\}$ will usually decrease monotonically; this indicates that K_i is close to the required K. If the sequence $\{\mathbf{y}^{(i)}(t_0)\}$ does not converge, we may find one or more K_i much larger than K_0. In general, both of these possibilities are found in practice.

To illustrate a more or less typical numerical experience, we list in Table 6.2 the Kantorovich norms, the initial velocity vector, and the final position for the two-body version of the problem of McGill and Kenneth [11, 28].

The two-body equations of motion are

$$\ddot{x}(t) = -\frac{kx(t)}{r^3}, \qquad \ddot{y}(t) = -\frac{ky(t)}{r^3}, \qquad \ddot{z}(t) = -\frac{kz(t)}{r^3}. \tag{6.8.1}$$

where

$$r = [x^2(t)+y^2(t)+z^2(t)]^{1/2},$$

$$k = 1.0, \text{ for canonical units.}$$

The boundary conditions for $t_0 = 0$ and $t_f = 2$ are

$$x(0) = 1.076000, \qquad x(2) = 0.000000,$$

$$y(0) = 0.000000, \qquad y(2) = 0.576000,$$

$$z(0) = 0.000000, \qquad z(2) = 0.997661.$$

The trial initial velocity vector is taken as the average velocity vector based on the difference between the final and initial position vectors divided by the time interval; that is, the initial conditions correspond to McGill and Kenneth's starting function. The values are

$$\dot{x}(0) = -0.538000,$$

$$\dot{y}(0) = 0.288000,$$

$$\dot{z}(0) = 0.498830.$$

In Table 6.2 the calculations were made for 20 integration steps over the interval (0,2), and the h_i were computed using β_i, η_i, K_i in (6.4.5)–(6.4.7). Fifty integration steps yielded essentially the same results. The h_i are monotone, except for the third iteration of the Goodman-Lance method. As stated above, this behavior can be traced directly to the evaluation of K_i. The 4th iteration of the Goodman-Lance method provides the 0th iteration for the application of the Kantorovich theorem. From the 4th iteration on, the h_i are less than 1/2, and the h_i and η_i are monotonic decreasing. We infer that the conditions of the Kantorovich theorem are satisfied, and expect convergence. A check on the computed terminal position vector confirms that the problem indeed converges.

As a matter of interest, we list in Table 6.3 the computed rate of convergence, the left-hand side of inequality (6.4.10), and the estimated rate of convergence, the right-hand side of (6.4.10). The calculations are based on using the 4th iteration of the Goodman-Lance method as the 0th iteration for the evaluation of the Kantorovich norms. The 6th iteration of the Goodman-Lance method is assumed to yield the "true" solution. We observe that the estimated rates of convergence are realistic.

Table 6.3. Comparison of Computed and Estimated Rates of Convergence for Problem Presented in Table 6.2.

Goodman-Lance Method Iteration Number	*Rates of Convergence*	
	Computed (*LHS* (6.4.10))	*Estimated* (*RHS* (6.4.10))
4	$2.2518(10^{-3})$	$4.500716(10^{-3})$
5	$2.8(10^{-6})$	$1.473389(10^{-4})$
6	0.0	$3.158037(10^{-7})$

Our computational experience suggests the following rules of thumb for similar problems. First, if the problem is going to converge, it will do so, in general, within 5–10 iterations of the Goodman-Lance method. Second, it is better to calculate with smaller time steps per iteration of the Goodman-Lance method, and employ fewer iterations than to calculate with larger time steps per iteration and employ more iterations. Another interesting computational observation is that, if the computed h_i remains less than $\frac{1}{2}$

Table 6.4. Modified Method of Adjoints Applied to Eq. (6.8.1)

Iteration No.	Kantorovich Norms h_i	β_i	η_i	K_i
0	6.763579(10²)	3.659561	1.069325	1.728374(10²)
1	6.536509(10¹)	3.444090	1.663449	1.140937(10¹)
2	3.397647(10²)	2.991697(10⁻²)	8.161847(10⁻²)	1.391465(10⁵)
3	1.484853(10⁻¹)	5.182107(10⁻¹)	3.235824	8.855075(10⁻²)
4	2.688236(10¹)	2.106461	2.846395	4.483515
5	1.662440	6.269471(10⁻¹)	1.432732	1.850759
6	3.217605(10¹)	2.340128(10⁻²)	6.246358(10⁻²)	2.201233(10⁴)
⋮	⋮	⋮	⋮	⋮
20	3.275644(10⁻²)	5.000002(10⁻¹)	4.869078(10³)	1.345487(10⁻⁵)

Iteration No.	Calculated Missing Initial Conditions $\dot{x}(0)$	$\dot{y}(0)$	$\dot{z}(0)$	Calculated Terminal Conditions $x(2)$	$y(2)$	$z(2)$
0	−5.3799999(10⁻¹)	2.8799999(10⁻¹)	4.9882999(10⁻¹)	6.0863367(10⁻¹)	−3.7337224(10⁻¹)	−6.4669886(10⁻¹)
1	−1.3763137	5.2427226(10⁻¹)	9.0806412(10⁻¹)	−2.2182091	−1.8377300(10⁻¹)	−3.1830344(10⁻¹)
2	−9.7733935(10⁻²)	8.1932398(10⁻²)	1.4190935(10⁻¹)	−7.0690255	−4.2945087	−7.4382167
3	2.9456488	−1.0847001	−1.8786805	6.5937129	−2.1226356	−3.6763658
4	−2.9420842	7.2204478(10⁻¹)	1.2508203	−5.2512476	3.9822786(10⁻¹)	6.8986231(10⁻¹)
5	7.6721351(10⁻¹)	−4.8602507(10⁻¹)	−8.4151227(10⁻¹)	1.7326841	−8.4152145(10⁻¹)	−1.4570249
6	−1.0781199	5.8827505(10⁻²)	1.0269023(10⁻¹)	−1.3777060(10¹)	−3.6528749	−6.3765168
⋮	⋮	⋮	⋮	⋮	⋮	⋮
20	−1.2999610	−4.8687901(10³)	2.8085323(10³)	−1.5527118	−9.7375762(10³)	5.6170623(10³)

Table 6.5. Example in Which Both the Method of Adjoints and the Modified Method of Adjoints Converge[a]

A. Method of Adjoints Iteration No.	*Kantorovich Norms*			
	h_i	β_i	η_i	K_i
0	4.2229972(10^{-1})	1.0824180	2.2228302(10^{-2})	1.7551712(10^{1})
1	1.9874074(10^{-2})	1.0151739	1.3795129(10^{-3})	1.4191250(10^{1})
2	1.8627939(10^{-5})	1.0135362	1.3020693(10^{-6})	1.4115341(10^{1})
3	2.2818950(10^{-8})	1.0135348	1.5950266(10^{-9})	1.4115265(10^{1})
4	1.8159086(10^{-11})	1.0135348	1.2693059(10^{-12})	1.4115263(10^{1})
5	1.5110349(10^{-14})	1.0135348	1.0562014(10^{-15})	1.4115265(10^{1})

	Calculated Missing Initial Conditions			*Calculated Terminal Conditions*		
	$\dot{x}(0)$	$\dot{y}(0)$	$\dot{z}(0)$	$x(2)$	$y(2)$	$z(2)$
0	8.6999999(10^{-2})	4.4999999(10^{-1})	7.9999999(10^{-1})	−6.5145038(10^{-2})	5.1882565(10^{-1})	9.2235671(10^{-1})
1	1.0182591(10^{-1})	4.7222478(10^{-1})	8.1663938(10^{-1})	−7.6940894(10^{-4})	5.7518715(10^{-1})	9.9469680(10^{-1})
2	1.0165788(10^{-1})	4.7228206(10^{-1})	8.1801915(10^{-1})	−2.5203506(10^{-6})	5.7599766(10^{-1})	9.9766041(10^{-1})
3	1.0165880(10^{-1})	4.7228336(10^{-1})	8.1801856(10^{-1})	−1.4735255(10^{-9})	5.7599999(10^{-1})	9.9766099(10^{-1})
4	1.0165880(10^{-1})	4.7228336(10^{-1})	8.1801856(10^{-1})	−1.8980234(10^{-12})	5.7599999(10^{-1})	9.9766099(10^{-1})
5	1.0165880(10^{-1})	4.7228336(10^{-1})	8.1801856(10^{-1})	−1.1379786(10^{-15})	5.7599999(10^{-1})	9.9766099(10^{-1})

B. Modified Method of Adjoints Iteration No.	*Kantorovich Norms*			
	h_i	β_i	η_i	K_i
0	4.2229972(10^{-1})	1.0824180	2.2228302(10^{-2})	1.7551712(10^{1})
1	1.9874074(10^{-2})	1.0151739	1.3795129(10^{-3})	1.4191250(10^{1})
2	5.9124365(10^{-4})	1.0136209	4.1311284(10^{-5})	1.4119594(10^{1})
3	5.2390109(10^{-5})	1.0135392	3.6619598(10^{-6})	1.4115463(10^{1})
4	1.9951430(10^{-6})	1.0135350	1.3945877(10^{-7})	1.4115277(10^{1})
5	1.4947761(10^{-7})	1.0135348	1.0448366(10^{-8})	1.4115266(10^{1})
6	6.6473046(10^{-9})	1.0135348	4.6464135(10^{-10})	1.4115265(10^{1})
7	4.4217505(10^{-10})	1.0135348	3.0907688(10^{-11})	1.4115264(10^{1})

	Calculted Missing Initial Conditions			*Calculated Terminal Conditions*		
	$\dot{x}(0)$	$\dot{y}(0)$	$\dot{z}(0)$	$x(2)$	$y(2)$	$z(2)$
0	8.6999999(10^{-2})	4.4999999(10^{-1})	7.9999999(10^{-1})	−6.5145038(10^{-2})	5.1882565(10^{-1})	9.2235671(10^{-1})
1	1.0182591(10^{-1})	4.7222478(10^{-1})	8.1663938(10^{-1})	−7.6940894(10^{-4})	5.7518715(10^{-1})	9.9469680(10^{-1})
2	1.0164567(10^{-1})	4.7224205(10^{-1})	8.1798921(10^{-1})	−7.9355623(10^{-5})	5.7590793(10^{-1})	9.9755301(10^{-1})
3	1.0165895(10^{-1})	4.7228369(10^{-1})	8.1801490(10^{-1})	−2.5247112(10^{-6})	5.7599841(10^{-1})	9.9765308(10^{-1})
4	1.0165877(10^{-1})	4.7228322(10^{-1})	8.1801847(10^{-1})	−2.0419541(10^{-7})	5.7599971(10^{-1})	9.9766070(10^{-1})
5	1.0165880(10^{-1})	4.7228336(10^{-1})	8.1801854(10^{-1})	−7.1454555(10^{-9})	5.7599999(10^{-1})	9.9766097(10^{-1})
6	1.0165880(10^{-1})	4.7228336(10^{-1})	8.1801856(10^{-1})	−5.5531839(10^{-10})	5.7599999(10^{-1})	9.9766099(10^{-1})
7	1.0165880(10^{-1})	4.7228336(10^{-1})	8.1801856(10^{-1})	−1.9003882(10^{-11})	5.7599999(10^{-1})	9.9766099(10^{-1})

[a] Same boundary conditions as for example in Table 6.2.

for two or three consecutive iterations, then the process will probably converge.

6.9. MODIFIED METHOD OF ADJOINTS [19]

In Section 6.7 we showed that the method of adjoints is a particular realization of the abstract Newton-Raphson method. This enabled us to state conditions under which the method of adjoints converges, and to estimate the rate of convergence and the accuracy of the solution. The identification also suggests alternative methods of solving two-point boundary value problems numerically, since every variant of the Newton-Raphson method can be translated into a numerical method for the solution of these problems. As an example of the process, we derive a modified method of adjoints (MMA for short), which under certain circumstances may require less computing time for the same accuracy than does the method of adjoints.

Let us first recall two methods of carrying out the solution of the operator equation

$$P(x) = 0 \tag{6.9.1}$$

by Newton-Raphson methods, where P is a mapping from the Banach space X to the Banach space X. The ordinary Newton-Raphson method gives the sequence $\{x^{(n)}\}$ according to the formula

$$x^{(n+1)} = x^{(n)} - [P'(x^{(n)})]^{-1}P(x^{(n)}), \tag{6.9.2}$$

in which a new operator, $[P'(x^{(n)})]^{-1}$ (the inverse of the Fréchet derivative $P'(x^{(n)})$ at the point $x^{(n)}$), is formed at each iteration. The modified Newton-Raphson method gives the sequence $\{x^{(n)}\}$ according to the formula

$$x^{(n+1)} = x^{(n)} - [P'(x^{(0)})]^{-1}P(x^{(n)}) \tag{6.9.3}$$

in which the same operator $[P'(x^{(0)})]^{-1}$ (the inverse of the Fréchet derivative $P'(x^{(0)})$ at the initial point $x^{(0)}$) is employed in each iteration. Since the method of adjoints is the usual Newton-Raphson method (6.9.2), there must be a modified method of adjoints corresponding to the modified Newton-Raphson method (6.9.3). With regard to convergence of the MMA, we can apply the appropriate Kantorovich theorem to state conditions under which it will converge, and to estimate the rate of convergence and the accuracy. As for the practical use of the MMA, it has the potential advantage of faster computation because the adjoint equations are integrated only once and the Jacobian matrix corresponding to $P'(x^{(0)})$ is inverted only once. This is a *potential* advantage in that it makes it possible that, for the same accuracy, the MMA will require less total computation time, since it requires less computation per iteration (with the exception of the first iteration, of course). However, the Kantorovich theorem predicts that the MMA will require more iterations. (We are assuming here that the conditions of applicability of the Kantorovich theorem are met by the initial vector $x^{(0)}$). The total computing time, which is the product of the time per iteration and the number of iterations, is dependent on the particular problem at hand, so that no general conclusion can be drawn. It can only be suggested that the MMA be considered, particularly if the matrices corresponding to $P'(x^{(n)})$ are becoming more singular with increasing n. A mixed strategy is also possible in which the method of adjoints is used at first and then, once the elements of the Jacobian matrix change only moderately from iteration to iteration, MMA is used.

Table 6.6. Comparison of Theoretical and Computed Error Bounds on the Initial Velocity Vector [a, b]

		Norm of Initial Velocity Vector	
	Iteration No.	*Theoretical*	*Computed*
Method of Adjoints	0	$4.5(10^{-2})$	$2.2(10^{-2})$
	1	$1.6(10^{-2})$	$1.4(10^{-3})$
	2	$3.8(10^{-3})$	$1.3(10^{-6})$
	3	$4.6(10^{-4})$	$1.6(10^{-9})$
	4	$1.3(10^{-5})$	$6.0(10^{-13})$
	5	$2.2(10^{-8})$	$1.0(10^{-16})$
Modified Method of Adjoints	0	$2.9(10^{-2})$	$2.2(10^{-2})$
	1	$1.3(10^{-2})$	$1.4(10^{-3})$
	2	$5.9(10^{-3})$	$4.0(10^{-5})$
	3	$2.7(10^{-3})$	$3.7(10^{-7})$
	4	$1.2(10^{-3})$	$1.4(10^{-7})$
	5	$5.4(10^{-4})$	$1.0(10^{-8})$

[a] This table was computed from data in Table 6.5 by using an adjusted value for h_0. The h_0 in Table 6.5 was calculated from $K_0 = 1.7551712(10^1)$, which is too high compared to other point estimates for K given in the table. Since a better estimate of the true K is given by $1.4115(10^1)$, a corrected value for h_0 is given by $4.2229972(10^{-1})\,(1.4115/1.7551712) = 3.4578(10^{-1})$. The value used for η_0 is $2.2228303(10^{-2})$.

[b] The bounds are given by the norm of the initial velocity vector, which is defined as Max $\{|\dot{x}_c - \dot{x}_t|, |\dot{y}_c - \dot{y}_t|, |\dot{z}_c - \dot{z}_t|\}$, where subscript c means calculated initial value and subscript t means true velocity. The "true" values were obtained from the tenth iterate of the method of adjoints.

To illustrate behavior that might be encountered in comparing the method of adjoints and the MMA, consider again problem (6.8.1) with the initial conditions as given. As Table 6.2 shows, the method of adjoints converges even though the first few iterates do not satisfy the conditions of the Kantorovich theorem. Table 6.4 for the same initial conditions shows that MMA does not converge for the same initial conditions even after 20 iterations. Here the method of adjoints worked but MMA did not; however, the theory developed for the methods was not applicable. Such examples lead us to the rule of thumb that the method of adjoints is preferred when the sufficiency conditions are not fulfilled; no theoretical justification is claimed for the rule.

In contrast, Table 6.5 gives an example in which the sufficiency conditions are fulfilled and both methods converge. Here the method of adjoints required half as many iterations as the MMA to achieve the same accuracy. Timing estimates for this particular case gave a time per iteration of 6.2 sec for the method of adjoints and 4.9 sec for the MMA, thus a total time of 18.6 sec for the method of adjoints and 29.4 sec for the MMA. Thus the method of adjoints is superior in this example; for another problem the conclusions could very well be reversed.

The error bounds were also calculated; they are given in Table 6.6. As in Section 6.8, the estimated errors are too large, but the Kantorovich estimates can be made fairly sharp by updating the quantities which enter into the computation of the estimate.

6.10. FRIEDRICHS' SOLUTION OF $P(x) = 0$

Section 6.9 indicates that, generally speaking, any method of approximating the solution to the abstract operator equation $P(x) = 0$ (where P maps the Banach space X into the Banach space Y) may be expected to lead to a concrete method for solving the two-point boundary value problem once the proper identifications have been made.

Friedrichs [29] is among the authors whose treatment of the solution of the operator equation $P(x)$ appears to be different from the Newton-Raphson-Kantorovich method, and we are therefore interested to see if his approach leads to another method of solving two-point boundary value problems. He assumes that $P(x)$ has a "first variation", that is, there is defined for each $x \in X$ a linear operator $P_1(x, h)$ from X to Y such that, if

$$R(x,h) = \begin{cases} \dfrac{P(x+h)-P(x)-P_1(x,h)}{\|h\|}, & \text{for } h \in X \neq 0, \\ 0 \qquad , & \text{for } h = 0, \end{cases} \tag{6.10.1}$$

then $\|R(x,h)\| \to 0$ as $\|h\| \to 0$. Now, if $x^{(0)}$ is an approximate solution of $P(x) = 0$, put $x = x^{(0)} + \Delta x$. By (6.10.1),

$$P(x) = P(x^{(0)}) + P_1(x^{(0)}, \Delta x) + \|\Delta x\| R(x^{(0)}, \Delta x); \tag{6.10.2}$$

thus the equation $P(x) = 0$ takes the form

$$P_1(x^{(0)}, \Delta x) = -P(x^{(0)}) - \|\Delta x\| R(x^{(0)}, \Delta x). \tag{6.10.3}$$

If the linear operator $P_1(x^{(0)}, \Delta x)$ has a bounded inverse $\Gamma_0 y \equiv \Gamma(x^{(0)}, y)$ (which is necessarily linear), that is, $\Gamma(x^{(0)}, P_1(x^{(0)}, \Delta x)) = \Delta x$ and $\|\Gamma_0 y\| \leq \|\Gamma_0\| \|y\|$, where the real number $\|\Gamma_0\| > 0$ is a bound for Γ_0, then Γ_0 can be applied to (6.10.3) to give

$$\Delta x = -\Gamma_0 P(x^{(0)}) - \|\Delta x\| \Gamma_0 R(x^{(0)}, \Delta x). \tag{6.10.4}$$

But (6.10.4) is of the form $z = H(z)$, and for such equations it is known that the sequence $\{z^{(n)}\}$, where $z^{(n+1)} = H(z^{(n)})$, converges to z provided that H is a contraction operator. Friedrichs then establishes conditions which guarantee that $H(z) = -\Gamma_0 P(x^{(0)}) - \|z\| \Gamma_0 R(x^{(0)}, z)$ will be a contraction operator with contraction constant θ, $0 < \theta < 1$. Under these conditions the sequence $\{\Delta x^{(n)}\}$, where

$$\Delta x^{(n+1)} = -\Gamma_0 P(x^{(0)}) - \|\Delta x^{(n)}\| \Gamma_0 R(x^{(0)}, \Delta x^{(n)}), \tag{6.10.5}$$

converges to the solution Δx of (6.10.4) which in turn furnishes the solution $x = x^{(0)} + \Delta x$ of $P(x) = 0$. Ficken [22] in his investigation of the continuation method for solving operator equation $P(x, \mu) = 0$, where μ is a scalar, takes Friedrichs'approach as a point of departure.

Friedrichs' results may be summed up in the following theorem, which is a restatement of his Theorem 10.1.

THEOREM 6.5 (Friedrichs). *Let $x^{(0)}$ be an approximate solution of $P(x) = 0$, and suppose the following assumptions hold.*

(i) *$P(x)$ has a first variation $P_1(x, \delta x)$ throughout a certain sphere $\|x - x^{(0)}\| \leq A$, and for each $\varepsilon > 0$ there exist numbers $\alpha(\varepsilon, x^{(0)}) > 0$ and $\beta(\varepsilon, x^{(0)}) > 0$ such that*

$$a. \quad \|x - x^{(0)}\| \leq \alpha(\varepsilon, x^{(0)}) \text{ and } \|\delta x\| \leq \delta(\varepsilon, x^{(0)})$$

imply

$$\|R(x, \delta x)\| \leq \varepsilon;$$

and

$$b. \quad \|x - x^{(0)}\| \leq \beta(\varepsilon, x^{(0)})$$

implies

$$\|P_1(x, \delta x) - P_1(x^{(0)}, \delta x)\| \leq \varepsilon \|\delta x\|.$$

(ii) $P_1(x^{(0)}, y)$ *has a bounded inverse* $\Gamma(x^{(0)}, y) = \Gamma_0$ *with bound*

$$\|\Gamma_0\| = \mu(x^{(0)}) > 0.$$

(iii) *Finally,* $x^{(0)}$ *is such that*

$$\|P(x^{(0)})\| \leq \frac{1-\mu\varepsilon}{\mu} \min\left(\alpha, \frac{\beta}{2}\right).$$

Choose θ *such that* $0 < \theta < 1$, *and define* $\varepsilon = \theta/2\mu(x^{(0)})$.

If assumptions (i) *and* (iii) *hold for* $x^{(0)}$ *and this value of* ε, *then the equation* $P(x) = 0$ *has a unique solution in the sphere*

$$\|x - x^{(0)}\| \leq \min(\alpha(\varepsilon, x^{(0)}), \tfrac{1}{2}\beta(\varepsilon, x^{(0)})),$$

to which the sequence $\{x^{(n)}\}$, *where* $x^{(n)} = x^{(0)} + \Delta x^{(n)}$ *and* $\Delta x^{(n)}$ *is given by* (6.10.5), *converges.*

However, although it is not explicitly stated in [29], it turns out that Friedrichs' method for solving $P(x)$ is simply the modified Newton-Raphson method, as can be easily shown. If we recognize that in (6.10.1) the first variation $P_1(x^{(0)}, h)$ is the Fréchet differential, we may write it in the more usual notation $P'(x^{(0)})h$. Then, using (6.10.1),

$$\|\Delta x^{(n)}\| R(x^{(0)}, \Delta x^{(n)}) = P(x^{(0)} + \Delta x^{(n)}) - P(x^{(0)}) - P'(x^{(0)}) \Delta x^{(n)}, \qquad (6.10.6)$$

which when substituted in (6.10.5) gives

$$\Delta x^{(n+1)} = \Delta x^{(n)} - \Gamma_0 P(x^{(0)} + \Delta x^{(n)}), \qquad (6.10.7)$$

since $\Gamma_0 P'(x^{(0)}) \Delta x^{(n)} = \Delta x^{(n)}$ by the definition of the operator Γ_0. Then adding $x^{(0)}$ to both sides of (6.10.7) and setting $x^{(0)} + \Delta x^{(n)} = x^{(n)}$, we get the sequence $\{x^{(n)}\}$, where

$$x^{(n+1)} = x^{(n)} - \Gamma_0 P(x^{(n)}). \qquad (6.10.8)$$

Equation (6.10.8) is the modified Newton-Raphson method for solving the operator equation $P(x) = 0$.

A direct comparison between Kantorovich's theorem (Theorem 6.2, of Section 6.3) and Friedrichs' theorem on the convergence of $\{x^{(n)}\}$ to the solution x does not seem possible, since Kantorovich requires that the second derivative $P''(x)$ be uniformly bounded on a certain sphere Ω_0 containing the initial approximation $x^{(0)}$, while Friedrichs requires that the first derivative $P'(x)$ satisfy a Lipschitz condition in a (possibly different) sphere around $x^{(0)}$. However, since a bound on the derivative of a function implies a Lipschitz condition on the function itself, the bounded derivative requirement is often made in theory, and is sometimes the only practical way to guarantee that a function will satisfy a Lipschitz condition. Therefore let us assume that the hypotheses of Kantorovich's theorem are satisfied and see what this implies with regard to Friedrichs' theorem. We have, then, that an $x^{(0)}$ has been found such that:

$$\|\Gamma_0\| \le B'; \quad \|P(x^{(0)})\| \le \eta'; \quad \|P''(x)\| \le K'$$

in a sphere $S(r_0, x^{(0)})$ of radius $r_0 = 2B'\eta'$ around $x^{(0)}$; and $h = K'B'^2\eta' \le \frac{1}{2}$. Under these conditions the modified Newton-Raphson method will converge to the solution x of $P(x) = 0$ which, moreover, will be in $S(r_0, x^{(0)})$.

Turning to Friedrichs' theorem, assumption (ii) is satisfied with $\mu(x^{(0)}) \equiv B'$. Assume that a θ (the contraction constant) has been picked with $0 < \theta < 1$, which then defines $\varepsilon = (\theta/2B')$. Assumption (i) is satisfied with $A = 2B'\eta'$, $\alpha(x^{(0)}, \varepsilon) = A = 2B'\eta'$, and $\beta(x^{(0)}, \varepsilon) = \varepsilon/K'$. Here the bound on $P''(x)$ has been used to verify the Lipschitz condition on the first derivative, (ib). Finally, assumption (iii) requires that $\eta' \le (2-\theta)/2B'$ $\min(2B'\eta', \varepsilon/2K')$. If $2B'\eta'$ is the minimum, the condition is essentially vacuous, reducing to $1 \le 2-\theta$ or $\theta \le 1$. So we may assume that $\varepsilon/2K'$ is the minimum; this leads to the requirement (since $\varepsilon = \theta/2B'$)

$$\eta' \le \frac{(2-\theta)\theta}{8B'^2K'}$$

or

$$h = \eta' B'^2 K' \le \frac{(2-\theta)\theta}{8}.$$

This in turn implies that h will be less than $1/8$ instead of $\frac{1}{2}$, as Kantorovich requires. Moreover, if it can be assumed that K' and B' are the same in both theorems so that only the value of η' is open, then the implication is that Friedrichs' theorem requires a better initial approximation $x^{(0)}$ in order to guarantee convergence.

Our conclusion is thus that Friedrichs' method is the modified Newton-Raphson method and that, for applications where a bounded second derivative may be assumed, the theorem of Kantorovich is stronger and more precise than the corresponding theorem of Friedrichs.

6.11. QUASILINEARIZATION AND NEWTON-RAPHSON METHOD

In this section we prove an assertion we have made a number of times, namely, that quasilinearization is also a realization of the abstract Newton-

Raphson method. This means that we again have available Kantorovich's theorem for the specification of conditions sufficient to ensure the convergence of quasilinearization, and to estimate the rate of convergence and the accuracy of the solution. Since we have already given a proof of convergence of quasilinearization (due to McGill and Kenneth; see Section 5.5, Chapter 5), we content ourselves here with sketching the ideas rather than giving all the details. A second reason for not giving the proof in detail is that, like the McGill and Kenneth theorem, the convergence theorem given here is not particularly strong in that it requires the interval $[a, b]$ over which the two-point boundary value problem is to be solved to be sufficiently small. In problems met in practice, we are usually given the interval $[a, b]$ so that theorems for "sufficiently small $[a, b]$" are not directly applicable. What the theorems furnish, in a sense, is the hope that the method (quasilinearization) will work for the interval $[a, b]$ given, since it is known to work for the interval $[a, b']$, $b' < b$.

Consider, then, the problem

$$\frac{dy_i}{dt} = g_i(y_1, \ldots, y_n, t), \qquad i = 1, 2, \ldots, n, \tag{6.11.1}$$

with the boundary conditions

$$\begin{aligned} y_i(t_0) &= c_i, \qquad & i &= 1, 2, \ldots, r, \\ y_{i_m}(t_f) &= c_{i_m}, \qquad & m &= 1, 2, \ldots, n-r. \end{aligned} \tag{6.11.2}$$

We would like to consider the space of all continuous n-vector functions $\mathbf{y}(t)$ on the closed interval $[t_0, t_f]$ with continuous first derivatives, such that $\mathbf{y}(t)$ satisfies the boundary conditions (6.11.2) and, since we are going to apply the abstract Newton-Raphson method, we require that the space be a Banach space. However, unless all the c_j in (6.11.2) are zero, this space is not a linear space (for if $y_j^{(1)}(\bar{t}) = c_j$ and $y_j^{(2)}(\bar{t}) = c_j$, where $\bar{t} = t_0$ or t_f, then $y_j^{(1)}(\bar{t}) + y_j^{(2)}(\bar{t}) = 2c_j$), and therefore cannot be a Banach space. But the given problem is readily transformed into one with zero boundary conditions by the transformations

$$\begin{aligned} z_i(t) &= y_i(t) - c_i, \qquad && \text{if } c_i \text{ is given at } t_0 \text{ but not } t_f, \\ z_j(t) &= y_j(t) - c_j, \qquad && \text{if } c_j \text{ is given at } t_f \text{ but not } t_0, \\ z_k(t) &= y_k(t) - \frac{c_k(t_f - t) - c_k'(t - t_0)}{t_f - t_0}, \qquad && \text{if } c_k \text{ is given at } t_0 \text{ and } c_k' \text{ at } t_f. \end{aligned} \tag{6.11.3}$$

Thus, without loss of generality, we can consider the problem

$$\frac{dz_i}{dt} = h_i(z_1, \ldots, z_n, t), \qquad i = 1, 2, \ldots, n, \tag{6.11.4}$$

with the zero boundary conditions

$$\begin{aligned} z_i(t_0) &= 0, \qquad & i &= 1, 2, \ldots, r, \\ z_{i_m}(t_f) &= 0, \qquad & m &= 1, 2, \ldots, n-r. \end{aligned} \tag{6.11.5}$$

The problem given by (6.11.4) and (6.11.5) can be considered to be one of solving the operator equation

$$P(\mathbf{z}) = 0, \tag{6.11.6}$$

where P is the mapping from the space Z of continuous n-vector functions on $[t_0, t_f]$ satisfying (6.11.5) and having continuous first derivatives, to the space W of continuous n-vector functions on $[t_0, t_f]$, given in component form by

$$(P(z))_i = \frac{dz_i}{dt} - h_i(z_1, \ldots, z_n, t), \qquad i = 1, 2, \ldots, n, \tag{6.11.7}$$

or in vector form by

$$P(\mathbf{z}) = \frac{d\mathbf{z}}{dt} - \mathbf{h}(\mathbf{z}, t). \tag{6.11.8}$$

The solution of (6.11.8) by the abstract Newton-Raphson method generates the sequence $\{\mathbf{z}^{(n)}\}$ according to the formula

$$\mathbf{z}^{(n+1)} = \mathbf{z}^{(n)} - [P'(\mathbf{z}^{(n)})]^{-1} P(\mathbf{z}^{(n)}) \tag{6.11.9}$$

or, equivalently, by

$$P'(\mathbf{z}^{(n)})\mathbf{z}^{(n+1)} = P'(\mathbf{z}^{(n)})\mathbf{z}^{(n)} - P(\mathbf{z}^{(n)}). \tag{6.11.10}$$

It is a straightforward matter to show that the Fréchet derivative $P'(\mathbf{z}^{(0)})$ operating on $\mathbf{z}$ is given in vector form by the expression

$$P'(\mathbf{z}^{(0)})\mathbf{z} = \frac{d\mathbf{z}}{dt} - \frac{\partial \mathbf{h}(\mathbf{z}^{(0)}, t)}{\partial \mathbf{z}} \mathbf{z}, \tag{6.11.11}$$

where $\partial\mathbf{h}/\partial\mathbf{z}$ is the $n \times n$ (Jacobian) matrix, whose entry in the ith row, jth column is $(\partial\mathbf{h}/\partial\mathbf{z})_{ij}$, is

$$\left(\frac{\partial \mathbf{h}}{\partial \mathbf{z}}\right)_{ij} = \frac{\partial h_i(z_1^{(0)}, \ldots, z_n^{(0)}, t)}{\partial z_j}. \tag{6.11.12}$$

It follows that (6.11.10) is the process of solving the sequence of linear two-point boundary value problems

$$\frac{d\mathbf{z}^{(n+1)}}{dt} = \frac{\partial \mathbf{h}(\mathbf{z}^{(n)}, t)}{\partial \mathbf{z}} (\mathbf{z}^{(n+1)} - \mathbf{z}^{(n)}) + \mathbf{h}(\mathbf{z}^{(n)}), \tag{6.11.13}$$

$$z_i^{(n+1)}(t_0) = 0, \qquad i = 1, \ldots, r, \qquad z_{i_m}^{(n+1)}(t_f) = 0, \qquad m = 1, \ldots, n-r.$$

But these are precisely the quasilinearization equations (see Eq. (5.2.6) of Chapter 5)! Thus quasilinearization is a realization of the abstract Newton-Raphson method.

In order to establish conditions under which the quasilinearization process (6.11.13) will converge, we now can use one of the forms of Kantorovich's theorem on the convergence of the Newton-Raphson method. For our purposes it is convenient to use Theorem 6.2 of Section 6.3. Thus we must estimate the bounds

$$\|\Gamma_0\| = \|P'(\mathbf{z}^{(0)})^{-1}\| \leq B',$$

$$\|P(\mathbf{z}^{(0)})\| \leq \eta',$$

$$\|P''(\mathbf{z})\| \leq K', \qquad \mathbf{z} \in \Omega_0,$$

where $\mathbf{z}^{(0)}$ is the initial approximation to the solution $\mathbf{z}^*$ of (6.11.8) and Ω_0 is the sphere in Z, $\|\mathbf{z} - \mathbf{z}^{(0)}\| \leq r$. Then, if the Kantorovich $h = \eta' B'^2 K' \leq \frac{1}{2}$ and $r \geq r_0 = (1 - \sqrt{1-2h}) B'\eta'/h$, (6.11.13) will converge to the solution $\mathbf{z}^*$ of (6.11.8) which will be in the sphere $\|\mathbf{z}^* - \mathbf{z}^{(0)}\| \leq r_0$, with a speed characterized by the inequality

$$\|\mathbf{z}^* - \mathbf{z}^{(n)}\| \leq \frac{1}{2^n} (2h)^{2^n} \frac{\eta'}{h}. \tag{6.11.14}$$

We must first set up norms in the spaces Z and W. For our purposes the norm in Z,

$$\|\mathbf{z}\| = \max_i \left\{ \max_{t_0 \leq t \leq t_f} |z_i(t)| + \lambda \max_{t_0 \leq t \leq t_f} \left| \frac{dz_i(t)}{dt} \right| \right\}, \tag{6.11.15}$$

where $\lambda > 0$ will be prescribed later, is suitable, since Z can be shown to be complete under this norm, which is necessary if Z is to be a Banach space. Furthermore convergence of $\mathbf{z}^{(n)}$ to $\mathbf{z}^*$ in this norm will ensure that $\mathbf{z}^{(n)}$ will be close to $\mathbf{z}^*$ in its derivative values as well as its function values. A suitable norm in W is

$$\|\mathbf{w}\| = \max_i \max_{t_0 \leq t \leq t_f} |w_i(t)|. \tag{6.11.16}$$

The linear operator L which is given by the matrix $\mathbf{D}(t)$ with the elements $d_{ij}(t)$ has the operator norm

$$\|L\| = \max_{t_0 \le t \le t_f} \left\{ \max_i \sum_{j=1}^{n} |d_{ij}(t)| \right\}, \qquad i = 1,2,\ldots,n. \tag{6.11.17}$$

The matrix $\partial \mathbf{h}(\mathbf{z}^{(0)}, t)/\partial \mathbf{z}$ in (6.11.11) is a linear operator.

The bilinear operator Q which is given by the matrix $\mathbf{E}(t)^{(k)}$ with the elements $e_{ij}^{(k)}(t)$ has the operator norm

$$\|Q\| = \max_{t_0 \le t \le t_f} \sum_{k=1}^{n} \sum_{i=1}^{n} \sum_{j=1}^{n} |e_{ij}{}^{(k)}(t)|. \tag{6.11.18}$$

The second derivative operator $P''(\mathbf{z})$ is a bilinear operator.

Next, assumptions have to be made about the vector function $\mathbf{h}(\mathbf{z}(t), t)$ which defines the differential equation (6.11.8). Since we are not attempting to prove the most general result possible, we assume for convenience that the functions $h_i(z_1, \ldots, z_n, t)$, $\partial h_i(z_1, \ldots, z_n, t)/\partial z_j$, and $\partial^2 h_i(z_1, \ldots, z_n, t)/\partial z_j \partial z_k$, $i,j,k = 1,2,\ldots,n$, are continuous (and therefore bounded) in a sufficiently large compact region of the cylinder $S^{n+1} = R^n \times [t_0, t_f]$ including the sphere Ω_0, where R^n is n-dimensional real Cartesian space, with uniform bounds M_0, M_1, and M_2 respectively. Fulfillment of these conditions will ensure that the initial value problem for Equation (6.11.8) has a solution, which is certainly necessary if the two-point boundary value problem is to be solvable.

Given an initial or starting profile for the quasilinearization process (6.11.13), we can estimate η'. If $\mathbf{u} = P(\mathbf{z}^{(0)})$, where $\mathbf{u} \in W$, then, by (6.11.7),

$$\max_i \max_{t_0 \le t \le t_f} |u_i(t)| = \max_i \max_{t_0 \le t \le t_f} \left| \frac{dz_i{}^{(0)}}{dt} - h_i(z_1{}^{(0)}, \ldots, z_n{}^{(0)}, t) \right| \tag{6.11.19}$$

$$\le \max_i \max_{t_0 \le t \le t_f} \left| \frac{dz_i{}^{(0)}}{dt} \right| + \max_{S^{n+1}} |h_i(z_1{}^{(0)}, \ldots, z_n{}^{(0)}, t)|$$

$$\le \|\mathbf{z}^{(0)}\| + M_0.$$

Thus, since $\mathbf{u} \in W$, by (6.11.16) we can estimate a bound on the norm in W as

$$\|\mathbf{w}\| = \|\mathbf{u}\| = \|P(\mathbf{z}^{(0)})\| \le \|\mathbf{z}^{(0)}\| + M_0, \tag{6.11.20}$$

and we can take

$$\eta' = \|\mathbf{z}^{(0)}\| + M_0. \tag{6.11.21}$$

Furthermore, since $P''(\mathbf{z})$ is the bilinear operator defined by the matrices with elements $\partial^2 h_i(z_1, \ldots, z_n, t)/\partial z_j \partial z_k$, by (6.11.18)

$$\|P''(\mathbf{z})\| \leq \max_{S^{n+1}} \sum_{k=1}^{n} \sum_{i=1}^{n} \sum_{j=1}^{n} \left| \frac{\partial^2 h_i}{\partial z_j \partial z_k} \right| \leq n^3 M_2, \tag{6.11.22}$$

so that we can take

$$K' = n^3 M_2. \tag{6.11.23}$$

Finally, we must show that $\Gamma_0 = [P'(\mathbf{z}^{(0)})]^{-1}$ exists, and estimate $\|\Gamma_0\|$. By (6.11.11), if $\mathbf{w} = P'(\mathbf{z}^{(0)})\mathbf{z}$, then $\mathbf{z} = [P'(\mathbf{z}^{(0)})]^{-1}\mathbf{w}$. Therefore $\mathbf{z}$ is the solution of the system of linear differential equations

$$\mathbf{w}(t) = \frac{d\mathbf{z}}{dt} - \frac{\partial \mathbf{h}(\mathbf{z}^{(0)}, t)}{\partial \mathbf{z}} \mathbf{z} \tag{6.11.14}$$

with boundary conditions (6.11.5), or in component form,

$$\frac{dz_i}{dt} = \sum_{j=1}^{n} \frac{\partial h_i(\mathbf{z}^{(0)}, t)}{\partial z_j} z_j + w_i, \qquad i = 1, 2, \ldots, n, \tag{6.11.25}$$

$$\begin{aligned} z_i(t_0) &= 0, \qquad & i &= 1, 2, \ldots, r, \\ z_{i_m}(t_f) &= 0, \qquad & m &= 1, 2, \ldots, n-r. \end{aligned} \tag{6.11.5}$$

Let us assume for the moment that there is a matrix function $\mathbf{G}(t, \tau)$ (a matrix "Green's function") such that the solution of (6.11.25) is

$$\mathbf{z}(t) = \int_{t_0}^{t_f} \mathbf{G}(t, \tau)\mathbf{w}(\tau)\, d\tau. \tag{6.11.26}$$

This will take care of the existence of Γ_0. Then

$$\begin{aligned} \max_i \max_{t_0 \leq t \leq t_f} |z_i(t)| &\leq \max_i \sum_{j=1}^{n} \int_{t_0}^{t_f} \max_{t_0 \leq t \leq t_f} |G_{ij}(t, \tau)| \max_{t_0 \leq \tau \leq t_f} |w_j(\tau)|\, d\tau \\ &\leq nG(t_f - t_0)\|\mathbf{w}\|, \end{aligned} \tag{6.11.27}$$

where G is the uniform bound for the elements $G_{ij}(t, \tau)$ of $\mathbf{G}(t, \tau)$ over the square $[t_0, t_f] \times [t_0, t_f]$, whose existence is yet to be established. Moreover, from (6.11.25),

$$\max_i \max_{t_0 \le t \le t_f} \left| \frac{dz_i}{dt} \right| \le \max_i \sum_{j=1}^{n} \max_{S^{n+1}} \left| \frac{\partial h_i}{\partial z_j} \right| \max_{t_0 \le t \le t_f} |z_j| + \max_{t_0 \le t \le t_f} |w_i|$$

$$\le n^2 GM_1(t_f - t_0) \|\mathbf{w}\| + \|\mathbf{w}\| \tag{6.11.28}$$

$$= (n^2 GM_1(t_f - t_0) + 1) \|\mathbf{w}\| = \theta \|\mathbf{w}\|.$$

It follows from (6.11.27), (6.11.28), and the norm of $\mathbf{z}$, defined in (6.11.15), that

$$\|\mathbf{z}\| \le (nG(t_f - t_0) + \lambda\theta) \|\mathbf{w}\|. \tag{6.11.29}$$

Therefore, by the definition of an operator norm,

$$\|\Gamma_0\| \le nG(t_f - t_0) + \lambda\theta = B'. \tag{6.11.30}$$

Combining (6.11.21), (6.11.23), and (6.11.30), we have that

$$h = K'B'^2\eta' = n^3 M_2(nG(t_f - t_0) + \lambda\theta)^2 (\|\mathbf{z}^{(0)}\| + M_0), \tag{6.11.31}$$

and it is clear from (6.11.31) that, if we make $t_f - t_0$ and λ sufficiently small, we will have $h \le \frac{1}{2}$. It will follow then that the quasilinearization process (6.11.13) with the initial profile $\mathbf{z}^{(0)}$ will converge for sufficiently small $t_f - t_0$, provided that we can establish (6.11.26) and estimate G.

To this end, we write (6.11.25) and its boundary conditions in the form

$$\frac{d\mathbf{z}}{dt} = \mathbf{\Phi}(t)\mathbf{z} + \mathbf{w}, \qquad \mathbf{B}_0 z(t_0) + \mathbf{B}_f z(t_f) = 0, \tag{6.11.32}$$

where $\mathbf{\Phi}(t)$ is an $n \times n$ matrix of functions continuous over $[t_0, t_f]$, and $\mathbf{B}_0$ and $\mathbf{B}_f$ are $n \times n$ constant matrices. Let $\mathbf{Z}(t)$ be the $n \times n$ matrix solution of the initial value problem

$$\frac{d\mathbf{Z}}{dt} = \mathbf{\Phi}(t)\mathbf{Z}, \qquad \mathbf{Z}(t_0) = \mathbf{I}, \tag{6.11.33}$$

where $\mathbf{I}$ is the $n \times n$ identity matrix. $\mathbf{Z}(t)$ is guaranteed to exist over the interval $[t_0, t_f]$ by the fundamental existence theorem. Likewise $\mathbf{Y}(t) = \mathbf{Z}^{-1}(t)$ exists over the same interval, since it is the solution of the problem adjoint to (6.11.33),

$$\frac{d\mathbf{Y}}{dt} = -\mathbf{Y}(t)\mathbf{\Phi}(t), \qquad \mathbf{Y}(t_0) = \mathbf{I} \tag{6.11.33a}$$

(see Lefschetz [26,] p. 54). Then it can be verified directly that the solution of the boundary value problem (6.11.32) can be expressed in the form

$$\mathbf{z}(t) = \int_{t_0}^{t_f} \mathbf{G}(t,\tau)\mathbf{w}(\tau)\,d\tau, \tag{6.11.34}$$

where

$$\begin{aligned} \mathbf{G}(t,\tau) &= \mathbf{Z}(t)\,[\mathbf{I}-\mathbf{V}^{-1}\mathbf{B}_f\mathbf{Z}(t_f)]\mathbf{Z}^{-1}(\tau), \qquad t_0 \leqslant \tau < t, \\ &= -\mathbf{Z}(t)\mathbf{V}^{-1}\mathbf{B}_f\mathbf{Z}(t_f)\mathbf{Z}^{-1}(\tau), \qquad t < \tau \leqslant t_f, \end{aligned} \tag{6.11.35}$$

and where $\mathbf{V} = \mathbf{B}_0+\mathbf{B}_f\mathbf{Z}(t_f)$. Now, if we confine ourselves to the class of problems in which $\mathbf{B}_0+\mathbf{B}_f$ is nonsingular (the class considered by Keller [9] and Osborne [14] to name but two authors), then $\mathbf{V}^{-1} = (\mathbf{B}_0+\mathbf{B}_f\mathbf{Z}(t_f))^{-1}$ will exist by continuity for $[t_0, t_f]$ sufficiently small, since $\mathbf{Z}(t_0) = \mathbf{I}$. Also, even if $\mathbf{B}_0+\mathbf{B}_f$ is singular, $(\mathbf{B}_0+\mathbf{B}_f\mathbf{Z}(t_f))^{-1}$ may still exist, in which case our results apply. In these cases it can also be verified directly that, as a function of t, $\mathbf{G}(t,\tau)$ satisfies the differential equation in (6.11.33) except at $t = \tau$, and that

$$\mathbf{G}(t,t^+)-\mathbf{G}(t,t^-) = \mathbf{I}, \tag{6.11.36}$$

the matrix analog of the usual properties of Green's function. Finally the the uniform bound G of $\mathbf{G}(t,\tau)$ can be estimated from (6.11.35), since all matrices occurring there are bounded over $[t_0, t_f]\times[t_0, t_f]$, either because they are constant or by virtue of the existence theorem for systems of linear differential equations.

In summary, the result of this section is that we have proved the following.

THEOREM 6.6. *Let there be given the two-point boundary value problem* (6.11.4) *and* (6.11.5) *over the interval* $[t_0, t_f]$. *Suppose that the functions* $h_i(z_1,\ldots,z_n,t), i=1,2,\ldots,n$, *are such that the functions themselves*, $\partial h_i(z_1,\ldots, z_n,t)/\partial z_j$, *and* $\partial^2 h_i(z_1,\ldots,z_n,t)/\partial z_j \partial z_k$, where $i,j,k = 1,2,\ldots,n$ *are all continuous over a sufficiently large compact region* S^{n+1} *of the cylinder* $R^n\times[t_0,t_f]$, *where* R^n *is n-dimensional Cartesian space. Suppose also that an initial estimate* $\mathbf{z}^{(0)}(t)$ *of the solution of* (6.11.4) *and* (6.11.5) (*if it exists*) *is given. Finally, suppose that* $(\mathbf{B}_0+\mathbf{B}_f)^{-1}$ *exists or, if not, that* $(\mathbf{B}_0+\mathbf{B}_f\mathbf{Z}(t_f))^{-1}$ *exists. Then the problem* (6.11.4) *and* (6.11.5) *has a solution* $\mathbf{z}^*(t)$ *to which the quasilinearization process* (6.11.13) *will converge, provided* t_f-t_0 *is sufficiently small.*

REFERENCES

1. H. A. Antosiewicz, Newton's Method and Boundary Value Problems, *J. of Computer and System Sci.*, (2) **2** (1968), 177–201.
2. R. G. Bartle, Newton's Method in Banach Spaces, *Proc. Amer. Math. Soc.*, **6** (1955), 827–831.
3. K. M. Brown and J. E. Dennis, On Newton-Like Iteration Functions: General Convergence Theorems and a Specific Algorithm, *Numer. Math.*, **12** (1968), 186–191.
4. L. Collatz, *Functional Analysis and Numerical Analysis*, English translation, Academic, New York, 1966.
5. J. E. Dennis, On Newton-Like Methods, *Numer. Math.*, **11** (1968), 324–330.
6. T. R. Goodman and G. N. Lance, The Numerical Solution of Two Point Boundary Value Problems, *MTAC*, **10** (1956), 82–86.
7. L. V. Kantorovich, Functional Analysis and Applied Mathematics, *Usepkh. Mat. Nauk*, (6) **3** (1948), 89–185. Translated by C. D. Benster, National Bur. Stds., 1952.
8. L. V. Kantorovich and G. P. Akilov, *Functional Analysis in Normed Spaces*, English Translation, Pergamon Press 1964.
9. H. B. Keller, *Numerical Methods for Two-Point Boundary Value Problems*, Blaisdell, Waltham, Mass., 1968.
10. P. Lancaster, Error Analysis for the Newton-Raphson Method, *Numer. Math.*, **9** (1966), 55–68.
11. R. McGill and P. Kenneth, Solution of Variational Problems by Means of a Generalized Newton-Raphson Operator, *AIAA J.*, (10) **2** (1964), 1761–1766.
12. R. H. Moore, Newton's Method and Variations, in *Nonlinear Integral Equations*, P. M. Anselone, ed., University of Wisconsin Press, Madison, Wisc., 1964.
13. R. H. Moore, Approximations to Nonlinear Operator Equations and Newton's Method, *Numer. Math.*, **12** (1968), 23–34.
14. M. R. Osborne, On Shooting Methods for Boundary Value Problems, *J. Math. Anal. Appl.*, (2) **27** (Aug. 1969), 417–433.
15. A. M. Ostrowski, *Solution of Equations and Systems of Equations*, 2nd ed., Academic Press, New York, 1966.
16. L. B. Rall, *Computational Solution of Nonlinear Operator Equations*, Wiley, New York, 1969.
17. S. M. Roberts and J. S. Shipman, The Kantorovich Theorem and Two-Point Boundary-Value Problems, *IBM J. Res. Develop.*, (5) **10** (Sept. 1966), 402–406.
18. S. M. Roberts and J. S. Shipman, Continuation in Shooting Methods for Two-Point Boundary Value Problems, *J. Math. Anal. Appl.*, (1) **18** (April 1967), 45–48.
19. S. M. Roberts and J. S. Shipman, Some Results in Two-Point Boundary Value Problems, *IBM J. Res. Develop.*, (4) **11** (July 1967), 383–388.
20. S. M. Roberts and J. S. Shipman, Justification for the Continuation Method in Two-Point Boundary Value Problems, *J. Math. Anal. Appl.*, (1) **21** (Jan. 1968), 23–30.
21. T. L. Saaty and J. Bram, *Nonlinear Mathematics*, McGraw-Hill, New York, 1964.
22. F. A. Ficken, The Continuation Method for Functional Equations, *Comm. Pure Appl. Math.*, **4** (1951), 435–456.
23. W. Fulks, *Advanced Calculus*, Wiley, New York, 1964.
24. C. Goffman, *Real Functions*, Holt, Rinehart, and Winston, New York, 1953.
25. C. Goffman and G. Pedrick, *First Course in Functional Analysis*, Prentice-Hall, Englewood Cliffs, N.J., 1965.
26. S. Lefschetz, *Lectures on Differential Equations*, Princeton University Press, Princeton, N.J., 1948.

27. P. Henrici, *Discrete Variable Methods in Ordinary Differential Equations*, Wiley, New York, 1962.
28. R. McGill and P. Kenneth, A Convergence Theorem on the Iterative Solution of Non-Linear Two-Point Boundary-Value Systems, *Proc. XIV Internat. Astronautical Congress*, Paris, **IV.12** (1963), 173–186.
29. K. O. Friedrichs, *Functional Analysis and Applications*, New York University Institute of Mathematical Sciences, 1953.

General References

P. M. Anselone and L. B. Rall, The Solution of Characteristic Value–Vector Problems by Newton's Method, *Numer. Math.*, **11**(1968), 38–45.

P. B. Bailey, L. F. Shampine, and P. E. Waltman, *Nonlinear Two Point Boundary Value Problems*, Academic, New York, 1968.

R. Bellman and R. Kalaba, *Quasilinearization and Nonlinear Boundary Value Problems*, American Elsevier, New York, 1965.

T. R. Goodman, System Identification and Prediction–An Algorithm Using a Newtonian Iteration Procedure, *Quart. Appl. Math.*, (3) **24** (Oct. 1966), 249–255.

T. R. Goodman, The Numerical Solution of Eigenvalue Problems, *Math. Comp.*, (91) **19** (July 1965), 462–466.

F. B. Hildebrand, *Introduction to Numerical Analysis*, Prentice-Hall, Englewood Cliffs, N.J., 1956.

D. D. Morrison, J. D. Riley, and J. F. Zancanaro, Multiple Shooting Method for Two-Point Boundary Value Problems, *Comm. ACM*, (12) **5** (Dec. 1962), 613–614.

S. M. Roberts, J. S. Shipman, and W. Ellis, A Perturbation Technique for Non-linear Two-Point Boundary Value Problems, *SIAM J. Numer. Anal.*, (3)**6** (Sept. 1969), 341-358.

V. E. Šamanskiĭ, Methods for the Numerical Solution of Boundary Value Problems. Part II: Nonlinear Boundary Value Problems and Eigenvalue Problems for Differential Equations [Russian], "Naukova Dumka", Kiev, 1966.

W. T. Whiteside, Patterns of Mathematical Thought in the Later Seventeenth Century, *Arch. History Exact Sci.*, **1** (1961).

Chapter 7

CONTINUATION

7.1. INTRODUCTION

Continuation refers to the situation in which a problem depends on a parameter and in which the solution to the problem for one value of the parameter is used to find the solution for a nearby value of the parameter. The word "continuation" suggests both that a given solution has been continued (in an appropriate sense) to a larger domain and that continuity plays a role.

In the typical case the solution to the operator equation

$$P(x, \mu) = 0 \tag{7.1.1}$$

is sought, where it is known that x is a continuous function of μ. The solution is known or has been found for $\mu = \mu_0$. This solution is then used to find the solution for $\mu = \mu_1$, where μ_1 is close to μ_0, by an iterative method using the solution for μ_0 to start the iteration. Often the parameter μ ranges over some interval, say $0 \leqq \mu \leqq 1$; the solution can be found for $\mu = 0$, and the solution for $\mu = 1$ is desired. From the solution for $\mu_0 = 0$ the solution for μ_1 is found, and so on for a finite sequence $\{\mu_i\}_0^N$ such that $\mu_i < \mu_{i+1}$ and $\mu_N = 1$. As we shall see, certain problems which seem to require the solution of the equation

$$P(x) = 0 \tag{7.1.2}$$

can, in practice, be more effectively solved by imbedding (7.1.2) in a series of problems (7.1.1), such that the solution of $P(x, 0) = 0$ is readily found and such that $P(x, 1) = P(x)$, and by applying a continuation technique. In transforming a problem which does not appear to depend on a parameter into one which depends on a parameter μ, the parameter may be chosen in a variety of ways, a few of which we will use in the course of our development.

In this chapter we are, of course, concerned with the application of continuation to the solution of two-point boundary value problems. We show:

(i) that continuation can be used as a theoretical tool to establish the existence of solutions to two-point boundary value problems;

(ii) how continuation can be used in conjunction with shooting methods in the practical solution of numerically sensitive problems; and

(iii) how continuation can be used in conjunction with quasilinearization in the practical solution of numerically sensitive problems.

Shooting methods have often been tried and rejected for two-point boundary value problems which are particularly sensitive to the initial conditions and therefore are troublesome numerically. The appeal of continuation is that, when it is used in conjunction with shooting methods, problems can be solved which could not be solved by shooting methods alone. As a consequence the analyst can avoid the lengthy problem preparation which may be required to solve these problems by finite difference methods.

7.2. CONTINUATION METHOD IN TWO-POINT BOUNDARY VALUE PROBLEMS [1]

Consider the set of n nonlinear differential equations

$$\dot{y}_i = g_i(y_1, y_2, \ldots, y_n, t), \qquad i = 1, 2, \ldots, n, \tag{7.2.1}$$

where the functions g_i are twice differentiable with respect to all the $y_j, j = 1, 2, \ldots, n$. The initial conditions are

$$y_i(t_0) = c_i, \qquad i = 1, 2, \ldots, r \tag{7.2.2}$$

and the terminal conditions are

$$y_{i_m}(t_f) = c_{i_m}, \qquad m = 1, 2, \ldots, n-r. \tag{7.2.3}$$

Let us assume that we will attack this problem using the method of adjoints of Goodman and Lance (Chapter 3, Section 3.6) [2].

Suppose we guess the missing initial conditions and attempt to integrate the equations (7.2.1) forward from t_0 to t_f. If the problem is very sensitive to the initial conditions, our initial guesses may be sufficiently in error that the equations can be integrated only from t_0 to t_1, where $t_1 < t_f$. Beyond t_1, the solution blows up. Faced with this difficulty, we can try other guesses for the missing initial conditions and try to integrate once again from t_0 to t_f. From a practical point of view it may not be possible to break out of the cycle of guessing the missing initial conditions and trying to integrate (7.2.1). It is exactly this situation which has caused many investigators to give up on shooting methods.

The continuation method picks up where conventional shooting methods fail. Briefly, the idea is to solve successively a sequence of two-point boundary value problems for a sequence of time spans $[t_0, t_1]$, $[t_0, t_2], \ldots,$ $[t_0, t_f]$, $t_0 < t_1 \ldots < t_f$. For each successive problem the boundary conditions are held constant, even though the terminal times are successively larger. The initial conditions determined by the solution of the qth successive two-point boundary value problem over the time span $[t_0, t_q]$ are taken as the initial guess for the initial values for the $(q+1)$st two-point boundary value problem for the extended time interval $[t_0, t_{q+1}]$, $t_q < t_{q+1}$.

In outline, the continuation method is the following:

1. If $y_i(t_0) = c_i$, $i = 1, 2, \ldots, r$, are the specified initial conditions at t_0, choose the remaining initial conditions $y_i(t_0) = c_i$, $i = r+1, \ldots, n$.

2. Integrate (7.2.1) forward until numerical difficulties or overflow problems appear. Choose as the final time t_1, a value at which the solutions still have good behavior.

3. Over the time span $[t_0, t_1]$ solve the two-point boundary value problem for the initial conditions $y_i(t_0) = c_i$, $i = 1, 2, \ldots, r$, and the terminal conditions $y_{i_m}(t_1) = c_{i_m}$, $m = 1, 2, \ldots, n-r$, using, say, the method of adjoints.

4. To extend the time interval from $[t_0, t_1]$ to $[t_0, t_2]$, let the initial conditions that were developed for the solution of the two-point boundary value problem over the interval $[t_0, t_1], y_i(t_0) = c_i^*, i = 1, 2, \ldots, n$, be the initial conditions for the first iteration of the two-point boundary value problem over $[t_0, t_2]$. Note that $y_i(t_0) = c_i^* = c_i$, $i = 1, \ldots, r$, are the specified initial conditions of the problem and $y_i(t_0) = c_i^*$, $i = r+1, \ldots, n$, are the guessed values for the missing initial conditions over the span $[t_0, t_2]$ for the first iteration of the shooting method. The terminal boundary conditions are the same as before, except that they are applied at t_2:

$$y_{i_m}(t_2) = c_{i_m}, \; m = 1, 2, \ldots, n-r.$$

Solve the two-point boundary value problem over the time span $[t_0, t_2]$.

5. Step 4 is repeated successively for the terminal times $t_1, t_2, \ldots$ until the terminal time equals t_f. At this time the original problem will have been solved.

7.3. DISCUSSION OF THE CONTINUATION METHOD

From the abstract point of view our continuation method is equivalent to what Moore [3] calls the construction of "curves of solutions" for the operator equation with a parameter $P(x, \mu) = 0$. Here, as we have shown [4],

x is a member of the $(n-r)$-dimensional vector space of missing initial conditions, P is the nonlinear vector function of "miss distances", while the sequence of values μ_i is to be identified with our sequence of final times $t_{i+1}, i = 0, 1, 2, \ldots$. Thus the theorem and observations cited by Moore apply. (See Section 6.7, Chapter 6.)

From the concrete or practical point of view we note that the continuation method uses the original nonlinear differential equations. This is in contrast to the quasilinearization of Bellman [5] and the finite difference methods of Fox [6], which deal with linear approximations to the original nonlinear differential equations. It is a matter of some importance whether the solution of the linearized set of equations does, in fact, produce the missing initial conditions that satisfy numerically the original nonlinear boundary value problem. To be more specific, we are concerned about whether the missing initial conditions when introduced into the original problem will result in the nonlinear equations being numerically integrable over $[t_0, t_f]$ and will result in satisfying the terminal boundary conditions. It is not clear whether various investigators have confirmed the results of their approximation techniques by such a check. With the continuation method, no such doubts exist since, as stated above, we deal directly with the original nonlinear differential equations.

For the continuation method the investigator has two devices to control the generation of a solution and the speed of producing a solution. These are: the incremental time span step $\Delta\tau_i = t_{i+1} - t_i$, to be added to the terminal time t_i; and the multiplier $F (0 < F \leq 1.0)$ applied to the corrections to the missing initial conditions $\delta y_i(t_0)$, $i = r+1, \ldots, n$, obtained by solving Eq. (3.6.16), Section 3.6, Chapter 3:

$$\delta y_i^*(t_0) = F \delta y_i(t_0),$$

where

$\delta y_i(t_0)$ = correction from (3.6.16), Section 3.6, Chapter 3,
$\delta y_i^*(t_0)$ = correction applied.

If the $\Delta\tau_i$ is set too small, it may take a long time to integrate over $[t_0, t_f]$. On the other hand, if $\Delta\tau_i$ is set too large, the forward integration of (7.2.1) may blow up. If $F = 1.0$, the full correction is taken. For some very delicate problems the full correction $\delta y_i(t_0)$ to $y_i(t_0)$ is too much, and (7.2.1) cannot be integrated. If F is too small, the Goodman-Lance method must go through many iterations to converge. While no general rule is applicable at this time, the best strategy appears to be at first to set $\Delta\tau_i$ to a modest size, perhaps less than 10% of the range $[t_0, t_i]$, and to set $F = 1.0$. (This conclusion is based on the fact that for each new continuation step, during an iteration of the shooting method, (7.2.1) must be integrated forward once and the adjoint equations must be integrated $(n-r)$ times. It is therefore worthwhile to make each extension produce equations which are integrable; otherwise, the problem has to be restarted for a smaller $\Delta\tau_i$.) Our computational experience has shown that the size of the $\Delta\tau_i$ and F generally must then be decreased as the overall time span $[t_0, t_i]$ increases. Eventually, in sensitive problems, we usually reach a point in time beyond which the continuation method fails.

One possible important advantage of the continuation method is that it does not require the analysis and development of new equations, such as that in the methods of Bellman and Kalaba [5], Fox [6], Henrici [7], Holt [8]. The creation of a suitable set of finite difference equations for example, may be a formidable job. The continuation method, on the other hand, can be programmed for a digital computer using more or less standard integration techniques, such as Runge-Kutta. It is believed the continuation method can be applied directly to sensitive two-point boundary value problems with a minimum of preparation. As a matter of fact, in our computer program, the only subroutines which need to be changed from problem to problem are those which specify the particular set of differential equations to be handled.

Having pointed out some of the potential advantages of the continuation method, we must now examine some of its possible disadvantages. The first one is that the selection of the first time span $[t_0, t_1]$ may be quite delicate. It cannot be too large, of course, because then it will not be possible to integrate the system of differential equations from t_0 to t_1. But it usually cannot be too small, either, for the original problem, assumed solvable over $[t_0, t_f]$, may not be solvable over the smaller interval. Although we do not know in general how small the interval $[t_0, t_1]$ may be relative to $[t_0, t_f]$, we have solved problems (see Section 7.4) in which the first time span was approximately 25% of the desired interval $[t_0, t_f]$. Secondly, while we may be able to integrate (7.2.1) forward over the interval $[t_0, t_i]$ using the initial guess of the missing initial conditions, it is quite possible that the Goodman-Lance method may generate corrections to the guessed initial conditions such that integration over the interval $[t_0, t_{i+1}]$ is not possible.

Finally, it can happen that, as the attempt at computing the solution proceeds, the necessary values of F and Δt_i continue to decrease to such an extent that the solution cannot be continued beyond $t_m < t_f$. In this case the continuation method of Section 7.2 has failed. It may be possible to use the solution obtained over $[t_0, t_m]$ in the combination of continuation and quasilinearization described in Section 7.7, or in the perturbation technique of Section 7.10. If not, the problem must be attacked by methods other than shooting, usually one of the finite difference methods (Chapter 8).

7.4. NUMERICAL RESULTS

As a rather difficult two-point boundary value problem, let us consider a problem proposed by Holt [8]:

Example 1.

$$
\begin{aligned}
\dot{y}_1 &= y_2, \\
\dot{y}_2 &= y_3, \\
\dot{y}_3 &= -\left(\frac{3-n}{2}\right) y_1 y_3 - n y_2^{\,2} + 1 - y_4^{\,2} + s y_2, \\
\dot{y}_4 &= y_5, \\
\dot{y}_5 &= -\left(\frac{3-n}{2}\right) y_1 y_5 - (n-1) y_2 y_4 + s(y_4 - 1).
\end{aligned}
\tag{7.4.1}
$$

The initial conditions are

$$y_1(0) = 0, \qquad y_2(0) = 0, \qquad y_4(0) = 0,$$

and the terminal conditions are

$$y_2(t_f) = 0, \qquad y_4(t_f) = 1.$$

Let us specify

$$n = -0.1, \qquad s = +0.2,$$

$$t_0 = 0.0, \qquad t_f = 11.3.$$

Holt claimed that this problem is not solvable by conventional shooting methods. He was able to solve it by using a combination of quasilinearization and finite difference methods.

In order to appreciate the difficulties of solving this two-point boundary value problem, let us examine Table 7.1. In the first two-columns we list the initial guesses of the missing $y_3(0)$ and $y_5(0)$. We tabulate in the third column the time interval over which these initial conditions led to the successful solution of the problem. We list in the fourth column a slightly larger time interval over which the problem cannot be solved for the same initial conditions, because the integration blows up. It is very instructive to observe the sensitivity of the solution to slight changes in the initial conditions and the time interval. From Table 7.1 it is quite obvious why conventional shooting methods fail.

Table 7.1. Example 1. Sensitivity Study of Initial Guess of Missing Initial Conditions for Various Terminal Times

Initial Guess Values of Missing Initial Conditions		*Time Span*	
$y_3(0)$	$y_5(0)$	*Successful Solution of Two-Point Boundary Value Problem*	*Unsuccessful Solution of Two-Point Boundary Value Problem*
−1.000000000	0.600000000	[0, 3.5]	[0, 4.0]
−0.978197694	0.646786696	[0, 4.0]	[0, 5.0]
−0.969167054	0.652538306	[0, 5.0]	[0, 6.0]
−0.966297293	0.652998781	[0, 6.0]	[0, 7.0]
−0.966304111	0.652913439	[0, 7.0]	[0, 8.0]
−0.966311455	0.652909499	[0, 8.5]	[0, 9.0]
−0.966311753	0.652909577	[0, 9.5]	[0, 10.0]

To show in detail the sequence of successful initial conditions and the corresponding intervals $[t_0, t_i]$ generated by the continuation method, Table 7.2 is presented. In this table we list in double precision in the first two columns the initial guess values for $y_3(0)$ and $y_5(0)$, and in the third and fourth columns the values of $y_3(0)$ and $y_5(0)$ determined by the Goodman-Lance shooting method. Using the values of $y_3(0)$ and $y_5(0)$ in the third and fourth columns, Eqs. (7.4.1) were integrated to satisfy each terminal boundary condition with an error less than 10^{-4}. As long as the continuation method worked for the chosen $\Delta\tau_i$ and F, the program automatically stored the entries in the third and fourth columns as the missing initial conditions $y_3(0)$ and $y_5(0)$ for the next continuation process. For example, in row 1, entries in the third and fourth columns for the time span [0, 10.0] became the entries in row 2, in the first and second columns for the time span [0, 10.2]. When the continuation method failed, owing to say, $\Delta\tau_i$ or F being too large, the program terminated. An example of this appears in row 5, where the two-point boundary value problem could not be solved over the time span [0, 10.8] although it was solved over [0, 10.6]. In row 6 the guessed missing initial conditions, which are the numbers taken from row 4, first and second columns, truncated to 10 decimal places, were used to initiate the two-point boundary value program over the time span [0, 10.7] and yielded the entries $y_3(0)$ and $y_5(0)$ in the third and fourth columns on convergence. One of the most interesting points about Table 7.2 is that, to extend the time interval, corrections to the missing initial conditions are in

Table 7.2. Example 1

Guessed Missing Initial Conditions		*Missing Initial Conditions Determined By Two-Point Boundary Value Program*			
$y_3(0)$	$y_5(0)$	$y_3(0)$	$y_5(0)$	t_i	F
−0.96631174099999990	0.65290958299999993	−0.96631173964458963	0.65290959353970734	10.0	1.0
−0.96631173964458963	0.65290959353970734	−0.96631173434973681	0.65209059476491492	10.2	1.0
−0.96631173434973681	0.65290959476491492	−0.96631172877727307	0.65209059605688981	10.4	1.0
−0.96631172877827307	0.65290959605688981	−0.96631172290039586	0.65290959742338757	10.6	1.0
−0.96631172290039586	0.65290959742338757	—	—	10.8	Did not solve TPBV
−0.9663117229	0.6529095974	−0.96631171983909825	0.65290959813601912	10.7	1.0
−0.96631171983909825	0.65290959813601912	−0.96631171669285489	0.65290959886874953	10.8	1.0
−0.96631171669285489	0.65290959886874953	−0.96631171345940601	0.65290959962187083	10.9	1.0
−0.96631171345940601	0.65290959962187083	—	—	11.1	Did not solve TPBV
−0.9663117134	0.6529095996	−0.96631171345940610	0.65290959962187074	10.9	1.0
−0.96631171345940610	0.65290959962187074	−0.96631171180931785	0.65290960000615792	10.95	1.0
−0.96631171180931785	0.65290960000615792	−0.96631171013665034	0.65290960039563508	11.00	1.0
−0.96631171013665034	0.65290960039563508	−0.96631170844115939	0.65290960079033038	11.05	1.0
−0.96631170844115939	0.65290960079033038	−0.96631170672260636	0.65290960119027126	11.10	1.0
−0.96631170672260636	0.65290960119027126	−0.96631170498075685	0.65209060159548526	11.15	1.0
−0.96631170498075685	0.65290960159548526	−0.96631170321538118	0.65290960200599910	11.20	1.0
−0.96631170321538118	0.65290960200599910	−0.96631170142625207	0.65290960242184060	11.25	1.0
−0.96631170142625207	0.65290960242184060	−0.96631169961314560	0.65290960284303701	11.30	0.5

Table 7.3. Example 1 [a]

Time	$y_1(t)$	$y_2(t)$	$y_3(t)$	$y_4(t)$	$y_5(t)$
0.0000	0.	0.	−9.66311693(10⁻¹)	0.	6.52909595(10⁻¹)
0.0565	−1.25670226(10⁻¹)	−3.97764826(10⁻¹)	−4.68124777(10⁻¹)	3.39400265(10⁻¹)	5.50824451(10⁻¹)
1.1300	−4.05545026(10⁻¹)	−5.62249017(10⁻¹)	−1.41289239(10⁻¹)	6.22806859(10⁻¹)	4.52044249(10⁻¹)
1.6950	−7.34100139(10⁻¹)	−5.82764459(10⁻¹)	5.08505023(10⁻²)	8.50075769(10⁻¹)	3.52721885(10⁻¹)
2.2600	−1.04866613	−5.20736521(10⁻¹)	1.57629611(10⁻¹)	1.02215350	2.57439291(10⁻¹)
2.8250	−1.31432994	−4.14716822(10⁻¹)	2.10480505(10⁻¹)	1.14245442	1.69924465(10⁻¹)
3.3900	−1.51374745	−2.89601451(10⁻¹)	2.27478865(10⁻¹)	1.21606904	9.24784923(10⁻²)
3.9550	−1.64121246	−1.62380719(10⁻¹)	2.19322291(10⁻¹)	1.24914077	2.66281226(10⁻²)
4.5200	−1.69914684	−4.51885200(10⁻²)	1.92945437(10⁻¹)	1.24856937	−2.64278278(10⁻²)
5.0850	−1.69583352	5.32016087(10⁻²)	1.53566974(10⁻¹)	1.22188050	−6.56703115(10⁻²)
5.6500	−1.64371775	1.26786053(10⁻¹)	1.05937897(10⁻¹)	1.17710450	−9.03708446(10⁻²)
6.2150	−1.55786994	1.72304597(10⁻¹)	5.51074070(10⁻²)	1.12252490	−1.00435726(10⁻¹)
6.7800	−1.45436738	1.89504519(10⁻¹)	6.65482533(10⁻³)	1.06618552	−9.68879545(10⁻²)
7.3450	−1.34852652	1.81335369(10⁻¹)	−3.38136074(10⁻²)	1.01514298	−8.22308147(10⁻²)
7.9100	−1.25314547	1.53654452(10⁻¹)	−6.18365341(10⁻²)	9.74629498(10⁻¹)	−6.03509825(10⁻²)
8.4750	−1.17711279	1.14222206(10⁻¹)	−7.52863276(10⁻²)	9.47445333(10⁻¹)	−3.58069664(10⁻²)
9.0400	−1.12473631	7.12452650(10⁻²)	−7.46437484(10⁻²)	9.33884835(10⁻¹)	−1.27087998(10⁻²)
9.6050	−1.09586686	3.21500003(10⁻²)	−6.19014412(10⁻²)	9.32339120(10⁻¹)	6.54088545(10⁻³)
10.1700	−1.08647805	3.28554133(10⁻³)	−3.85182148(10⁻²)	9.40719199(10⁻¹)	2.30777159(10⁻²)
10.7350	−1.08909157	−9.30043542(10⁻³)	−4.16728240(10⁻³)	9.59557080(10⁻¹)	4.63308668(10⁻²)
11.3000	−1.09281866	3.96523276(10⁻⁷)	3.65269014(10⁻²)	9.99999583(10⁻¹)	1.08170825(10⁻¹)

[a] $n = -0.1$, $s = 0.2$.

Table 7.4. Example 1 [a]

Time	$y_1(t)$	$y_2(t)$	$y_3(t)$	$y_4(t)$	$y_5(t)$
0.0000	0.	0.	−1.02023558	0.	6.74866498(10^{-1})
0.6540	−1.72857293(10^{-1})	−4.62741297(10^{-1})	−4.19004709(10^{-1})	4.36543137(10^{-1})	6.47774827(10^{-1})
1.3080	−5.31902182(10^{-1})	−5.89983714(10^{-1})	−3.29417863(10^{-3})	8.27050793(10^{-1})	5.32197446(10^{-1})
1.9620	−8.98872149(10^{-1})	−5.06845415(10^{-1})	2.30456927(10^{-1})	1.11915350	3.55014971(10^{-1})
2.6160	−1.17233741	−3.19541830(10^{-1})	3.21895772(10^{-1})	1.28864871	1.64746051(10^{-1})
3.2700	−1.31183192	−1.08498976(10^{-1})	3.09173396(10^{-1})	1.34023449	−3.26135677(10^{-4})
3.9240	−1.32151081	7.00392377(10^{-2})	2.28482434(10^{-1})	1.29917751	−1.15596882(10^{-1})
4.5780	−1.23460446	1.83378385(10^{-1})	1.15755863(10^{-1})	1.20236725	−1.70185173(10^{-1})
5.2320	−1.09808709	2.22038794(10^{-1})	5.52625269(10^{-3})	1.08917131	−1.67452204(10^{-1})
5.8860	−9.58101368(10^{-1})	1.97308880(10^{-1})	−7.44553864(10^{-2})	9.92303133(10^{-1})	−1.23863727(10^{-1})
6.5400	−8.48369372(10^{-1})	1.34321609(10^{-1})	−1.10684562(10^{-1})	9.30863976(10^{-1})	−6.32013565(10^{-2})
7.1940	−7.84523052(10^{-1})	6.13948858(10^{-2})	−1.06658912(10^{-1})	9.08502972(10^{-1})	−7.41737777(10^{-3})
7.8480	−7.65388489(10^{-1})	3.59175217(10^{-4})	−7.72418386(10^{-2})	9.17142081(10^{-1})	3.02757448(10^{-2})
8.5020	−7.79000807(10^{-1})	−3.78416961(10^{-2})	−3.94259354(10^{-2})	9.43398917(10^{-1})	4.66237807(10^{-2})
9.1560	−8.09642708(10^{-1})	−5.22046965(10^{-2})	−5.95852125(10^{-3})	9.74269271(10^{-1})	4.53788698(10^{-2})
9.8100	−8.43234098(10^{-1})	−4.80683237(10^{-2})	1.65246694(10^{-2})	1.00041194	3.33864078(10^{-2})
10.4640	−8.70207083(10^{-1})	−3.33278206(10^{-2})	2.65576443(10^{-2})	1.01712300	1.75943986(10^{-2})
11.1800	−8.86205399(10^{-1})	−1.56790704(10^{-2})	2.58497462(10^{-2})	1.02377261	3.30593202(10^{-3})
11.7720	−8.91434228(10^{-1})	−1.27545387(10^{-3})	1.70305368(10^{-2})	1.02236339	−6.92323339(10^{-3})
12.4260	−8.89608049(10^{-1})	5.19079322(10^{-3})	1.72950268(10^{-3})	1.01518992	−1.53172398(10^{-2})
13.0800	−8.87215376(10^{-1})	−2.33526736(10^{-5})	−1.77311935(10^{-2})	9.99985147(10^{-1})	−3.53176868(10^{-2})

[a] $n = -0.5$, $s = 0.0$.

the lower half of the double precision word. In other words, the first 7 or 8 decimal digits in $y_3(0)$ and $y_5(0)$ are essentially the same throughout Table 7.2.

Table 7.3 lists in single precision $y_1, \ldots, y_5$ profiles over the span $[0, 11.3]$, where the integration step size was 0.0565. Table 7.4 lists the profiles for the same problem, except $n = -0.5$, $s = 0$, and the time span is $[0, 13.08]$. In both Tables 7.3 and 7.4 we held the number of integration steps equal to 200 for all the time spans considered. For the Runge-Kutta integration method we employed and for this problem, the success of the integration was quite dependent on the number of integration steps.

Example 2. As a second example consider the equations (Holt [8])

$$\dot{y}_1 = -\frac{y_1}{y_5}[y_6+2(y_3+y_5)],$$

$$\dot{y}_2 = y_1 y_3(y_3+y_5),$$

$$\dot{y}_3 = y_4,$$

$$\dot{y}_4 = \frac{R_e y_1}{y_7}[-2\sigma A y_7{}^2+y_4 y_5+y_3(y_3+y_5)]+\frac{y_4 y_5 y_6}{2Ay_7{}^2}+\frac{2R_e y_2}{y_7}$$

$$\dot{y}_5 = y_6,$$

$$\dot{y}_6 = \frac{y_1}{(4/3R_e)y_7}\left[-\frac{\sigma A y_7{}^2}{y_5}\{y_6+2(y_3+y_5)\}+y_5 y_6(1-\sigma)\right] - \frac{y_3 y_5 y_6}{2Ay_7{}^2}-\frac{y_4}{2}-\frac{y_5}{2A}\left(\frac{y_6}{y_7}\right)^2,$$

$$\dot{y}_7 = -\frac{y_5 y_6}{2Ay_7},$$

with the initial conditions

$$y_1(t_0) = 0.9617, \qquad y_5(t_0) = -0.0212,$$
$$y_2(t_0) = -0.1018, \qquad y_7(t_0) = 0.9998,$$
$$y_3(t_0) = 0.4078,$$

and the terminal conditions

$$y_3(t_f) = 1.0, \qquad y_5(t_f) = -1.0,$$

where

$$R_e = 100.0,$$
$$A = 0.515,$$
$$\sigma = 0.400,$$
$$t_0 = 0.05,$$
$$t_f = 0.3303.$$

The set of equations are equivalent to Holt's Example 3, while the boundary conditions differ, in part because we have used as initial conditions Holt's tabulated solution at $t = 0.05$ rather than Holt's initial conditions at $t_0 = 0.0$. This was done to avoid the further analysis that is required by the fact that his v, our y_5, which is specified to be equal to zero at time zero, occurs in the denominator in our equations for $\dot{y}_1$ and $\dot{y}_6$.* The problem in either form is a difficult one to solve. Our results over the time span [0.05, 0.3303] are given in Table 7.5. It is significant that, on originally checking Holt's Table 3, we found that his results given to four decimal digits were not accurate enough to integrate the original differential equations.

7.5. JUSTIFICATION FOR THE CONTINUATION METHOD IN TWO-POINT BOUNDARY VALUE PROBLEMS [9]

In Sections 7.2–7.4 we gave a continuation process to be used in conjunction with shooting methods (in particular, the method of adjoints of Goodman and Lance [5]) for the numerical solution of the two-point boundary value problem

$$\dot{y}_i = g_i(y_1, \ldots, y_n, t), \qquad i = 1, 2, \ldots, n, \tag{7.5.1}$$

where r initial conditions are specified at $t = t_0$ and $n-r$ final conditions at $t = t_f$. The continuation process is intended to be used when numerical sensitivity to initial values does not allow the equations to be integrated over the entire interval $[t_0, t_f]$ with the $n-r$ guessed initial values. The process consists in solving the given boundary value problem, but for the interval $[t_0, t_1]$ over which the integration can be accomplished, then using

* The reader interested in making a direct comparison with Holt may want to verify that the appropriate boundary conditions for our equations are

$$y_1(0) = 1, \quad y_3(0) = 0, \quad y_5(0) = 0, \quad y_6(0) = 0, \quad y_7(0) = 1,$$
$$y_3(t_f) = 1, \quad y_5(t_f) = -1.$$

Table 7.5. Example 2

Time	$y_1(t)$	$y_2(t)$	$y_3(t)$	$y_4(t)$	$y_5(t)$	$y_6(t)$	$y_7(t)$
0.05000	$9.6170000(10^{-1})$	$-1.0180000(10^{-1})$	$4.0779999(10^{-1})$	7.5249155	$-2.1200000(10^{-2})$	$-7.8877221(10^{-1})$	$9.9979999(10^{-1})$
0.07002	$9.4829479(10^{-1})$	$-9.7660192(10^{-2})$	$5.4666740(10^{-1})$	6.3611086	$-3.9607126(10^{-2})$	-1.0409544	$9.9925650(10^{-1})$
0.09004	$9.3499551(10^{-1})$	$-9.1284603(10^{-2})$	$6.6317369(10^{-1})$	5.2971326	$-6.2592025(10^{-2})$	-1.2539909	$9.9811478(10^{-1})$
0.11006	$9.0290583(10^{-1})$	$-8.2881929(10^{-2})$	$7.5965267(10^{-1})$	4.3651288	$-9.0905873(10^{-2})$	-1.7045214	$9.9599854(10^{-1})$
0.13008	$7.3291338(10^{-1})$	$-7.3674509(10^{-2})$	$8.3888886(10^{-1})$	3.5712532	$-1.4388608(10^{-1})$	-4.1096237	$9.8988986(10^{-1})$
0.15010	$4.6138256(10^{-1})$	$-6.6703075(10^{-2})$	$9.0304049(10^{-1})$	2.8473898	$-2.6291316(10^{-1})$	-7.4934846	$9.6584861(10^{-1})$
0.17012	$3.0624987(10^{-1})$	$-6.2604296(10^{-2})$	$9.5336767(10^{-1})$	2.1939093	$-4.2537168(10^{-1})$	-8.3483231	$9.0794784(10^{-1})$
0.19015	$2.3081467(10^{-1})$	$-6.0193573(10^{-2})$	$9.9138643(10^{-1})$	1.6148050	$-5.8602411(10^{-1})$	-7.5236221	$8.1646830(10^{-1})$
0.21017	$1.9133763(10^{-1})$	$-5.8718343(10^{-2})$	1.0183191	1.0809043	$-7.2229869(10^{-1})$	-6.0361307	$7.0250827(10^{-1})$
0.23019	$1.6927998(10^{-1})$	$-5.7793414(10^{-2})$	1.0347897	$5.6661892(10^{-1})$	$-8.2704952(10^{-1})$	-4.4408558	$5.7961212(10^{-1})$
0.25021	$1.5656761(10^{-1})$	$-5.7210668(10^{-2})$	1.0411079	$6.8432246(10^{-2})$	$-9.0135639(10^{-1})$	-3.0285629	$4.5963694(10^{-1})$
0.27023	$1.4921390(10^{-1})$	$-5.6852602(10^{-2})$	1.0378205	$-3.8522080(10^{-1})$	$-9.5038139(10^{-1})$	-1.9274821	$3.5091709(10^{-1})$
0.29025	$1.4499429(10^{-1})$	$-5.6652282(10^{-2})$	1.0263311	$-7.3898528(10^{-1})$	$-9.8070732(10^{-1})$	-1.1586936	$2.5749735(10^{-1})$
0.31027	$1.4255023(10^{-1})$	$-5.6571129(10^{-2})$	1.0091106	$-9.4791616(10^{-1})$	$-9.9863503(10^{-1})$	$-6.8017305(10^{-1})$	$1.7852702(10^{-1})$
0.33030	$1.4100281(10^{-1})$	$-5.6585727(10^{-2})$	$9.8878235(10^{-1})$	-1.0534341	-1.0093838	$-4.4308019(10^{-1})$	$1.0450045(10^{-1})$

the converged initial values as the first guess of the missing initial values for the same boundary value problem over the interval $[t_0, t_2]$, $t_2 > t_1$, etc. The process continues until t_f is reached.

It was noted in Section 7.3 that, in view of our identification of the Goodman-Lance method of adjoints with the abstract Newton-Raphson-Kantorovich method [4, 10], the continuation method could be justified by appealing to theorems on the solution by the Newton-Raphson-Kantorovich method of operator equations, which depend on a parameter as discussed, for example, by Moore [3]. In this section we give an explicit justification of our continuation method.

Recall the sufficient conditions for the convergence of the Goodman-Lance method of adjoints to the solution of the two-point boundary value problem (See Sections 6.3 and 6.7, Chapter 6). Define the miss distances

$$\phi_m(\mathbf{y}(t_0); t) = Y_m - y_{i_m}(\mathbf{y}(t_0); t), \qquad m = 1, 2, \ldots, n-r, \tag{7.5.2}$$

where the Y_m are the $n-r$ given final values and the $y_{i_m}(\mathbf{y}(t_0), t)$ are the calculated values of the solutions $\mathbf{y}(t)$ of (7.5.1) at t with initial values $\mathbf{y}(t_0) = \{y_1(t_0), \ldots, y_r(t_0), y_{r+1}(t_0), \ldots, y_n(t_0)\}$, and let $\mathbf{A}(\mathbf{y}(t_0))$ be the $(n-r)\times(n-r)$ matrix $(\partial\phi_m/\partial y_{r+j}(t_0))$. If a first guess of the missing initial conditions $y_{r+j}^{(0)}(t_0) = z_j^{(0)}, j = 1, 2, \ldots, n-r$, is known such that

(i) $\mathbf{\Gamma}_0 = \mathbf{A}^{-1}(\mathbf{z}^{(0)})$ exists and $\|\mathbf{\Gamma}_0\| \le B'$, where the norm is the maximum row sum norm;

(ii) $\|\boldsymbol{\phi}(\mathbf{z}^{(0)})\| \le \eta'$, where the norm is the maximum component norm;

(iii) $\sum_{j,k=1}^{n-r} \partial^2\phi_i/\partial z_j\,\partial z_k \le K'$ for $i = 1, 2, \ldots, n$ in the region defined by inequality (7.5.3) below;

(iv) $h = K'B'^2\eta' \le \frac{1}{2}$;

then (by the Kantorovich sufficiency theorem for the Newton-Raphson method, Theorem 6.2, Chapter 6) the Goodman-Lance method of adjoints will converge to a solution of the two-point boundary value problem. The missing initial values $\mathbf{z}$ for this solution will lie in the region

$$\|\mathbf{z} - \mathbf{z}^{(0)}\| \le \frac{1-\sqrt{1-2h}}{h} B'\eta', \tag{7.5.3}$$

and the rate of convergence can be estimated by the inequalities

$$\|\mathbf{z}^{(k)} - \mathbf{z}\| \le \frac{1}{2^k}(2h)^{2^k}\frac{\eta'}{h}, \qquad k = 0, 1, \ldots. \tag{7.5.4}$$

It can be seen that, in order to guarantee convergence of the Goodman-Lance method of adjoints (which, incidentally, then establishes the existence of solutions of the two-point boundary value problem), two kinds of conditions must be met. First, the solution to the initial value problem (7.5.1), along with its first and second partial derivatives with respect to the initial values, must exist and be bounded; second,an approximation to the missing initial values must be known, good enough to ensure that $h \leq \frac{1}{2}$.

For a wide class of two-point boundary value problems met in practice it is reasonable to assume that a solution to the two-point boundary value problem is being sought in a subregion of the closed and bounded region D^n. The classical existence theorem guarantees that a unique solution of the initial value problem exists for each initial point in D^n. More precisely, if I is a closed and bounded interval of the real line $-\infty < t < \infty$ and R^n is a closed and bounded subset of the n-dimensional space of the reals such that the $g_i(y_1, \ldots, y_n; t), i = 1, 2, \ldots, n$, have continuous partial derivatives up through third order in $R^n \times I$, then there is a closed subregion $D^n \subset R^n \times I$ such that, for any point $(y_1^{(0)}, y_2^{(0)}, \ldots, y_n^{(0)}, t_0) \in D^n$, Eq. (7.5.1) has a unique solution; it passes through $(y_1^{(0)}, y_2^{(0)}, \ldots, y_n^{(0)}; t_0)$; has continuous first and second partial derivatives with respect to the $y_i^{(0)}$; and remains in D^n.

We assume that the projection of D^n on the real line includes the closed and bounded interval $[t_0, t_f]$ and that R^{n-r}, the projection of D^n on the $(n-r)$-dimensional space containing the missing initial conditions, is a closed and bounded set.

Since, for the reals, a closed and bounded set is compact, by the Tychonoff theorem the Cartesian product spaces D^n and D^{n-r}, where

$$D^{n-r} = R^{n-r} \times [t_0, t_f] \subset D^n \subset R^n \times I$$

are compact. Since a continuous function defined on a compact set is uniformly continuous and uniformly bounded, the functions $\boldsymbol{\phi}(\mathbf{z}^{(0)})$, and matrices $\mathbf{A} \equiv (\partial\phi_i/\partial z_j), i, j = 1, 2, \ldots, n-r; \partial^2\phi_i/\partial z_j \partial z_k, i, j, k = 1, 2, \ldots, n-r; \mathbf{A}^{-1}$ are uniformly continuous and uniformly bounded. Consequently the bounds in (i), (ii), (iii) of the Kantorovich theorem (Theorem 6.2, Chapter 6) can be replaced by uniform bounds over D^{n-r}: $\bar{B}, \bar{\eta}, \bar{K}$. It should be mentioned that the continuity of $\boldsymbol{\phi}(\mathbf{z}^{(0)})$, $\partial\phi_i/\partial z_j$, $\partial^2\phi_i/\partial z_j \partial z_k$, and the existence and continuity of $\mathbf{A}^{-1}$ follow from the unique existence and continuity of the solution to (7.5.1).

Now suppose initial values $\mathbf{z}_1$ are known such that the boundary value problem has been solved over the interval $[t_0, t_1]$. Then there is a $t_2 > t_1$ such that the Goodman-Lance method will converge with the initial approximation $\mathbf{z}^{(0)} = \mathbf{z}_1$. For, by the mean value theorem, the miss distance at t_2

can be evaluated in terms of the miss distance at t_1:

$$\phi_m(\mathbf{z}^{(0)};t_2) = Y_m - y_{i_m}(\mathbf{z}^{(0)},t_2) = Y_m - y_{i_m}(\mathbf{z}_1,t_1)$$

$$-(t_2-t_1)\,\dot{y}_{i_m}(\mathbf{z}_1,t_1+\theta(t_2-t_1)), \qquad 0 \le \theta \le 1, \tag{7.5.5a}$$

$$= -(t_2-t_1)g_{i_m}(\mathbf{y}(\mathbf{z}_1;t_1+\theta(t_2-t_1));t_1+\theta(t_2-t_1)),$$

$$m = 1,2,\ldots,n-r, \tag{7.5.5b}$$

where $\mathbf{z}_1$ solves the two-point boundary value problem over $[t_0,t_1]$. Thus $\|\boldsymbol{\phi}(\mathbf{z}^{(0)})\| = (t_2-t_1)\,\|\mathbf{g}\| \le (t_2-t_1)\overline{M}$; where $\overline{M}$ may be taken as the uniform bound of $\|\mathbf{g}(y_1,\ldots,y_n,t)\|$ over D^{n-r}. Since $\|\boldsymbol{\Gamma}_0\| = \|\mathbf{A}^{-1}(\mathbf{z}^{(0)})\| < \bar{B}$ and $\sum_{j,k=1}^{n}(\partial^2\phi_i)/\partial z_j\,\partial z_k) < \bar{K}$, where $\bar{B}$ and K are uniform bounds over D^{n-r},

$$h \le \overline{K}\overline{B}^2\eta' \le \overline{K}\overline{B}^2(t_2-t_1)\overline{M}$$

and h will be less than $\frac{1}{2}$ provided

$$t_2-t_1 \le \frac{1}{2\overline{M}\overline{K}\overline{B}^2}. \tag{7.5.6}$$

Hence the Goodman-Lance method of adjoints will converge to a solution to the boundary value problem over the interval $[t_0,t_2]$ with initial values $\mathbf{z}_2$. The argument may be repeated for $t_3 > t_2$, etc. But, since the increment t_i-t_{i-1} is independent of t_{i-1}, the solution of the boundary value problem over the interval $[t_0,t_f]$ will be reached in a finite number of steps of the continuation process.

Our discussion may be summarized in the following theorem:

THEOREM 7.1. (i) *If for the set of n nonlinear differential equations the g_i's have continuous partial derivatives with respect to the y_j up through third order over the region $R^n \times I$, where R^n is a closed and bounded subset of the n-dimensional space of the reals and I is a closed and bounded interval of the real line* $-\infty < t < \infty$;

(ii) *if there is a region $D^n \subset R^n \times I$ containing the r given initial values and the interval $[t_0,t_f]$ such that the solutions of* (7.5.1) *passing through these initial values remains in D^n*;

(iii) *if the conditions of the Kantorovich theorem for the two-point boundary value problem are satisfied in a region containing the Cartesian product of R^{n-r} (the projection of D^n onto the $(n-r)$-dimensional space of the missing initial conditions) and the interval* $[t_0,t_1]$;

then the two-point boundary value problem can be solved by the continuation method in a finite number of extensions Δt_i; from t_1 to t_f. The Δt_i can be at least as large as

$$\Delta t_i \leq \frac{1}{2\bar{M}\bar{K}\bar{B}^2}, \tag{7.5.7}$$

where

$\bar{M}$ = *uniform bound on* $\|\mathbf{g}\|$ *over* D^{n-r},

$\bar{K}$ = *uniform bound on* $\sum_{j,k=1}^{n-r} (\partial^2\phi_i/\partial z_j \partial z_k)$ *over* D^{n-r},

$\bar{B}$ = *uniform bound on* $\|\mathbf{A}^{-1}(\mathbf{z})\|$ *over* D^{n-r},

$\mathbf{z}$ = *set of* $n-r$ *missing initial conditions,*

$\mathbf{A}(\mathbf{z}) \equiv \partial\phi_i/\partial z_j$ *evaluated at* $(\mathbf{z}, t)$ *in* D^{n-r},

$D^{n-r} = R^{n-r} \times [t_0, t_f]$.

7.6. CONTINUATION METHOD OF FICKEN†

In [11] Ficken developed a continuation method for functional equations. He considered solving the nonlinear operator equation $T(x) = 0$ by imbedding $T(x)$ in a family of nonlinear operators $T(s, x)$ such that a solution x^* of the equation $T(0, x) = 0$ is known and such that $T(1, x) = T(x)$. Using the pair $0, x^*$ to initiate the process, a sequence of equations,

$$T(s_i, x) = 0, \qquad 0 = s_0 < s_1 < \ldots < s_n = 1, \tag{7.6.1}$$

is solved until the solution of $T(1, x) = T(x) = 0$ is obtained. Ficken's assumptions and proof that the solution of $T(x) = 0$ will be reached after solving a finite number of equations (7.6.1) are quite involved, since he deals with the general case of an infinite dimensional Banach space S. In this section we show that our continuation method satisfies Ficken's assumptions, and therefore that his conclusions furnish an alternative convergence proof for our continuation method. The principal reasons why we were able to give the simple convergence proof of the preceding section are that we employ a finite dimensional Banach space, and that we use the Kantorovich theorem.

We briefly review the assumptions under which Ficken is able to guarantee the existence, continuity, and uniqueness in $S^N = \{x \in S | \; \|x\| \leq N\}$ of the solution $x(s)$ of $T(s, x) = 0$, $0 \leq s \leq 1$, and that the solution of $T(1, x) = T(x) = 0$ can be obtained in a finite number of extensions from the solution x^* of $T(0, x) = 0$. Here $N > 0$ is a suitably chosen constant, and $T(s, \cdot)$ is a

† In this section we use the notation of Ficken.

family of operators mapping S into S. First it is assumed that, for each $s \in J$, where $J = \{0 \le s \le 1\}$, and $x \in S^N$,

$$T(s, x+u) - T(s, x) = L(s, x, u) + \|u\| \; R(s, x, u), \tag{7.6.2}$$

where

$L(s, x, u)$ = the differential, which is linear in u, and has domain S and range $V \subset S$;

$R(s, x, u)$ = the remainder, with the property that $R(s, x, u) \to 0$ as $u \to 0$.

The remaining assumptions are (using Ficken's nomenclature)

LBCU1 : $\exists$ a constant $\mu > 0 \ni s \in J$ and $x \in S^N \Rightarrow \forall u$, $\|L(s, x, u)\| \ge \|u\|/\mu$.

LBCU2 : $\forall \varepsilon_2 > 0$, $\exists$ scalars $\delta_2(\varepsilon_2) > 0$ and $\alpha_2(\varepsilon_2) > 0 \ni \forall s, s' \in J$, $\forall x, x' \in S^N$, $|s' - s| \le \delta_2$, and $\|x' - x\| \le \alpha_2$, $\Rightarrow \forall u$, $\|L(s', x', u) - L(s, x, u)\| \le \varepsilon_2 \|u\|$

RU : As $u \to 0$, $R(s, x, u) \to 0$, uniformly for $s \in J$, $x \in S^N$

TV : $\forall s \in J, \forall x \in S^N, T(s, x) \in V$

TCU : $T(s, x)$ is continuous in s uniformly for $s \in J$ and $x \in S^N$

O*G : For $\|x\| \le N$ (that is, $x \in S^N$), the equation $T(0, x) = T^*x = 0$ has a unique solution x^* and $\|x^*\| < N$

BS : $\exists$ scalar $\rho > 0 \ni$ if $x \in \Sigma_N$ (where Σ_N is the set of those x in S^N such that $T(s, x) = 0$ for some s in J), then $\|x\| < N - \rho$.

Ficken's main theorem, his Theorem 4.1, is now given.

THEOREM 7.2 (Ficken). *Let $x \to T(x)$ be a transformation of a Banach space into itself. Let $x \to T^*(x)$ be a transformation with property O^*G. Suppose that a one-parameter family of transformations $x \to T(s, x)$, $0 \le s \le 1$, can be found such that $T(0, x) \equiv T^*(x)$ and $T(1, x) \equiv T(x)$, and such that the requirements of assumptions* LBCU, RU, TV, TCU, *and* BS *are met. Then a continuous globally unique solution $x(s)$ can be generated for* $0 \le s \le 1$, *in a finite number of steps from the solution x^* of $T^*(x) = 0$. In particular, the solution $x(1)$ of the equation $T(x) \equiv T(1, x) = 0$ can be constructed by the continuation method.*

The relationship between Ficken's nomenclature and ours may be found in the table of correspondence, Table 7.6. With this identification it can be verified that the assumptions of Ficken's main theorem 4.1 are also satisfied in our case, either explicitly or as conclusions from other assumptions.

Table 7.6. Correspondence between Ficken's Nomenclature and that of Roberts and Shipman

Ficken	Roberts and Shipman
1. s, scalar, $0 \leq s \leq 1$	$s = \dfrac{t-t_1}{t_f-t_1}$, $t_1 \leq t \leq t_f$, where $[t_0, t_1]$ is first interval over which two-point boundary value problem is solved
2. S infinite dimensional Banach Space	R^n, n-dimensional Cartesian product space of the reals, a Banach space
3. x, $u \in S$, u is increment to x	$\mathbf{y}(t_0) \in R^n$, $\mathbf{z} \in R^{n-r}$, $\mathbf{z} = (y_{r+1}(t_0), \ldots, y_n(t_0)) =$ set of missing initial conditions
4. T, a nonlinear operator, $T: S \rightarrow S$	ϕ, a nonlinear operator, $\phi: R^{n-r} \rightarrow R^{n-r}$, for $\phi(\mathbf{z}) = 0$, or $\phi: R^{n-r} \times [t_0, t] \rightarrow R^{n-r}$, for $\phi(t, \mathbf{z}) = 0$
5. Operator equations	
(i) $T(x) = 0$,	(i) $\phi(\mathbf{z}) = 0$, where $\phi(\mathbf{z}) =$ miss distance,
(ii) $T(s,x)$,	(ii) $\phi(t,\mathbf{z}) = \mathbf{Y} - \mathbf{y}(\mathbf{y}(t_0), t)$, where $\mathbf{Y} =$ specified terminal conditions,
(iii) $T(s, x+u) - T(s,x) = L(s,x,u) + \|u\| R(s,x,u)$	(iii) $\phi(t, \mathbf{z}^{(k+1)}) - \phi(t, \mathbf{z}^{(k)}) = \mathbf{A}(t, \mathbf{z}^{(k)}) \Delta z^{(k)} + + \dfrac{\|\Delta \mathbf{z}\|}{2} \left(\sum_{j,m} \dfrac{\partial^2 \phi_l}{\partial z_j \partial z_m} \Delta z_j \right) \dfrac{\Delta \mathbf{z}^{(k)}}{\|\Delta \mathbf{z}\|}$
6. (i) $L(s,x,.)$ derivative,	(i) $\mathbf{A}(t,\mathbf{z})$, Jacobian matrix with elements $\dfrac{\partial \phi_l}{\partial z_j}$
(ii) $L(s,x,u) =$ differential, linear in u,	(ii) $\mathbf{A}(t,\mathbf{z}) \Delta \mathbf{z}$
(iii) $L(s,x,u)^{-1} = M(s,x,u)$	(iii) $\mathbf{A}^{-1}(t,\mathbf{z})$
7. S^N domain of $T(s,x)$, $\{x \in S \mid \|x\| \leq N\}$	$R^{n-r} \times [t_0, t_f]$ domain of $\phi(t,\mathbf{z})$ and $\mathbf{A}(t,\mathbf{z})$
8. $V \subset S$, range of $L(s,x,u)$	Range of $\mathbf{A}(t,\mathbf{z}) \Delta \mathbf{z} \subset R^{n-r} \times [t_0, t_f]$

To begin with, Ficken deals with an infinite dimensional Banach space S and with the closed and bounded domain $S^N \subset S$, where $S^N = \{x \in S \mid \|x\| \leq N\}$, and the closed and bounded interval for s, $J = [0,1]$. Since S^N is assumed closed and bounded, but not totally bounded, S^N is not necessarily compact, and therefore Ficken could not use the Tychonoff theorem for $S^N \times J$ and take advantage of compactness. In contrast, we consider the closed and bounded space R^{n-r} of the $n-r$ missing initial conditions and the closed and bounded interval $[t_0, t_f]$. Since each of these spaces is compact, the Cartesian product space $R^{n-r} \times [t_0, t_f]$ is compact by the Tychonoff theorem. In view of our assumptions (see Section 7.5) concerning the initial value problem, the continuity of $\mathbf{y}$, $\mathbf{g}(\mathbf{y})$, $\boldsymbol{\phi}(\mathbf{y})$, $\partial\phi_i/\partial z_j$, and $\partial^2\phi_i/\partial z_j \partial z_k$, $i,j,k = 1,2,\ldots,n-r$, is established on the domain $R^{n-r} \times [t_0, t_f]$. Since a function continuous on a compact space is uniformly continuous and uniformly bounded, these functions are uniformly continuous and uniformly bounded. In addition the $\|\mathbf{A}\|$ and $\|\mathbf{A}^{-1}\|$ (occurring in the Kantorovich theorem) are uniformly continuous and uniformly bounded. This replaces Ficken's assumption LBCU1. In place of the Lipschitz-type condition LBCU2, we assume uniformly bounded second derivatives $\partial^2\phi_i/\partial z_j \partial z_k$ as required by the Kantorovich theorem. Superficially, our assumptions are stronger than Ficken's, but the loss of generality does not seem to be significant. RU holds by virtue of the compactness of the region D^{n-r}. Assumption TV is made in our case too, while, as we have seen above, assumption TCU also holds by virtue of compactness.

The assumption O*G requires that a unique solution to $T(0,x) = 0$, namely x^*, exists and that $\|x^*\| < N$. Our assumption that the two-point boundary value problem has been solved over the interval $[t_0, t_1]$ corresponds to solving $T(0,x) = 0$. The uniqueness of our assumed solution of the two-point boundary value problem and the fact that the solution lies in the specified domain can be deduced from the theorem of Kantorovich on the abstract Newton-Raphson method.

The assumption BS is designed to prevent penetration of the boundary of S^N. As mentioned above under O*G, Kantorovich showed that for our case the solution to the abstract Newton-Raphson problem always lies within the specified domain.

We may therefore conclude that Ficken's theorem applies to our continuation method for solving two-point boundary value problems; that is, under the assumptions made in Section 7.5 it is possible to solve the sequence of problems over the intervals $[t_0, t_i]$, $t_1 < t_2 < \ldots < t_n = t_f$ and thus to obtain a solution of the original problem over $[t_0, t_f]$ in a finite number of steps.

7.7. CONTINUATION IN QUASILINEARIZATION [12]

The solution of nonlinear two-point boundary value problems by quasilinearization was discussed in Chapter 5. In this method the original differential equations are linearized around a nominal solution, that is, a function which satisfies the boundary conditions but need not satisfy the differential equations exactly. A sequence of linear two-point boundary value problems results, the solutions of which converge under appropriate conditions to the solution of the original nonlinear problem.

One of the advantages of quasilinearization over "shooting" methods for solving nonlinear two-point boundary value problems is that the resulting linear equations are often numerically more stable than the original nonlinear equations.

However, despite the generally improved stability of the linear equations resulting from quasilinearization, there are problems in which even these equations are unstable, as our examples in Section 7.9 show. The first purpose of this section is to describe a continuation procedure which extends the applicability of quasilinearization to problems in which the resulting linear equations are unstable.

Quasilinearization is an iterative method which requires an initial "point" or nominal solution to start the process. As a matter of fact, since it can be shown that quasilinearization is a realization of the abstract Newton-Raphson method in Banach space, it may be expected that the selection of the initial point will be crucial for the convergence of the method. (See Section 6.11, Chapter 6.) The practical application of any Newton-Raphson method usually involves a number of trials until a satisfactory initial point is found which starts the iterations converging to the solution.

In the practical application of quasilinearization, we might expect particular difficulties in supplying the initial trial "points" (that are actually functions over the specified interval) which satisfy the prescribed boundary conditions. The question of generating suitable nominal profiles (those which ultimately lead to a solution of the problem) has not received much attention in the literature. The second purpose of this section is to suggest a practical technique for finding suitable nominal profiles.

The combination of the proposed method for generating nominal profiles together with the continuation technique has enabled us to solve, by quasilinearization and standard programs for numerical integration of ordinary differential equations, two-point boundary value problems that previously had required applications of time consuming finite difference methods, either to the original equations or to the linear equations resulting from quasilinearization. Numerical experience with the continuation method and the technique for generating nominal profiles is illustrated by several examples.

We again consider the two-point boundary value problem (7.2.1)–(7.2.3), but now we assume that it is being solved by quasilinearization and that the resulting linear equations are unstable and thus it is not possible to integrate them forward over the interval $[t_0, t_f]$. The continuation technique applied to quasilinearization follows.

1. Use the solution of the nonlinear two-point boundary value problem over $[t_0, t_1]$ as the nominal profile for an application of quasilinearization

over $[t_0, t_1]$. This will usually supply a suitable nominal profile, that is, one that ultimately leads to a solution of the original problem. Note that t_1, where $t_1 < t_f$, is a point at which the solution exhibits good behavior.

2. Solve the resulting linear two-point boundary value problem over the interval $[t_0, t_1]$. Theoretically this solution should be nothing but the nominal profile; in practice this is not always the case.

3. Form the nominal profiles $y_i^{(0)}(t)$ for an application of quasilinearization over the extended interval $[t_0, t_2]$, where $t_1 < t_2$, by combining the solution of the linearized problem over $[t_0, t_1]$ with the function constant over the interval $[t_1, t_2]$ whose value is $y_i^{(0)}(t_1)$.

4. Repeat step 2 for the extended time interval $[t_0, t_2]$.

5. Repeat steps 2 through 4, advancing the subscripts on the right-hand end points t_i until $t_i = t_f$, at which point the original problem will be solved.

7.8. DISCUSSION OF THE PROCEDURE

In Section 7.5 the theoretical justification for continuation of the solution by the "nonlinear process" was established.* Here we use the quasilinearization approach and present what we believe to be a practical way of starting on the interval $[t_0, t_1]$, and then continuing on the intervals $[t_0, t_2]$, $[t_0, t_3]$, ...

As a practical computation procedure, a number of questions rise:

1. How is $\Delta t_i = (t_{i+1} - t_i)$ chosen?

2. How far can a problem be extended?

3. What determines the extent of continuation?

4. If the method fails, how does it fail?

On the basis of a limited computing experience using the quasilinearization process for numerically very unstable nonlinear differential equations, we can make the following statements which, we believe, have a wider validity.

1. In Section 7.5 for the nonlinear differential equations a theoretical value of the minimum size of the continuation step, Δt_i, was established on the basis of the norms in the Kantorovich sufficiency theorem for the abstract

* By "nonlinear process" we mean the application of the method of adjoints to the original nonlinear differential equations.

Newton-Raphson method. From a practical computing point of view, we never in fact determine these norms. In practice we choose the Δt_i rather intuitively after examining current results and then try it out. We have used Δt_i as small as several integration steps and as large as the original time span $[t_1 - t_0]$.

2. In our problems the original time span has been extended incrementally anywhere from 2 to 10 times.

3. The continuation capability of a given set of linearized equations seems to depend in a complex way on:

the number of initial and terminal conditions specified;
the specific variables chosen for the initial and terminal conditions;
the numerical values specified for the initial and terminal variables;
the size of the Δt_i;
the numerical procedures used to integrate the differential equations.

The one place where the effects of all these items (except possibly the last one) and their interrelations come together is in the matrix **A**, whose i,j element is $\partial\phi_i/\partial y_{r+j}(t_0)$, where ϕ_i is the miss distance of the ith terminal condition and $y_{r+j}(t_0)$ is the jth missing initial condition. In quasilinearization the matrix **A** is the matrix of coefficients of the set of linear equations whose solution is the set of missing initial conditions. The elements of **A** (or $\mathbf{A}^{-1}$) are of course crucial in determining the size of the $y_{r+j}(t_0)$. An examination of the relative size of the elements of the **A** (or $\mathbf{A}^{-1}$) matrix is the clue to explain numerical difficulties or to predict them.

4. When continuation fails, it does so in two ways. Generally we find that, when computing with a fixed precision, the boundary conditions at t_n are satisfied to less and less accuracy as t_n increases–until rather gross errors occur. Next we find that the missing initial conditions generated lead to overflow in numerical integration. In some problems we have been able to determine a limiting value of t_n beyond which we cannot continue, for the reasons above. In other problems we have been able to extend our intervals many times larger than the original interval and, within the computer time allotted, have not found a limiting value of t_n.

A topic of interest is the relative merit of continuation in quasilinearization and continuation in the method of adjoints [1, 9]. For the problems we have considered, the quasilinearization process permits continuation over a longer interval than does the method of adjoints. Quasilinearization is continued, however, with a loss in accuracy. To understand this we recall that the quasilinearization process fixes the solution at the specified initial and terminal conditions and determines the profiles of the variables over the given interval.

At each iteration of the quasilinear process, a set of missing initial conditions is found for the linearized differential equations which, when integrated forward, theoretically generate the specified terminal conditions and also the nominal profiles for the next iteration. In practice we find that the set of missing initial conditions determined by the quasilinear process may not satisfy the specified terminal conditions. In fact, the error between the calculated terminal conditions and the specified terminal conditions for the quasilinear process may be larger than the error for the method of adjoints at its limiting time. Then the error grows progressively larger as the quasilinear process is continued beyond the limiting time for the nonlinear process. Therefore in our programs each nominal profile is adjusted at its final point to the specified terminal condition before the next iteration is begun.

Another item of interest is that in our problems the initial conditions found by quasilinearization, when used in the original nonlinear differential equations, do not always produce solutions which satisfy numerically the original nonlinear two-point boundary value problem. Instead, forward integration of the nonlinear equations using the initial conditions found by quasilinearization sometimes causes machine overflow.

While this phenomenon needs more study, it does raise the interesting problem of whether either quasilinearization or the method of adjoints does in fact furnish a solution of the original problem, as we have claimed. In one sense we can explain the apparent failure of the methods by recognizing that the convergence proofs of the shooting methods as well as of the quasilinear method are based on limiting processes and on theoretical calculations without round-off. In numerical work, of course, programs work with rounded off numbers and finite approximations, and iteration schemes are always terminated within a finite and often a rather small number of passes. In other words, numerical iterative processes may differ considerably in their convergence properties from theoretical iterative processes, and the results of the two processes which ideally converge to the same result may differ considerably in practice. From our experience, it appears that inputs from the nonlinear process (nominal profiles) to the quasilinear process are usually acceptable, but inputs from the quasilinear process (missing initial conditions) to the nonlinear process can be unsatisfactory.

7.9. NUMERICAL EXAMPLES

In order to support the statements made in the preceding section, we direct our attention to two numerically very unstable problems, which are solved by a combination of continuation and quasilinearization with nominal profiles obtained from the nonlinear differential equations.

Example 3 is the quasilinear problem corresponding to Example 1 of Section 7.4.

Example 3. The differential equations to be integrated are

$$\dot{y}_1^{(k+1)} = y_2^{(k+1)},$$

$$\dot{y}_2^{(k+1)} = y_3^{(k+1)},$$

$$\dot{y}_3^{(k+1)} = ay_3^{(k)}y_1^{(k+1)} + (s - 2ny_2^{(k)})y_2^{(k+1)} + ay_1^{(k)}y_3^{(k+1)} - 2y_4^{(k)}y_4^{(k+1)} - ay_1^{(k)}y_3^{(k)} + ny_2^{(k)2} + y_4^{(k)2} + 1,$$

$$\dot{y}_4^{(k+1)} = y_5^{(k+1)},$$

$$\dot{y}_5^{(k+1)} = ay_5^{(k)}y_1^{(k+1)} + by_4^{(k)}y_2^{(k+1)} + (by_2^{(k)} + s)y_4^{(k+1)} + ay_1^{(k)}y_5^{(k+1)} - (s + ay_1^{(k)}y_5^{(k)} + by_2^{(k)}y_4^{(k)}),$$

where $a = -(3-n)/2$ and $b = -(n-1)$.
The initial conditions are

$$y_1^{(k)}(0) = 0, \qquad y_2^{(k)}(0) = 0, \qquad y_4^{(k)}(0) = 0, \qquad k = 0, 1, \ldots,$$

and the terminal conditions are

$$y_2^{(k)}(t_f) = 0, \qquad y_4^{(k)}(t_f) = 1, \qquad k = 0, 1, \ldots .$$

Let us specify

$$n = -0.1, \qquad s = +0.2, \qquad t_0 = 0.0.$$

Let t_f be as large as possible.

In Figure 7.1 are plotted for Example 3 the nominal profiles $y_i^{(0)}(t)$ which are the result of solving the original nonlinear two-point boundary value problem over [0, 5.3]. The 5th iteration, $y_i^{(5)}(t)$, of quasilinearization is also plotted in Figure 7.1. In Figure 7.2 the problem is extended from [0, 5.3] to [0, 6.6]. The initial profile consists of $y_i^{(5)}(t)$ from Figure 7.1 over [0, 5.3] with the values of the variables at $t = 5.3$ held constant over the interval [5.3, 6.6]. The 5th profile over [0, 6.6] is also plotted on Figure 7.2. Continuing in this manner, Figure 7.3 exhibits $y_i^{(5)}(t)$ for [0, 13.3], which is as far as we could extend by quasilinearization.

It is interesting to note that, under conditions of identical boundary conditions and comparable step size, the solution by the method of adjoints of the nonlinear problem could be extended only to $t_f = 11.3$ [1]. See Section 7.4.

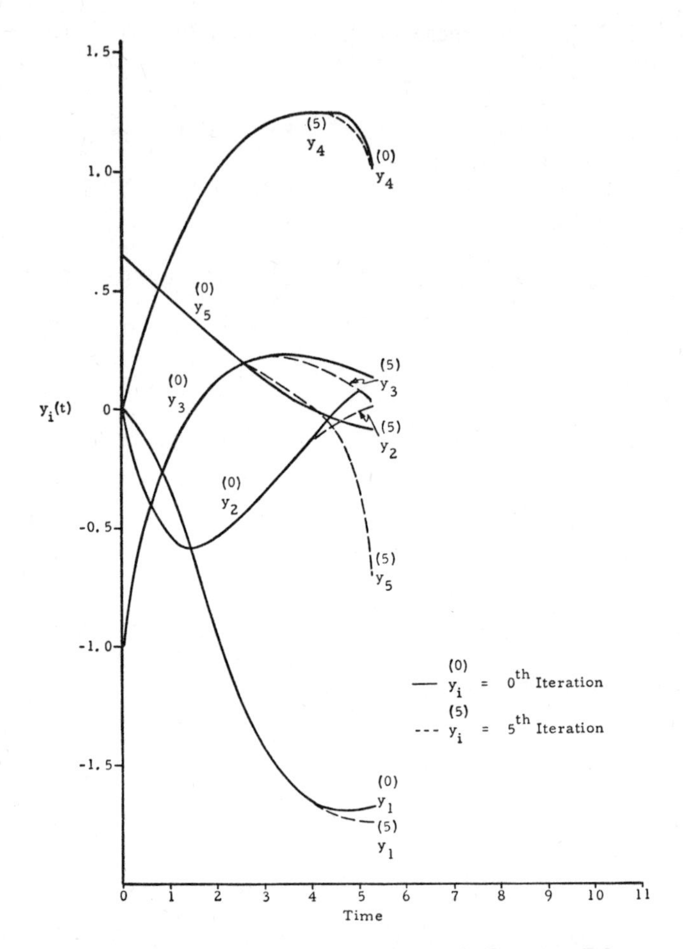

Figure 7.1. $y_i(t)$ **vs. time. Example 3.** $t_f = 5.3$**.**

To determine how large an extension can be made in a single step, we exhibit Figures 7.4 and 7.5. In these figures the nominal profile was taken as the solution over [0, 5.3] by the method of adjoints with the values of the variables at $t = 5.3$ held constant until $t = 8.0$ (Figure 7.4) and until $t = 10.6$ (in Figure 7.5). The $y_i^{(5)}(t)$ for the quasilinearization process are plotted in each figure. The problem could not be extended in one step from $t = 5.3$ beyond $t = 10.6$. In other words, for this problem the maximum single-step extension possible was equal in size to the interval over which the nonlinear problem was solved. The solutions obtained using these large extensions agreed with those obtained by smaller extensions.

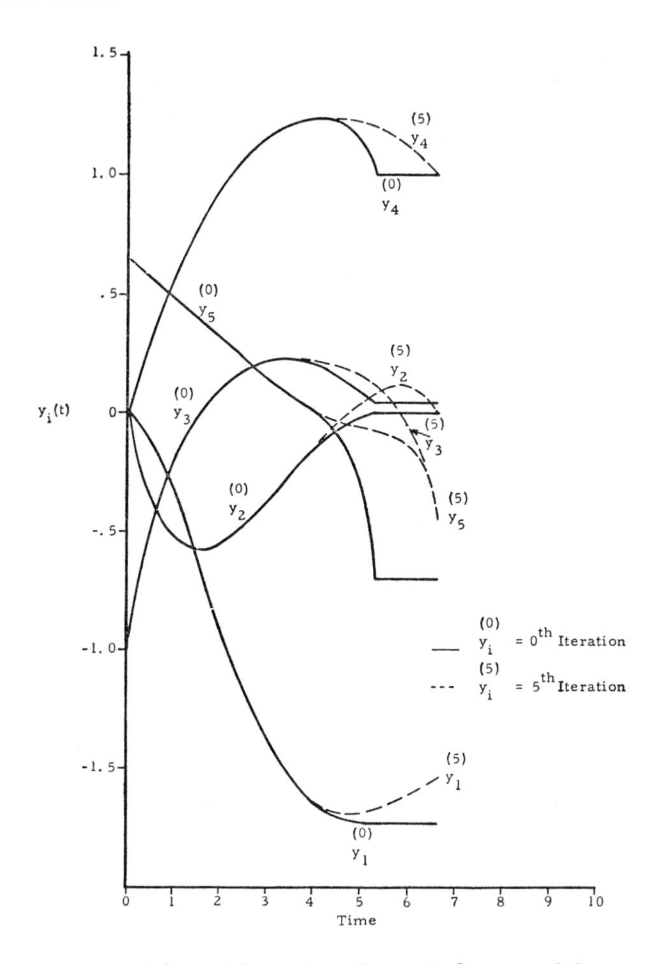

Figure 7.2. $y_i(t)$ **vs. time. Example 3.** $t_f = 6.6$.

Using the initial conditions found by a combination of quasilinearization and continuation over the interval [0.0, 13.3] with the integration step size $h = 0.05$, namely,

$$y_1^{(0)} = 0.0,$$
$$y_2^{(0)} = 0.0,$$
$$y_3^{(0)} = -0.966311765959431,$$
$$y_4^{(0)} = 0.0,$$
$$y_5^{(0)} = 0.652909585951888,$$

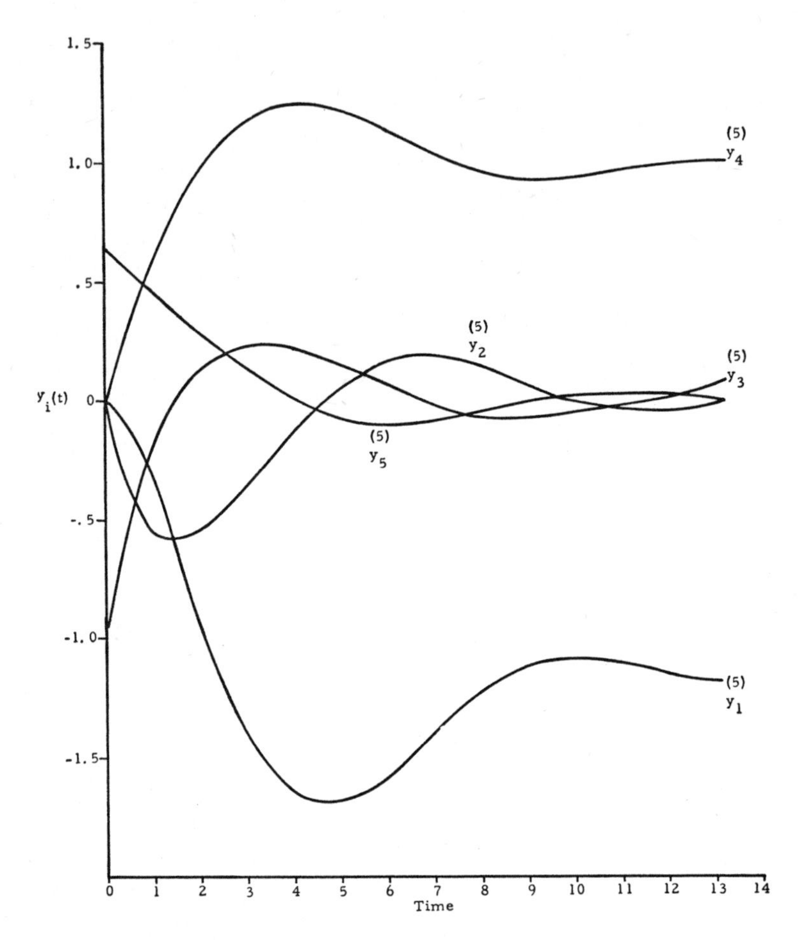

Figure 7.3. $y_i(t)$ **vs. time. Example 3.** $t_f = 13.3$.

we found that the original nonlinear differential equations could not be integrated forward for $h = 0.05$ or, for that matter, for any h in the range $[0.02, 0.10]$.

As another example of continuation and quasilinearization, consider Example 4, which is the linear problem corresponding to Example 2 of Section 7.4 or, again more precisely, the sequence of linear two-point boundary value problems that result when quasilinearization is applied to Example 2.

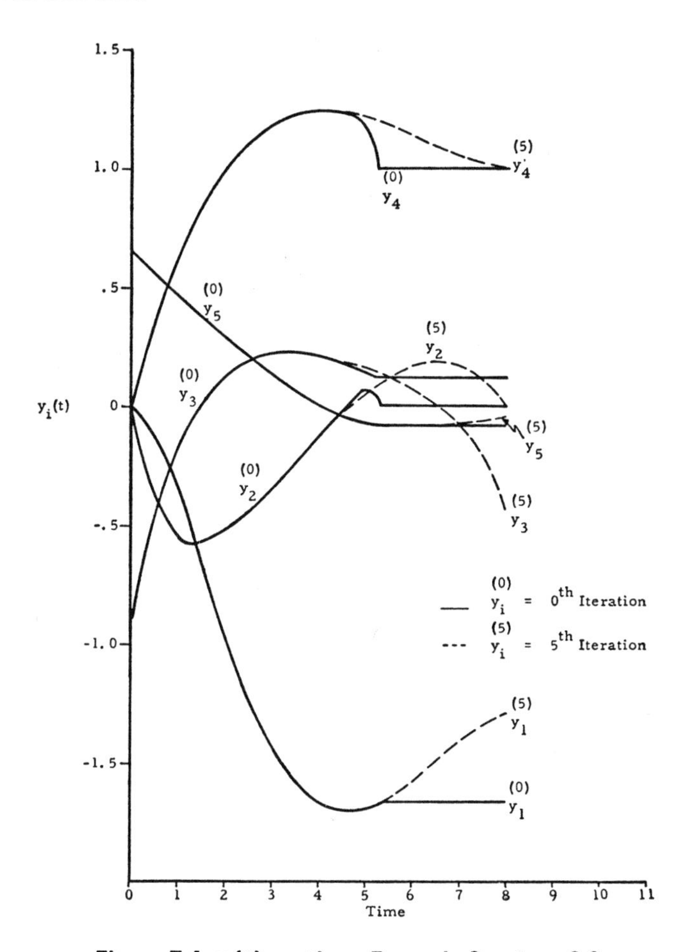

Figure 7.4. $y_i(t)$ vs. time. Example 3. $t_f = 8.0$.

Example 4. The differential equations to be integrated are

$$\dot{y}_1{}^{(k+1)} = -\left[\frac{y_6{}^{(k)}}{y_5{}^{(k)}} + \frac{2}{y_5{}^{(k)}}(y_3{}^{(k)}+y_5{}^{(k)})\right]y_1{}^{(k+1)} - \frac{2y_1{}^{(k)}}{y_5{}^{(k)}}y_3{}^{(k+1)}$$

$$+\frac{1}{y_5{}^{(k)2}}(y_1{}^{(k)}y_6{}^{(k)}+2y_1{}^{(k)}y_3{}^{(k)})y_5{}^{(k+1)} - \frac{y_1{}^{(k)}}{y_5{}^{(k)}}y_6{}^{(k+1)},$$

$$\dot{y}_2{}^{(k+1)} = y_3{}^{(k)}(y_3{}^{(k)}+y_5{}^{(k)})y_1{}^{(k+1)} + (2y_1{}^{(k)}y_3{}^{(k)}+y_1{}^{(k)}y_5{}^{(k)})y_3{}^{(k+1)}$$

$$+(y_1{}^{(k)}y_3{}^{(k)})y_5{}^{(k+1)} - 2(y_1{}^{(k)}y_3{}^{(k)}+y_1{}^{(k)}y_5{}^{(k)})y_3{}^{(k)},$$

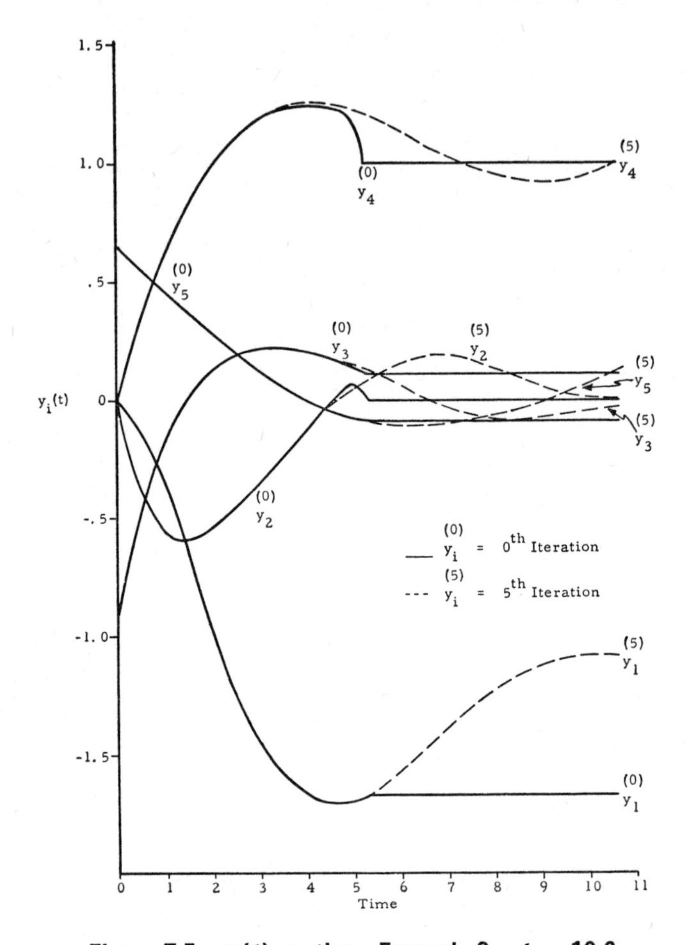

Figure 7.5. $y_i(t)$ **vs. time. Example 3.** $t_f = 10.6$.

$$\dot{y}_3^{(k+1)} = y_4^{(k+1)},$$

$$\begin{aligned}\dot{y}_4^{(k+1)} &= \frac{R_e}{y_7^{(k)}}\left(-2\sigma A y_7^{(k)2} + y_4^{(k)} y_5^{(k)} + y_3^{(k)}(y_3^{(k)} + y_5^{(k)})\right) y_1^{(k+1)} \\ &+ \frac{2R_e}{y_7^{(k)}} y_2^{(k+1)} + \frac{R_e y_1^{(k)}}{y_7^{(k)}}(2y_3^{(k)} + y_5^{(k)}) y_3^{(k+1)} \\ &+ \left[\frac{R_e y_1^{(k)} y_5^{(k)}}{y_7^{(k)}} + \frac{y_5^{(k)} y_6^{(k)}}{2A y_7^{(k)2}}\right] y_4^{(k+1)}\end{aligned}$$

$$+\left[\frac{R_e y_1^{(k)}}{y_7^{(k)}}(y_4^{(k)}+y_3^{(k)})+\frac{y_4^{(k)}y_6^{(k)}}{2Ay_7^{(k)2}}\right]y_5^{(k+1)}+\frac{y_4^{(k)}y_5^{(k)}}{2Ay_7^{(k)2}}y_6^{(k+1)}$$

$$-\left[2\sigma AR_e y_1^{(k)}+\frac{R_e y_1^{(k)}}{y_7^{(k)2}}\{y_5^{(k)}y_4^{(k)}+y_3^{(k)}(y_3^{(k)}+y_5^{(k)})\}\right.$$

$$\left.+\frac{y_4^{(k)}y_5^{(k)}y_6^{(k)}}{Ay_7^{(k)3}}+\frac{2R_e y_2^{(k)}}{y_7^{(k)2}}\right]y_7^{(k+1)}$$

$$+\frac{R_e y_1^{(k)}}{y_7^{(k)}}\left[-y_3^{(k)2}-(y_4^{(k)}+y_3^{(k)})y_5^{(k)}+2\sigma Ay_7^{(k)2}+\frac{2y_2^{(k)}}{y_1^{(k)}}\right],$$

$$\dot{y}_5^{(k+1)}=y_6^{(k+1)},$$

$$\dot{y}_6^{(k+1)}=\frac{1}{(4/3R_e)y_7^{(k)}}\left[-\frac{\sigma Ay_7^{(k)2}}{y_5^{(k)}}\{y_6^{(k)}+2(y_3^{(k)}+y_5^{(k)})\}\right.$$

$$\left.+y_5^{(k)}y_6^{(k)}(1-\sigma)\right]y_1^{(k+1)}$$

$$+\left[\frac{2}{4/3R_e}\left(-\frac{\sigma Ay_1^{(k)}y_7^{(k)}}{y_5^{(k)}}\right)-\frac{y_5^{(k)}y_6^{(k)}}{2Ay_7^{(k)2}}\right]y_3^{(k+1)}-\frac{1}{2}y_4^{(k+1)}$$

$$+\left[\frac{1}{4/3R_e}\frac{y_1^{(k)}}{y_7^{(k)}}\left\{\frac{\sigma Ay_7^{(k)2}}{y_5^{(k)2}}(y_6^{(k)}+2y_3^{(k)})+y_6^{(k)}(1-\sigma)\right\}\right.$$

$$\left.-\frac{y_3^{(k)}y_6^{(k)}}{2Ay_7^{(k)2}}-\frac{y_6^{(k)2}}{2Ay_7^{(k)2}}\right]y_5^{(k+1)}+\left[\frac{1}{4/3R_e}\frac{y_1^{(k)}}{y_7^{(k)}}\left\{-\frac{\sigma Ay_7^{(k)2}}{y_5^{(k)}}\right.\right.$$

$$\left.\left.+y_5^{(k)}(1-\sigma)\right\}-\frac{y_3^{(k)}y_5^{(k)}}{2Ay_7^{(k)2}}-\frac{y_5^{(k)}y_6^{(k)}}{Ay_7^{(k)2}}\right]y_6^{(k+1)}$$

$$+\left[\frac{1}{4/3R_e}\left(-\frac{\sigma Ay_1^{(k)}}{y_5^{(k)}}\right)\{y_6^{(k)}+2(y_3^{(k)}+y_5^{(k)})\}\right.$$

$$\left.+\frac{y_5^{(k)}y_6^{(k)}}{Ay_7^{(k)3}}(y_3^{(k)}+y_6^{(k)})\right]y_7^{(k+1)}$$

$$+\frac{\sigma Ay_1^{(k)}}{4/3R_e}\frac{\{y_6^{(k)}+2(y_3^{(k)}+y_5^{(k)})\}y_7^{(k)}}{y_5^{(k)}}-\frac{2y_1^{(k)}}{4/3R_e}\frac{(1-\sigma)y_5^{(k)}y_6^{(k)}}{y_7^{(k)}},$$

$$\dot{y}_7^{(k+1)}=\left(-\frac{y_6^{(k)}}{2Ay_7^{(k)}}\right)y_5^{(k+1)}-\left(\frac{y_5^{(k)}}{2Ay_7^{(k)}}\right)y_6^{(k+1)}+\left(\frac{y_5^{(k)}y_6^{(k)}}{2Ay_7^{(k)2}}\right)y_7^{(k+1)}.$$

The initial conditions are

$$\left.\begin{aligned} y_1^{(k)}(t_0) &= 0.9617, & y_5^{(k)}(t_0) &= -0.0212 \\ y_2^{(k)}(t_0) &= -0.1018, & y_7^{(k)}(t_0) &= 0.9998 \\ y_3^{(k)}(t_0) &= 0.4078, \end{aligned}\right\}, \qquad k = 0, 1, \ldots,$$

and the terminal conditions are

$$\left.\begin{aligned} y_3^{(k)}(t_f) &= 1.0 \\ y_5^{(k)}(t_f) &= -1.0 \end{aligned}\right\}, \qquad k = 0, 1, \ldots,$$

where

$$\begin{aligned} R_e &= 100.0 \\ A &= 0.515, \\ \sigma &= 0.400, \\ t_0 &= 0.05. \end{aligned}$$

Let t_f be as large as possible.

The solution by the method of adjoints over the interval [0.05, 0.3303] was used as the nominal profile for quasilinearization. Using an integration step size $h = 0.02$, and an extension of $\Delta t_i = 0.12$, the problem was extended from [0.05, 0.3303] to [0.05, 1.0510]. Beyond $t_f = 1.0510$, the terminal conditions were not satisfied and, in succeeding iterations, overflow occurred. In this example we once again were able to extend the solution by quasilinearization beyond that obtained by the nonlinear process [1]. See Section 7.4

On taking the initial conditions developed by quasilinearization for the interval [0.05, 0.3303], namely,

$$\begin{aligned} y_1(.05) &= 0.9617, \\ y_2(.05) &= -0.1018, \\ y_3(.05) &= 0.4078, \\ y_4(.05) &= 7.492778389, \\ y_5(.05) &= -0.0212, \\ y_6(.05) &= -0.7886885879, \\ y_7(.05) &= 0.9998, \end{aligned}$$

and using them in the original nonlinear equations, we could not integrate the equations forward using a step size $h = 0.02$, because of overflow.

7.10. A PERTURBATION TECHNIQUE

In the preceding sections we have described continuation methods for the method of adjoints and for quasilinearization. In each case the original two-point boundary value problem for the interval $[t_0, t_f]$ was solved for a sequence of intervals $[t_0, t_N]$, $i = 1, 2, \ldots, N$, with $t_N = t_f$. And in each case, if the original problem were put in operator form, that is, solve the operator equation $P(x) = 0$, then it could be embedded in a sequence of problems depending on a parameter μ, $P(x, \mu) = 0$. The μ is a parameter that defines the end point of the interval $[t_0, t_i]$ over which the two-point boundary value problem is to be solved, namely,

$$t_i = t_1 + \mu(t_f - t_1),$$

so that $P(x, 0) = 0$ corresponds to solution over $[t_0, t_1]$, while $P(x, 1) = P(x) = 0$ corresponds to solution over the original interval $[t_0, t_f]$.

In this section we give a continuation method in which the parameter μ now occurs in the differential equation itself. When $\mu = 0$, the problem $P(x, 0) = 0$ is a readily solved two-point boundary value problem over the original interval $[t_0, t_f]$ with the original boundary values, while $P(x, 1) = P(x) = 0$ is again the original problem. Since varying μ in effect perturbs the given system of differential equations, we call our method a perturbation technique.

We might add that, in the familiar perturbation technique as it is applied, for example, to the equation $\ddot{y} + y = \mu f(y, \dot{y}, t)$, a solution is assumed to be expressed as a power series in μ, the perturbation parameter, namely,

$$y = y_0 + \mu y_1 + \mu^2 y_2 + \ldots, \tag{7.10.1}$$

where y_0 is the unperturbed solution, that is, the solution of $\ddot{y} + y = 0$. Substituting the assumed power series in the differential equation and equating like powers of μ gives rise to a sequence of differential equations in $y_0, y_1, y_2, \ldots$ where the equation for y_k depends on $y_{k-1}, y_{k-2}, \ldots, y_0$ [13]. Here, too, we have a sequence of problems, but now μ is considered fixed (and usually small) so the method can be thought of as successive perturbations of the solution.

Returning to our perturbation technique, in the following section we describe it in detail, give its theoretical justification, and present some numerical experience with it. We also show how it can be used to generate the initial profiles required for quasilinearization. Consider again the situation in which the two-point boundary value problem

$$\dot{y}_i = g_i(y_1, y_2, \ldots, y_n, t), \qquad i = 1, 2, \ldots, n \tag{7.10.2}$$

$$\begin{aligned} y_i(t_0) &= c_i, & i &= 1, 2, \ldots, r, \\ y_{i_m}(t_f) &= c_{i_m}, & m &= 1, 2, \ldots, n-r, \end{aligned} \tag{7.10.3}$$

is to be solved by the method of adjoints, and suppose that Eqs. (7.10.2) are so unstable that they cannot be integrated from t_0 to t_f with the available initial conditions. The perturbation technique has successfully been used to overcome this difficulty.

We begin the application of the perturbation technique by writing the right-hand side of (7.10.2) in the form $g_i = p_i + q_i$ in such a way that the two-point boundary value problem

$$\dot{y}_i = p_i(y_1, \dots, y_n, t), \qquad i = 1, 2, \dots, n, \tag{7.10.4}$$

with boundary conditions (7.10.3) is solvable numerically. Then a sequence of two-point boundary value problems,

$$\dot{y}_i = p_i(y_1, \dots, y_n, t) + \mu_k q_i(y_1, \dots, y_n, t), \qquad i = 1, 2, \dots, n, \qquad k = 0, 1, 2, \dots, \tag{7.10.5}$$

with boundary conditions (7.10.3) is solved, where $\mu_0 = 0$, $\mu_{k+1} = \mu_k + \Delta\mu_k$, and the missing initial conditions for the solution of the kth problem are taken as the initial conditions for the shooting method solution of the $(k+1)$st problem. If a sequence $\Delta\mu_k$ can be formed such that each problem (7.10.5), (7.10.3) can be solved and such that $\mu_N = 1$ for some N, then the required problem (7.10.2), (7.10.3) will have been solved.

The partitioning of the g_i is not unique. We may expect that a variety of partitions will lead to the solution of the original problem. It is usually convenient, however, to choose when possible the p_i to be linear functions of the y_i, so that (7.10.5) takes the form

$$\dot{y}_i = \sum_{j=1}^{n} b_{ij} y_j + \mu_k q_i(y_1, \dots, y_n, t), \qquad i = 1, 2, \dots, n, \tag{7.10.6}$$

for in that case the problem $k = 0$ can be solved without iteration (if it is not too unstable) and without the assumption of trial values for the missing initial conditions. In this manner we solve first a linear two-point boundary value problem ($\mu_0 = 0$) and then solve sequentially two-point boundary value problems which become increasingly "more" nonlinear as $\mu_k \to 1.0$.

7.11. JUSTIFICATION OF THE PERTURBATION METHOD

Since the present perturbation technique is a particular realization of the continuation method, a theoretical justification could be given, as in Section 7.6., by making the appropriate identifications and applying Ficken's theorems [11]. A direct proof is possible along the lines of Section

7.5, and this is what we now give. We show that the problem (7.10.2)–(7.10.3) can be considered the problem of solving the operator equation $P(y) = 0$, where P is a nonlinear operator from the Banach space Y into itself. Then we show that this problem can be imbedded in a sequence of problems $P(y, \mu) = 0$, and that, beginning with the problem $P(y, 0) = 0$, the solution of $P(y, 1) = P(y) = 0$ can be reached in a finite number of increments to $\mu_k, \mu_{k+1} = \mu_k + \Delta\mu_k$.

Accordingly, we take as the Banach space Y the $(n-r)$-dimensional real Cartesian space R^{n-r}, and as $P(y, \mu)$ the "miss distance" vector for the system (7.10.5), (7.10.3) whose mth component is

$$(P(y, \mu))_m = \varphi_m[\mathbf{y}(t_0), \mu] = Y_m - y_{i_m}[\mathbf{y}(t_0), \mu], \qquad m = 1, 2, \ldots, n-r \tag{7.11.1}$$

where

$Y_m = c_{i_m}$, the specified terminal values, $m = 1, 2, \ldots, n-r$,

$y_{i_m}[\mathbf{y}(t_0), \mu] =$ calculated terminal values obtained by using the initial conditions $\mathbf{y}(t_0) = \{c_1, \ldots, c_r, y_{r+1}(t_0), \ldots, y_n(t_0)\}$ and by integrating (7.10.5), $m = 1, 2, \ldots, n-r$.

We assume further that the system (7.10.5), (7.10.3) is to be solved by the shooting method of Goodman and Lance. As we pointed out in [1], the discussion for other shooting methods would be similar.

Now let $\mathbf{A}[\mathbf{y}(t_0), \mu]$ be the $(n-r)\times(n-r)$ matrix $\{\partial\varphi_i/\partial y_{r+j}(t_0)\}$. If a first guess of the missing initial condition $y_{r+j}(t_0) = z_j^{(0)}$, $j = 1, 2, \ldots, n-r$ is known such that:

$\boldsymbol{\Gamma}_0 = \mathbf{A}^{-1}(\mathbf{z}^{(0)}, \mu)$ exists and $\|\boldsymbol{\Gamma}_0\| \leq B'$, where the norm is the maximum row sum norm,

$\|\boldsymbol{\varphi}(\mathbf{z}^{(0)}, \mu)\| \leq \eta'$, where the norm is the maximum component norm,

$\sum_{j,s=1}^{n-r} |\partial^2\varphi_i/\partial z_j \partial z_s| \leq K'$ for $i = 1, 2, \ldots, n$ in the region defined by the inequality (7.11.2),

$h = K'B'^2\eta' \leq \frac{1}{2}$,

then, by the Kantorovich theorem for the Newton-Raphson method (Theorem 6.2, Chapter 6), the Goodman-Lance method of adjoints will converge to the solution of the two-point boundary value problem (7.10.5), (7.10.3). The missing initial values $\mathbf{z}$ for this solution will lie in the region

$$\|\mathbf{z} - \mathbf{z}^{(0)}\| \leq \frac{1-\sqrt{1-2h}}{h} B'\eta', \tag{7.11.2}$$

and the rate of convergence can be estimated by

$$\|\mathbf{z}^{(k)} - \mathbf{z}\| \leq \frac{1}{2^k}(2h)^{2^k}\frac{\eta'}{h}, \qquad k = 0, 1, \ldots. \tag{7.11.3}$$

As we pointed out in Section 7.5, in order to guarantee the convergence of the method of adjoints, two kinds of conditions must be met. First, the solution of the initial value problem, Eq. (7.10.2) along with its first and second partial derivatives with respect to the initial values, must exist and be bounded. Second, an approximation of the missing initial values, good enough to ensure $h \leq 1/2$, must be known.

Accordingly, we now define

I = closed and bounded interval $[t_0, t_f]$,
J = $\{\mu_k \mid \mu_k \in [0, 1]\}$,
μ_k = scalar perturbation parameter, kth perturbation of μ, $0 \leq \mu_k \leq 1.0$,
R^n = closed and bounded subset of the n-dimensional space of the reals such that $g_i(\mathbf{y}, \mu) = p_i + \mu q_i$, $i = 1, 2, \ldots, n$, have continuous partial derivatives up through the third order in $R^n \times I$,
R^{n-r} = the closed and bounded set of the missing initial conditions $\mathbf{z}$,
D^n = closed and bounded set, $D^n \subset R^n \times I \times J$,
D^{n-r} = $R^{n-r} \times I \times J \subset D^n \subset R^n \times I \times J$.

Once again we assume that the solution to the two-point boundary value problem lies in a subset of the closed and bounded set D^n, which we take to be the case for (7.10.5), (7.10.3). The classical existence theorem then guarantees that a unique solution of the initial value problem exists for each point in D^n. That is, if $g_i(\mathbf{y}, \mu)$, $i = 1, 2, \ldots, n$, have continuous partial derivatives through the third order in $R^n \times I \times J$, then there is a closed bounded set $D^n \subset R^n \times I \times J$ such that, for any point $(y_1^{(0)}, y_2^{(0)}, \ldots, y_n^{(0)}, t_0) \in D^n$, Eq. (7.10.5) has a unique solution which passes through $(y_1^{(0)}, y_2^{(0)}, \ldots, y_n^{(0)}, t_0)$, which has continuous first and second partial derivatives with respect to the $y_i^{(0)}$, and which remains in D^n.

Finally, we assume that R^{n-r}, the projection of D^n on the $(n-r)$-dimensional space of the missing initial conditions, is a closed and bounded set. Since, for the reals, a closed and bounded set is compact, by the Tychonoff theorem the Cartesian product spaces D^n and D^{n-r} are compact. Since a continuous function defined on a compact set is uniformly continuous and uniformly bounded, the functions $\boldsymbol{\varphi}(\mathbf{z}^{(0)}, \mu)$, the matrix $\mathbf{A}(\mathbf{z}^{(0)}, \mu) = (\partial\varphi_i/\partial z_j)$, $i, j = 1, \ldots, n-r$, the inverse $\mathbf{A}^{-1}(\mathbf{z}^{(0)}, \mu)$, and the second partials $\partial^2\varphi_i/\partial z_j\,\partial z_k$, $i, j, k = 1, 2, \ldots, n-r$, are uniformly continuous and uniformly bounded. As a consequence, the bounds (i), (ii), (iii) in the

Kantorovich theorem (Theorem 6.2, Chapter 6) can be replaced by uniform bounds over D^{n-r}; namely, $\bar{B}, \bar{\eta}, \bar{K}$. It should be mentioned that the continuity of $\boldsymbol{\varphi}(\mathbf{z}^{(0)}, \mu)$, $\partial \varphi_i(\mathbf{z}^{(0)}, \mu)/\partial z_j$, $\partial^2 \varphi_i(\mathbf{z}^{(0)}, \mu)/\partial z_j \partial z_k$, and the existence and continuity of $\mathbf{A}^{-1}$ follow from the unique existence and continuity of the solution of (7.10.5).

Now suppose that, for μ_0 the two-point boundary value problem has been solved. Then there is $\mu_1 > \mu_0$ such that the missing initial values for the μ_0 case found by the method of adjoints, namely, $\mathbf{z}^{(0)}$, can be used as the trial missing initial conditions for the μ_1 case, which will lead to a solution of (7.10.5) with $\mu = \mu_1$. For, by the mean value theorem, the miss distance vector $P(y, \mu_1) = \boldsymbol{\varphi}(\mathbf{y}(t_0), \mu_1)$ can be evaluated in terms of the miss distance at $\mu_0, \boldsymbol{\varphi}(\mathbf{y}(t_0), \mu_0)$, namely,

$$\begin{aligned}
\varphi_m(\mathbf{z}^{(0)}, \mu_1) &= Y_m - y_{i_m}(\mathbf{z}^{(0)}, \mu_1), \\
&= Y_m - y_{i_m}(\mathbf{z}_1, \mu_0) - (\mu_1 - \mu_0)\left(\frac{\partial y_{i_m}(\mathbf{z}_1, \mu_0 + \theta(\mu_1 - \mu_0))}{\partial \mu}\right),
\end{aligned}$$

$$0 \le \theta \le 1, \qquad m = 1, 2, \ldots, n-r,$$

where $\mathbf{z}_1 = \mathbf{z}^{(0)}$. But, since

$$Y_m - y_{i_m}(\mathbf{z}^{(0)}, \mu_0) = Y_m - y_{i_m}(\mathbf{z}_1, \mu_0) = 0, \qquad m = 1, 2, \ldots, n-r,$$

and since, from (7.10.5),

$$\frac{d}{dt}\left(\frac{\partial y_i}{\partial \mu}\right) = q_i(\mathbf{y}, t),$$

it follows that

$$\varphi_m(\mathbf{z}^{(0)}, \mu_1) = -(\mu_1 - \mu_0) \int_{t_0}^{t_f} q_{i_m}(\mathbf{y}(\mathbf{z}_1, \mu_0 + \theta(\mu_1 - \mu_0)))\, dt,$$

$$m = 1, 2, \ldots, n-r.$$

Thus, taking absolute values,

$$|\varphi_m(\mathbf{z}^{(0)}, \mu_1)| \le |\mu_1 - \mu_0|\, |q_{i_m}|\, |t_f - t_0| \le |\mu_1 - \mu_0|\, |t_f - t_0| \bar{M},$$

where $\bar{M}$ is the uniform bound on $\|\mathbf{g}(y_1, y_2, \ldots, y_n, t)\|$, $\|\mathbf{p}(y_1, y_2, \ldots, y_n, t)\|$, and $\|\mathbf{q}(y_1, y_2, \ldots, y_n, t)\|$, over D^{n-r}. Since $\|\boldsymbol{\Gamma}\| = \|\mathbf{A}^{-1}(\mathbf{z}^{(0)}, \mu)\| \le \bar{B}$ and

$$\sum_{j,s=1}^{n-r} \left|\frac{\partial^2 \varphi_i}{\partial z_j \partial z_s}\right| \le \bar{K}$$

where $\bar{B}$ and $\bar{K}$ are uniform bounds over D^{n-r},

$$h = K'B'^2\eta' \leq \bar{K}\bar{B}^2 |\mu_1 - \mu_0| |t_f - t_0| \bar{M}.$$

Then h will be less than $\frac{1}{2}$ if

$$|\mu_1 - \mu_0| \leq \frac{1}{2\bar{K}\bar{B}^2\bar{M}|t_f - t_0|}.$$

Hence, under the condition on μ_1, the method of Goodman and Lance will converge to a solution of the boundary value problem, Eq. (7.10.5) with $\mu = \mu_1$ and with the boundary conditions (7.10.3). The argument may be repeated for $\mu_2 > \mu_1$, etc. Since the increment $\mu_k - \mu_{k-1}$ is independent of μ_{k-1}, the solution of the boundary value problem for $\mu_0, \mu_1, \ldots$, etc. until $\mu_k = 1.0$ will be reached in a finite number of steps.

Our discussion may be summarized in the following theorem.

THEOREM 7.3. (i) *If for the set of n nonlinear differential equations* (7.10.2) *the* g_i *have continuous partial derivatives with respect to the* y_j *up through the third order over the region* $R^n \times I \times J$, *where* R^n, I, *and* J *are defined above;*

(ii) *if there is a region* $D^n \subset R^n \times I \times J$ *containing the r given initial values such that the solutions of* (7.10.5) *passing through these initial points lie in* D^n,

(iii) *if the conditions of the Kantorovich theorem for the two-point boundary value problem are satisfied in* $R^{n-r} \times I$ *for* $\mu_0 = 0$,
then the two-point boundary value problem can be solved by the perturbation method in a finite number of extensions $\Delta\mu_i$ *from* $\mu_0 = 0$ *to* $\mu_i = 1.0$. *The* $\Delta\mu_i$ *can be at least as large as*

$$\frac{1}{2\bar{K}\bar{B}^2\bar{M}|t_f - t_0|},$$

where

$\bar{M}$ = *uniform bound on* $\|\mathbf{g}\|$, $\|\mathbf{p}\|$, $\|\mathbf{q}\|$ *over* D^{n-r},

$\bar{K}$ = *uniform bound of* $\sum_{j,s=1}^{n-r} |\partial^2\varphi_i/\partial z_j \partial z_s|$ *over* D^{n-r},

$\bar{B}$ = *uniform bound on* $\|\mathbf{A}^{-1}(\mathbf{z}, \mu)\|$ *over* D^{n-r},

$\mathbf{z}$ = *set of* $(n-r)$ *missing initial conditions,*

$\mathbf{A}(\mathbf{z}, \mu)$ = $\{\partial\varphi_i/\partial z_j\}$, *evaluated at* $(\mathbf{z}, \mu)$ *in* D^{n-r},

D^{n-r} = $R^{n-r} \times I \times J$.

7.12. NUMERICAL EXPERIENCE

To illustrate computational experience with the perturbation method we present some numerical results on the two numerically sensitive nonlinear boundary value problems which we previously discussed in Section 7.4.

Example 5. This example is the same problem as Example 1 in Section 7.4, except $t_f = 3.5$. The initial conditions are

$$y_1(0) = 0, \qquad y_2(0) = 0, \qquad y_4(0) = 0,$$

and the terminal conditions are

$$y_2(3.5) = 0, \qquad y_4(3.5) = 1.0.$$

We specify

$$n = -0.1, \qquad s = 0.2, \qquad t_0 = 0.0, \qquad t_f = 3.5.$$

The equations were partitioned into a linear and nonlinear parts as follows:

$$\begin{bmatrix} \dot{y}_1 \\ \dot{y}_2 \\ \dot{y}_3 \\ \dot{y}_4 \\ \dot{y}_5 \end{bmatrix} = \begin{bmatrix} 0 & 1 & 0 & 0 & 0 \\ 0 & 0 & 1 & 0 & 0 \\ 0 & s & 0 & 0 & 0 \\ 0 & 0 & 0 & 0 & 1 \\ 0 & 0 & 0 & s & 0 \end{bmatrix} \begin{bmatrix} y_1 \\ y_2 \\ y_3 \\ y_4 \\ y_5 \end{bmatrix} + \mu_k \begin{bmatrix} 0 \\ 0 \\ -\left(\dfrac{3-n}{2}\right) y_1 y_3 - n y_2{}^2 + 1 - y_4{}^2 \\ 0 \\ -\left(\dfrac{3-n}{2}\right) y_1 y_5 - (n-1) y_2 y_4 - s \end{bmatrix}.$$

Starting with $\mu_0 = 0.0$, using $\Delta\mu_k = 0.05$, and taking the fraction of the full correction to the missing initial conditions $= 1.0$, we solved a sequence of two-point boundary value problems. In Table 7.7 are tabulated the μ_k, the corresponding missing initial conditions, $y_3(0)$ and $y_5(0)$, and the corresponding terminal vector. For each μ_k the calculation was terminated either when the absolute value of the largest miss distance was less than 10^{-16} or when the number of iterations exceeded 10, whichever occurred first. We note that, for $\mu_0 = 0$, the terminal conditions are satisfied exactly; this is usually the case for linear two-point boundary value problems. As $\mu_k \to 1.0$ the absolute values of the maximum miss distance increase from $|10^{-18}|$ to $|10^{-7}|$. This suggests that to improve accuracy we should have increased the upper bound on the maximum iteration count as $\mu_k \to 1.0$.

Table 7.7. Example 5. Calculated Missing Initial Conditions and Terminal Vector. Case 4 [a–d]

μ	$y_3(0)$	$y_5(0)$	$y_1(3.5)$	$y_2(3.5)$	$y_3(3.5)$	$y_4(3.5)$	$y_5(3.5)$
0.00	0.0000000	1.9551078(10⁻¹)	0.0000000	0.0000000	0.0000000	1.0000000	4.8808244(10⁻¹)
0.05	−6.5879963(10⁻²)	2.0985476(10⁻¹)	−1.1295672(10⁻¹)	−3.1983964(10⁻¹⁸)	4.1522835(10⁻²)	1.0000000	4.7573132(10⁻¹)
0.10	−1.3135557(10⁻¹)	2.2374507(10⁻¹)	−2.2542775(10⁻¹)	2.4286128(10⁻¹⁷)	8.3807213(10⁻²)	1.0000000	4.6758426(10⁻¹)
0.15	−1.9666381(10⁻¹)	2.3735741(10⁻¹)	−3.3904668(10⁻¹)	−1.0842021(10⁻¹⁶)	1.2881783(10⁻¹)	1.0000000	4.6324185(10⁻¹)
0.20	−2.6199837(10⁻¹)	2.5088468(10⁻¹)	−4.5535742(10⁻¹)	1.1819191(10⁻¹³)	1.7868912(10⁻¹)	1.0000000	4.6240855(10⁻¹)
0.25	−3.2748821(10⁻¹)	2.6455721(10⁻¹)	−5.7578850(10⁻¹)	2.2476150(10⁻¹¹)	2.3588033(10⁻¹)	1.0000000	4.6479685(10⁻¹)
0.30	−3.9315706(10⁻¹)	2.7866840(10⁻¹)	−7.0153514(10⁻¹)	1.1476365(10⁻⁹)	3.0328781(10⁻¹)	1.0000000	4.6998385(10⁻¹)
0.35	−4.5885422(10⁻¹)	2.9360578(10⁻¹)	−8.3328308(10⁻¹)	2.5473335(10⁻⁸)	3.8423093(10⁻¹)	1.0000000	4.7717080(10⁻¹)
0.40	−5.2414959(10⁻¹)	3.0988258(10⁻¹)	−9.7068536(10⁻¹)	2.9733560(10⁻⁷)	4.8211475(10⁻¹)	1.0000000	4.8478489(10⁻¹)
0.45	−5.8820485(10⁻¹)	3.2814886(10⁻¹)	−1.1115433	1.9924282(10⁻⁶)	5.9942210(10⁻¹)	1.0000001	4.8990330(10⁻¹)
0.50	−6.4968516(10⁻¹)	3.4913543(10⁻¹)	−1.2508798	7.9406353(10⁻⁶)	7.3569199(10⁻¹)	1.0000007	4.8768204(10⁻¹)
0.55	−7.0685475(10⁻¹)	3.7346587(10⁻¹)	−1.3806226	1.8953780(10⁻⁵)	8.8482058(10⁻¹)	1.0000019	4.7142830(10⁻¹)
0.60	−7.5799721(10⁻¹)	4.0133505(10⁻¹)	−1.4910092	2.6961536(10⁻⁵)	1.0336861	1.0000031	4.3427821(10⁻¹)
0.65	−8.0205807(10⁻¹)	4.3223591(10⁻¹)	−1.5739069	2.2703322(10⁻⁵)	1.1649730	1.0000030	3.7246524(10⁻¹)
0.70	−8.3907139(10⁻¹)	4.6502164(10⁻¹)	−1.6260363	1.1105332(10⁻⁵)	1.2638256	1.0000016	2.8787049(10⁻¹)
0.75	−8.7002368(10⁻¹)	4.9832891(10⁻¹)	−1.6495435	2.8036373(10⁻⁶)	1.3232486	1.0000004	1.8730338(10⁻¹)
0.80	−8.9633214(10⁻¹)	5.3102765(10⁻¹)	−1.6499230	1.1705843(10⁻⁷)	1.3443500	1.0000000	7.9270076(10⁻²)
0.85	−9.1936583(10⁻¹)	5.6241679(10⁻¹)	−1.6334364	−1.5584084(10⁻⁸)	1.3330505	1.0000000	−2.8906685(10⁻²)
0.90	−9.4021398(10⁻¹)	5.9218413(10⁻¹)	−1.6055389	1.0410011(10⁻⁷)	1.2966790	1.0000000	−1.3221824(10⁻¹)
0.95	−9.5965233(10⁻¹)	6.2027759(10⁻¹)	−1.5703815	−8.2692603(10⁻⁸)	1.2419980	9.9999994(10⁻¹)	−2.2781004(10⁻¹)
1.00	−9.7819707(10⁻¹)	6.4678660(10⁻¹)	−1.5308940	−4.0183900(10⁻⁷)	1.1744953	9.9999986(10⁻¹)	−3.1437074(10⁻¹)

[a] $\Delta\mu = 0.05$.

[b] Integration step size = 0.0875.

[c] For each μ, criteria for terminating run were: $|\text{largest miss distance}| < 10^{-16}$, or maximum number of iterations = 10.

[d] Fraction of full correction to missing initial conditions = 1.0.

The analyst has at his disposal two adjustable parameters, appropriate values of which may materially assist him in the search for the solution of perturbed two-point boundary value problems. These parameters are: the size of the $\Delta \mu_k$ which may be changed for each μ_k; and the fraction of the full correction taken to the set of missing initial conditions. The choice of values for each of these parameters is a matter of experimentation, and in general the combination which leads to a solution is not unique.

Table 7.8 is presented to illustrate some of the combinations of $\Delta \mu_k$ and the fraction of full correction to the set of missing initial conditions which were used to solve Example 5. In each of the four cases given, the perturbation method was initiated with $\mu_0 = 0.0$, and with $\Delta \mu_k$ and the fraction of the full correction to the set of missing initial conditions held constant throughout the computation. For each k the process was terminated either when the absolute value of the largest miss distance was less then 10^{-16} or when the number of iterations exceeded 10, whichever occurred first. For the four cases cited, all calculations were in fact terminated by the maximum iteration count. In the last column of Table 7.8 are listed the absolute values of the maximum miss distance at $\mu = 1.0$ for each case. While the accuracy of Cases 1, 3, and 4 is approximately the same, case 2 yields inferior results. Case 4 has been presented in Table 7.7.

Table 7.8. Example 5

Case	*Initial* μ	*Final* μ	$\Delta \mu_k$	*Fraction of Full Correction to Missing Initial Conditions*	*No. of Iterations*	*Maximum Miss Distance at* $\mu = 1.0$
1	0.0	1.0	.10	1.0	10	$\|1.8(10^{-6}\|$
2	0.0	1.0	.05	0.5	10	$\|5.3(10^{-3}\|$
3	0.0	1.0	.025	1.0	10	$\|1.2(10^{-7})\|$
4	0.0	1.0	.05	1.0	10	$\|4.1(10^{-7})\|$

Example 6. As a second problem demonstrating the perturbation method, consider this example, which is the same problem as Example 2 in Section 7.4. For some problems, such as this one, after a while the perturbation technique requires such a very small $\Delta \mu_k$ that the method may not be practical. To cope with such a situation we may employ a mixed strategy of perturbation and quasilinearization, in which the perturbation technique is used as a device to supply the nominal profiles for the quasilinear process. This approach is used to solve the example.

The equations are partitioned into linear and nonlinear parts as follows:

$$
\begin{bmatrix} \dot{y}_1 \\ \dot{y}_2 \\ \dot{y}_3 \\ \dot{y}_4 \\ \dot{y}_5 \\ \dot{y}_6 \\ \dot{y}_7 \end{bmatrix}
=
\begin{bmatrix}
0 & 0 & 0 & 0 & 0 & 0 & 0 \\
0 & 0 & 0 & 0 & 0 & 0 & 0 \\
0 & 0 & 0 & 1 & 0 & 0 & 0 \\
0 & 0 & 0 & 0 & 0 & 0 & 0 \\
0 & 0 & 0 & 0 & 0 & 1 & 0 \\
0 & 0 & 0 & -\frac{1}{2} & 0 & 0 & 0 \\
0 & 0 & 0 & 0 & 0 & 0 & 0
\end{bmatrix}
\begin{bmatrix} y_1 \\ y_2 \\ y_3 \\ y_4 \\ y_5 \\ y_6 \\ y_7 \end{bmatrix}
+ \mu_k
\begin{bmatrix}
-\dfrac{y_1}{y_5}[y_6 + 2(y_3 + y_5)] \\
y_1 y_3 (y_3 + y_5) \\
0 \\
\dfrac{R_e y_1}{y_7}[-2\sigma A y_7^2 + y_4 y_5 + y_3(y_3 + y_5)] + \dfrac{y_4 y_5 y_6}{2A y_7^2} + \dfrac{2R_e y_2}{y_7} \\
0 \\
\dfrac{y_1}{(4/3R_e)y_7}\left[-\dfrac{\sigma A y_7^2}{y_5}\{y_6 + 2(y_3 + y_5)\} + y_5 y_6 (1 - \sigma)\right] - \dfrac{y_3 y_5 y_6}{2A y_7^2} - \dfrac{y_5}{2A}\left(\dfrac{y_6}{y_7}\right)^2 \\
-\dfrac{y_5 y_6}{2A y_7}
\end{bmatrix}
$$

Table 7.9. Example 6. Calculated Missing Initial Conditions and Terminal Vector

μ	$y_4(0.05)$	$y_6(0.05)$	$y_1(0.3303)$	$y_2(0.3303)$	$y_3(0.3303)$	$y_4(0.3303)$	$y_5(0.3303)$	$y_6(0.3303)$	$y_7(0.3303)$
0.00	2.11273635	−3.34392289	0.96170000	−0.10180000	1.00000000	2.11273635	−1.00000000	−3.64002289	0.99980000
0.05	2.59181747	−2.51032587	0.81705726	−0.10068131	0.93716109	0.92826250	−1.10775954	−3.72914284	0.96956338
0.10	3.10218036	−1.67754212	0.70520649	−0.09918668	0.93150996	0.38893675	−1.11654003	−3.51545684	0.93734178
0.15	3.53149893	−1.13147817	0.62661665	−0.09689841	0.93976942	0.10812770	−1.09860004	−3.31950417	0.90768952
0.20	3.88785235	−0.89928618	0.57216939	−0.09395605	0.94923138	−0.06679672	−1.07856836	−3.16431778	0.87966008
0.25	4.19832377	−0.83003351	0.52782163	−0.09078838	0.95560665	−0.19481164	−1.06737702	−3.01568259	0.85040069
0.30	4.48000516	−0.80976846	0.48671326	−0.08768816	0.95907092	−0.29442483	−1.06387838	−2.85391183	0.81857762
0.35	4.74244717	−0.80243868	0.44709876	−0.08479326	0.96060680	−0.37464614	−1.06550901	−2.67458854	0.78356085
0.40	4.99105597	−0.79891284	0.40896315	−0.08214380	0.96105964	−0.44105403	−1.06987210	−2.47959445	0.74515708
0.45	5.22827869	−0.79674789	0.37324784	−0.07965906	0.96125358	−0.49603257	−1.07398427	−2.27702800	0.70417679
0.50	5.45418507	−0.79518922	0.34076420	−0.07723034	0.96153571	−0.54143301	−1.07575880	−2.07664689	0.66184803
0.55	5.66635807	−0.79394781	0.31254994	−0.07466481	0.96210127	−0.57675633	−1.07296734	−1.89152303	0.62055657

Table 7.10. Example 6. Perturbation Method Profiles [a]

Time	$y_1(t)$	$y_2(t)$	$y_3(t)$	$y_4(t)$	$y_5(t)$	$y_6(t)$	$y_7(t)$
0.05000	9.6170000(10⁻¹)	−1.0180000(10⁻¹)	4.0780000(10⁻¹)	5.2282787	−2.1200000(10⁻²)	−7.9674739(10⁻¹)	9.9980000(10⁻¹)
0.07002	9.5174604(10⁻¹)	−1.0010029(10⁻¹)	5.0751814(10⁻¹)	4.7349766	−3.9061560(10⁻²)	−9.8982145(10⁻¹)	9.9956481(10⁻¹)
0.09004	9.3352692(10⁻¹)	−9.7726600(10⁻²)	5.9750069(10⁻¹)	4.2565289	−6.1454890(10⁻²)	−1.2943147	9.9907271(10⁻¹)
0.11006	8.8148704(10⁻¹)	−9.4777930(10⁻²)	6.7806595(10⁻¹)	3.7932050	−9.4095410(10⁻²)	−2.0894895	9.9796090(10⁻¹)
0.13008	7.8443582(10⁻¹)	−9.1579220(10⁻²)	7.4938365(10⁻¹)	3.3286306	−1.4941648(10⁻¹)	−3.4998367	9.9500516(10⁻¹)
0.15010	6.7890149(10⁻¹)	−8.8538720(10⁻²)	8.1118398(10⁻¹)	2.8400887	−2.3396093(10⁻¹)	−4.8831225	9.8786432(10⁻¹)
0.17012	5.9336438(10⁻¹)	−8.5897950(10⁻²)	8.6293339(10⁻¹)	2.3266744	−3.4131031(10⁻¹)	−5.7410174	9.7411465(10⁻¹)
0.19015	5.3037511(10⁻¹)	−8.3735120(10⁻²)	9.0430777(10⁻¹)	1.8077892	−4.5999808(10⁻¹)	−6.0292505	9.5254934(10⁻¹)
0.21017	4.8466843(10⁻¹)	−8.2051450(10⁻²)	9.3544830(10⁻¹)	1.3084107	−5.7975718(10⁻¹)	−5.8725388	9.2355208(10⁻¹)
0.23019	4.5119032(10⁻¹)	−8.0815740(10⁻²)	9.5697336(10⁻¹)	8.5042597(10⁻¹)	−6.9316966(10⁻¹)	−5.4188646	8.8874945(10⁻¹)
0.25021	4.2630955(10⁻¹)	−7.9982380(10⁻²)	9.6988131(10⁻¹)	4.4969193(10⁻¹)	−7.9564086(10⁻¹)	−4.7790794	8.5042496(10⁻¹)
0.27023	4.0754371(10⁻¹)	−7.9500090(10⁻²)	9.7542061(10⁻¹)	1.1529273(10⁻¹)	−8.8492201(10⁻¹)	−4.1157839	8.1096844(10⁻¹)
0.29025	3.9318574(10⁻¹)	−7.9317160(10⁻²)	9.7495330(10⁻¹)	−1.5046529(10⁻¹)	−9.6051498(10⁻¹)	−3.4413745	7.7247809(10⁻¹)
0.31027	3.8203890(10⁻¹)	−7.9384790(10⁻²)	9.6982466(10⁻¹)	−3.5140596(10⁻¹)	−1.0230893	−2.8208132	7.3654213(10⁻¹)
0.33030	3.7324784(10⁻¹)	−7.9659060(10⁻²)	1.0000000	−4.9603257(10⁻¹)	−1.0000000	−2.2770280	7.0417679(10⁻¹)

[a] $\mu = 0.45$.

Table 7.11. Example 6. Quasilinearization Process Profiles, 10th Iteration

Time	$y_1(t)$	$y_2(t)$	$y_3(t)$	$y_4(t)$	$y_5(t)$	$y_6(t)$	$y_7(t)$
0.05000	$9.61700000(10^{-1})$	$-1.0180000(10^{-1})$	$4.07800000(10^{-1})$	7.49272108	$-2.12000000(10^{-2})$	$-7.88688556(10^{-1})$	$9.99300000(10^{-1})$
0.07002	$9.48368090(10^{-1})$	$-9.7665862(10^{-2})$	$5.46035113(10^{-1})$	6.33017934	$-3.95844060(10^{-2})$	-1.03943121	$9.99257534(10^{-1})$
0.09004	$9.35400980(10^{-1})$	$-9.1309480(10^{-2})$	$6.61933916(10^{-1})$	5.26732249	$-6.25113290(10^{-2})$	-1.24726273	$9.98119348(10^{-1})$
0.11006	$9.10480800(10^{-1})$	$-8.2919238(10^{-2})$	$7.57825008(10^{-1})$	4.33625977	$-9.01014490(10^{-2})$	-1.58976285	$9.96069849(10^{-1})$
0.13008	$7.83574148(10^{-1})$	$-7.3422958(10^{-2})$	$8.36524641(10^{-1})$	3.54779530	$-1.35243809(10^{-1})$	-3.37322179	$9.91088861(10^{-1})$
0.15010	$5.21921380(10^{-1})$	$-6.5589040(10^{-2})$	$9.00390452(10^{-1})$	2.84263197	$-2.37543145(10^{-1})$	-6.74680558	$9.72213662(10^{-1})$
0.17012	$3.44895863(10^{-1})$	$-6.0739745(10^{-2})$	$9.50696541(10^{-1})$	2.19485507	$-3.90030413(10^{-1})$	-8.08746650	$9.23208729(10^{-1})$
0.10015	$2.55978327(10^{-1})$	$-5.7855486(10^{-2})$	$9.88761450(10^{-1})$	1.61934328	$-5.48839079(10^{-1})$	-7.56859711	$8.41162317(10^{-1})$
0.21017	$2.09437193(10^{-1})$	$-5.6078436(10^{-2})$	1.01588292	1.09725782	$-6.87747640(10^{-1})$	-6.23446519	$7.35377890(10^{-1})$
0.23019	$1.83435696(10^{-1})$	$-5.4952404(10^{-2})$	1.03290362	$6.07614318(10^{-1})$	$-7.97213533(10^{-1})$	-4.70152754	$6.18839153(10^{-1})$
0.25021	$1.68356634(10^{-1})$	$-5.4230124(10^{-2})$	1.04039869	$1.46936108(10^{-1})$	$-8.76884651(10^{-1})$	-3.29648305	$5.03460082(10^{-1})$
0.27023	$1.59509069(10^{-1})$	$-5.3770879(10^{-2})$	1.03911974	$-2.63076535(10^{-1})$	$-9.31053092(10^{-1})$	-2.16922813	$3.97990590(10^{-1})$
0.29025	$1.54328301(10^{-1})$	$-5.3492668(10^{-2})$	1.03046267	$-5.81725959(10^{-1})$	$-9.65786819(10^{-1})$	-1.35517518	$3.07307057(10^{-1})$
0.31027	$1.51283296(10^{-1})$	$-5.3346797(10^{-2})$	1.01660850	$-7.76283297(10^{-1})$	$-9.87106522(10^{-1})$	$-8.20710731(10^{-1})$	$2.32424709(10^{-1})$
0.33030	$1.49419339(10^{-1})$	$-5.3302963(10^{-2})$	1.00000000	$-8.57590516(10^{-1})$	$-9.99999999(10^{-1})$	$-5.02460606(10^{-1})$	$1.70735638(10^{-1})$

As we have seen, this is a very difficult problem to solve. It is somewhat remarkable that the linear problem with its very sparse matrix would generate numerically sufficiently accurate missing initial conditions to be used in the first nonlinear problem. In Table 7.9 are tabulated the μ_k and corresponding missing initial conditions and the terminal vector. Table 7.9 is a result of experimentation with the relative size of $\Delta\mu_k$ and the fraction of the full correction taken to the set of missing initial conditions. The $\Delta\mu_k$ was set equal to 0.05, and the fraction of the full correction was set at 0.10. For each μ_k the calculation was terminated either when the absolute value of the largest miss distance was less than 10^{-6} or when the number of iterations exceeded 10. While the terminal conditions for each μ were satisfied on the average to within only 5%, smaller miss distances could have been achieved by more iterations or by adjusting the fraction of the full correction taken to the set of missing initial conditions. In this problem, $\Delta\mu_k = 0.05$ was successful in the range from $\mu = 0.0$ to $\mu = 0.55$. Beyond $\mu = 0.55$, the $\Delta\mu_k$ had to be reduced to very small numbers. While the μ was edged up to 0.91, the $\Delta\mu$ became so small ($\Delta\mu = 5(10^{-4})$) that the perturbation method alone for this problem really did not appear practical.

However, a mixed strategy of the perturbation and quasilinearization techniques is practical. To employ the mixed strategy we merely take the calculated profiles for a certain μ_k and use the profiles as the nominal profiles for the quasilinearization process applied to the original nonlinear problem, that is, when $\mu_k = 1.0$. While not all the profiles developed for μ_k, $k = 0, 1, 2, \ldots,$ will lead to convergence in the quasilinear process, there seems to be a "threshold" μ_k such that all μ_k greater than the "threshold" μ_k will produce successful profiles. In Example 6 the "threshold" μ_k was 0.45.

In Table 7.10 are listed the profiles for $\mu_k = 0.45$. In Table 7.11 are the profiles for the 10th iteration of the quasilinear process, which employed as nominal profiles those for $\mu_k = 0.45$.

7.13. DISCUSSION

In practice the perturbation technique requires experimental determination of combinations of $\Delta\mu_k$, fraction of full correction to the set of missing initial conditions, and an upper bound on the number of iterations to progress from $\mu_0 = 0$ to $\mu = 1.0$. A small $\Delta\mu_k$ and the fraction of full correction to the set of missing initial conditions which is set equal to 1.0 are probably the best combination to employ. If $\Delta\mu_k$ is too small, then we must of course solve a sequence of many two-point boundary value problems. If $\Delta\mu_k$ is too large, overflow problems will occur and the problem must be restarted at μ_k with a smaller $\Delta\mu_k$. In our numerical experience a suitably small $\Delta\mu_k$ to try at first is approximately 0.05; however, much depends on the problem.

In a sophisticated computer program, logic can be built in to adjust $\Delta\mu_k$ automatically as a function of k, the size of the miss distance, and the number of iterations to converge, perhaps as an adaptive learning scheme.

As a potential scheme to speed the process, we may plot each missing initial condition as a function of μ_k and extrapolate to $\mu_k = 1.0$. As with all extrapolation procedures, this may not give sufficiently accurate results.

In our discussion we have only considered μ_k as a scalar. We may, however, consider μ_k as a vector with components $\mu_{j,k}$, $j = 1, 2, \ldots, n$, where j refers to the equation number and k refers to the kth two-point boundary value problem. Corresponding to each $\mu_{j,k}$ is a $\Delta\mu_{j,k}$. The advantage of the vector perturbation parameters is that certain equations could be treated more individually. The disadvantage, of course, is the complication of choosing the components of the correction terms $\Delta\mu_{j,k}$.

Nonlinear two-point boundary value problems for which the analyst may not have any idea of guess values for the set of missing initial conditions, or must have very good guess values because of the sensitivity of the problem, are attractive candidates for perturbation. This is true since the perturbation method, if it starts with a linear two-point boundary value problem, requires no input information from the analyst in the way of guess values for the missing initial conditions. The perturbation method automatically generates the set of missing initial conditions as a function of μ_k.

The perturbation method may also be used in conjunction with the continuation technique [1]. For example, we may use the perturbation method to get started over the initial interval $[t_0, t_1]$. For that matter, we may use the perturbation technique at any stage of the continuation process where there are difficulties in finding the missing initial conditions.

REFERENCES

1. S. M. Roberts and J. S. Shipman, Continuation in Shooting Methods for Two-Point Boundary Value Problems, *J. Math. Anal. Appl.*, (1) **18** (April 1967), 45–58.
2. T. R. Goodman and G. N. Lance, The Numerical Solution of Two Point Boundary Value Problems, *MTAC*, **10** (1956), 82–86.
3. R. H. Moore, Newton's Method and Variations in *Nonlinear Integral Equations*, P. M. Anselone, ed., University of Wisconsin Press, 1964.
4. S. M. Roberts and J. S. Shipman, The Kantorovich Theorem and Two-Point Boundary-Value Problems, *IBM J. Res. and Develop.*, (5) **10** (Sept. 1966), 402–406.
5. R. Bellman and R. Kalaba, *Quasilinearization and Nonlinear Boundary Value Problems*, American Elsevier, New York, 1965.

6. L. Fox, *Numerical Solution of Two-Point Boundary Problems in Ordinary Differential Equations*, Oxford, London, 1957.
7. P. Henrici, *Discrete Variable Methods in Ordinary Differential Equations*, Wiley, New York, 1962.
8. J. F. Holt, Numerical Solution of Nonlinear Two-Point Boundary Value Problems by Finite Difference Methods, *Comm. ACM*, (6) **7** (1964), 363–373.
9. S. M. Roberts and J. S. Shipman, Justification for the Continuation Method in Two-Point Boundary Value Problems, *J. Math. Anal. Appl.*, (1) **21** (Jan. 1968), 23–30.
10. L. V. Kantorovich and G. P. Akilov, *Functional Analysis in Normed Spaces*, English Translation, Pergamon Press, 1964.
11. F. A. Ficken, The Continuation Method for Functional Equations, *Comm. Pure Appl. Math.*, **4** (1951), 435–456.
12. S. M. Roberts, J. S. Shipman, and C. V. Roth, Continuation in Quasilinearization, *J. Optimization Theory Appl.*, (3) **2** (May 1968), 164–178)
13. R. Bellman, *Perturbation Techniques in Mathemat'cs, Physics, and Engineering*, Holt, Rinehart, and Winston, New York, 1963.

General References

K. O. Friedrichs, *Functional Analysis and Applications*, New York University Institute of Mathematical Sciences, 1953.

D. D. Morrison, J. D. Riley, and J. F. Zancanaro, Multiple Shooting Method for Two-Point Boundary Value Problems, *Comm. ACM*, (12) **5**(Dec. 1962), 613–614.

S. M. Roberts and J. S. Shipman, Some Results in Two-Point Boundary Value Problems, *IBM J. Res. and Develop.*, (4) **11** (July 1967), 383–388.

S. M. Roberts, J. S. Shipman, and W. Ellis, A Perturbation Technique for Nonlinear Two-Point Boundary Value Problems, *SIAM J. Numer. Anal.*, (3) **6** (Sept. 1969), 347-358.

V. E. Šamanskĭ, Methods for the Numerical Solution of Boundary Value Problems. Part II: Nonlinear Boundary Value Problems and Eigenvalue Problems for Differential Equations [Russian], "Naukova Dumka", Kiev, 1966.

Chapter 8

FINITE DIFFERENCE METHODS AND RELATED TOPICS

8.1. INTRODUCTION

Our attention in this book so far has been devoted to the solution of two-point boundary value problems by shooting and related methods which employ numerical integration. In order to give the reader a broader view of numerical methods of solution of two-point boundary value problems, and because shooting is inadequate to handle certain problems, this chapter deals with the important and widely used finite difference method. The finite difference method of solving two-point boundary value problems converts the set of ordinary differential equations into a finite set of algebraic or transcendental equations. The solution of the set of algebraic or transcendental equations yields approximations to the solution of the original differential equations at discrete points. If the original ordinary differential equations are linear, the finite difference equations will be linear algebraic equations. If the ordinary differential equations are nonlinear, the resulting finite difference equations will be nonlinear algebraic or transcendental equations. The resulting nonlinear equations may then be attacked by any numerical technique available for solving such equations. For example, the nonlinear equations may be solved by the Newton-Raphson method, or the equations put in the form symbolized by $x = F(x)$ and solved by the iteration scheme $x^{(n+1)} = F(x^{(n)})$ [1–10].

Finite difference methods have proved to be particularly useful techniques for solving numerically sensitive two-point boundary value problems. As will be shown later, the finite difference equations incorporate both specified initial and terminal conditions in the finite set of equations; thus the resulting solution of these equations is constrained to satisfy these boundary conditions. Quasilinearization shares this feature with finite difference methods, since the prescribed boundary conditions are required to be satisfied at each iteration.

Shooting methods, on the other hand, take trial values for the set of missing initial conditions and integrate the equations forward. The specified terminal conditions are in effect an "afterthought", since they do not demand to be satisfied for each iteration. For shooting methods the specified terminal conditions exercise influence only after the "miss distances" have been found. At that time appropriate adjustments are made in the corrections to the set of missing initial conditions.

We may characterize the difference in approach between finite difference methods and shooting methods as the difference between solving the problem in "parallel" or in "series". In finite difference methods we would not expect the solution at one point to be determined with an accuracy significantly different from that for the solution at another point, since all points of the solution are, in effect, produced simultaneously.

In contrast, the shooting method, since it employs a step-by-step numerical integration scheme, generates the solution sequentially. We may thus expect the solution at points near the final point to be determined with less accuracy than at points near the initial point because of propagation of round-off errors. This expectation seems to be borne out by numerical experience.

As we shall see, most applications of the finite difference method ultimately reduce to the solution of large systems of linear algebraic equations with sparse matrices having a well-defined structure, usually banded matrices such as tridiagonal. The theory of such matrices, which is highly developed, can then be brought to bear to determine under what conditions the problem has a solution, and to estimate the rate of convergence.

Finally, it should be noted that, while we are concerned with the application of finite difference methods to the numerical solution of ordinary differential equations, these methods are equally important, if not more so, in the numerical treatment of partial differential equations.

8.2. FINITE DIFFERENCE METHOD

To illustrate the finite difference method we consider the following, simple, two-point boundary value problem as an example:

$$\frac{d^2y}{dt^2} = y+t, \qquad a \leqq t \leqq b, \tag{8.2.1}$$

$$y(a) = c_1, \qquad y(b) = c_2. \tag{8.2.2}$$

The interval $[a, b]$ is divided into $(N+1)$ evenly spaced intervals of length h, where N is the number of internal points:

$$h = \frac{(b-a)}{N+1}. \tag{8.2.3}$$

The discrete points are given by

$$t_j = t_0+j\,\frac{(b-a)}{N+1}, \qquad j = 0,\ldots,N+1, \tag{8.2.4}$$

where $t_0 = a$ and $t_{N+1} = b$.
The second derivative at t_j may be approximated by the second central difference,

$$\frac{d^2y}{dt^2} = \frac{y_{j+1}-2y_j+y_{j-1}}{h^2} + O(h^2). \tag{8.2.5}$$

where $y_j = y(t_j)$. The discrete approximation to (8.2.1) is then

$$-y_{j-1}+(2+h^2)y_j-y_{j+1} = -h^2t_j, \qquad j = 1,2,\ldots,N. \tag{8.2.6}$$

The two-point boundary value problem given by (8.2.1) and (8.2.2) may now be approximated by the N equations of (8.2.6), where the boundary conditions are included. To be specific we write the finite difference two-point boundary value problem:

$$\begin{aligned}
&(2+h^2)y_1-y_2 &&= -h^2t_1+y_0,\\
&-y_1+(2+h^2)y_2-y_3 &&= -h^2t_2,\\
&\qquad -y_2+(2+h^2)y_3-y_4 &&= -h^2t_3,\\
&\qquad\qquad \vdots && \quad\vdots\\
&\qquad\qquad -y_{N-1}+(2+h^2)y_N &&= -h^2t_N+y_{N+1},
\end{aligned} \tag{8.2.7}$$

where $y_0 = y(a) = c_1$ and $y_{N+1} = y(b) = c_2$.

In matrix form the equations appear as

$$\begin{bmatrix} 2+h^2 & -1 & & \\ -1 & 2+h^2 & -1 & \\ & \ddots & \ddots & \ddots \\ & & \ddots & -1 \\ & & -1 & 2+h^2 \end{bmatrix} \begin{bmatrix} y_1 \\ y_2 \\ \vdots \\ y_{N-1} \\ y_N \end{bmatrix} = \begin{bmatrix} -h^2t_1+y_0 \\ -h^2t_2 \\ \vdots \\ -h^2t_{N-1} \\ -h^2t_N+y_{N+1} \end{bmatrix} \tag{8.2.8}$$

This is a set of N linear algebraic equations in the N unknowns $y_1, y_2, \ldots, y_N$ whose solution is an approximation to the solution of (8.2.1) at the points t_j. Whether or not (8.2.8) has a solution, as well as the efficiency of numerical methods of solution, depends on the nature of the matrix of the equations. In Section 8.5, we present briefly some matrix theory and two important theorems called on frequently in this connection. In practice, the solution of (8.2.8) is usually found by Gaussian elimination or by the Gauss-Seidel or Jacobi iterative methods.

8.3. DISCUSSION OF FINITE DIFFERENCE METHOD

The example in Section 8.2, while very simple, displays a number of features typical of finite difference methods applied to two-point boundary value problems. The finite difference method may be characterized by the following:

1. Finite difference approximation. The differential equation (8.2.1) in Section 8.2 is approximated by a finite difference equation. A differential equation may be approximated by many finite difference formulas, the choice of which is a compromise among many factors including simplicity, size of the truncation error, and the requirements of stability and consistency. A description of and the relative merits of various formulas are discussed in a number of standard books on numerical analysis; for example, see [1–10].

2. Banded matrix. The linear or linearized finite difference equations exhibit banded matrix structure. The band width depends on the number of points in the numerical differentiation formula. For second-order linear differential equations the matrix is commonly tridiagonal and diagonally dominant. As will be discussed below, matrices with diagonally dominant terms are particularly attractive because they are guaranteed to have inverses.

3. Problem size. The size of the problem depends on N, the number of discrete points considered. The choice of N (or the choice of h) is a compromise between accuracy and time for solution. As one check on the validity of his results, the investigator may solve the problem for several step sizes, and then may combine the solutions to obtain a still more accurate solution (Richardson extrapolation).

4. Boundary values. The boundary values specified at t_0 and t_{N+1}, namely, y_0 and y_{N+1}, are included in the set of equations and appear on the right-hand side.

5. Solution of the linear finite difference equations. The linear finite difference equations which are linear algebraic equations may be solved by a variety of techniques. These techniques include methods such as Gaussian elimination or any of its variants, direct matrix inversion, relaxation techniques, and indirect iterative methods such as Gauss-Seidel or Jacobi. Finally, for simple finite difference equations, analytical solutions may be obtained.

6. Nonlinear differential equations. Linear differential equations give rise to linear finite difference equations, while nonlinear differential equations

give rise to nonlinear finite difference equations. As an example, take the nonlinear differential equations

$$-\frac{d^2y}{dt^2}+y^2=t, \qquad y(a)=c_1, \qquad y(b)=c_2.$$

A finite difference approximation is

$$-y_{j-1}+2y_j+h^2y_j{}^2-y_{j+1}=-h^2t_j, \qquad j=1,2,\ldots,N,$$

which is a nonlinear algebraic expression since y_j appears to the second power. Nonlinear problems are usually solved by the Newton-Raphson method or some other linearizing technique.

8.4. PROBLEMS IN SETTING UP THE FINITE DIFFERENCE EQUATIONS

So far in our discussion we have examined only second-order ordinary differential equation boundary value problems. While the system of finite difference equations approximating such a boundary value problem displays many features common to boundary value problems, we do not want to give the impression that setting up the finite difference equations is a trivial problem. For systems of nonlinear ordinary differential equations in which the derivatives of the various variables may appear to several orders and in which cross products of various variables appear, the formulation of the finite difference equations can be a difficult and time-consuming task.

In this section we mention some of the problems and difficulties in the formation of the finite difference equations.

1. Finite difference formulas. The choice of the finite difference formulas to approximate the differential equations is of course very important. As with numerical integration formulas, attention must be paid not only to accuracy but also to stability. A good account of the stability theory for finite difference methods, which parallels that for numerical integration, can be found in Issacson and Keller [8]. In most cases the derivatives appearing in the differential equations will be approximated by a central difference formula of the same order as the derivative, in particular,

$$\frac{dy}{dt}=\frac{y(t+h)-y(t-h)}{2h}+\mathrm{O}(h^2),$$

$$\frac{d^2y}{dt^2}=\frac{y(t+h)-2y(t)+y(t-h)}{h^2}+\mathrm{O}(h^2).$$

At initial and terminal points where this is not possible, forward and backward difference formulas, respectively, are frequently used (but see item 3).

The problem of solving a two-point boundary value problem by finite differences is really two problems. The first problem is to solve the set of linear or linearized algebraic finite difference equations by various direct methods such as Gaussian elimination or iterative methods such as Gauss-Seidel or Jacobi. Matrix theory is particularly helpful here because we can specify the conditions under which the set of linear equations can be solved (see Section 8.5).

The second problem is to show that, in the case of linear problems, the solution of the finite difference equations y_k approaches the solution of the differential equations $y(t_k)$ at the point t_k and, in the case of nonlinear problems, that the solution of the sequence of linear problems converges to the solution of the nonlinear problem. See Henrici [4, 5] and Forsythe and Wasow [2].

2. Step size. As in the case of numerical integration, the choice of step size is a compromise between accuracy and speed of solution. If the step size is very small, a large number of equations must be solved with the attendant problem of longer time for solution and increased round-off error. If the step size is too large, the size of the problem is reduced but the solution of the finite difference equation may not be close enough to the solution of the differential equation. Between these extremes of step size there is a range of step sizes which is satisfactory. One practical way to ascertain the adequacy of the step size is to solve the problem for two different step sizes and compare the solutions. Furthermore, the solution for the two step sizes can be combined to give a solution of greater accuracy than either of them (Richardson extrapolation and related methods; see Pereyra [11, 12]).

The choice of step size and the choice of finite difference approximations are related. Some investigators use finite difference equations with relatively large truncation errors but solve the equations with a small step size by iterative procedures. The matrix iterative schemes are particularly well suited for digital computers. Fox [3], on the other hand, prefers finite difference equations with relatively small truncation errors of $O(h^3)$ or $O(h^4)$ and uses a larger step size. He then calculates the truncation error from the difference table associated with his solution and corrects his solution for the truncation error. Fox's procedure is also iterative but is better suited for hand computation with a desk calculator than for digital computers.

3. Initial and terminal conditions. Since the finite difference formulas commonly involve three or more points, we may find that we have more variables than equations if we use central difference formulas. For we may

have variables such as y_{-2}, y_{-1}, preceding the initial point y_0 and such variables as y_{N+2}, y_{N+3} beyond the terminal point y_{N+1}. Additional equations therefore must be introduced so that the number of variables equals the number of equations. For points such as y_{-2}, y_{-1} preceding the initial point, forward difference formulas may be used while, for points such as y_{N+2}, y_{N+3}, backward difference formulas may be employed. These "extra" equations should be so chosen that their truncation error is approximately the same as that for the main body of equations. The extra formulas are commonly another finite difference approximation to the differential equations. They all have the common feature of adding another equation without introducing more variables. As an example, see Section 8.9, where an initial value problem is converted into a boundary value problem. Another way to introduce an extra equation is to use a Taylor's series approximation to the differential equation.

A third possibility for obtaining the extra equation is to differentiate the differential equation and approximate it with a finite difference expression. See the example in item 6 in this section where this is carried out.

Although theoretically all these schemes do the same job, in practice the choice of the form of the extra equation and where the extra equations are applied may have significant impact. Holt [13], for example, cites a case in which the finite difference equation obtained from the differentiation of the differential equation did not produce a numerical solution when the finite difference equation was applied to the terminal point. When applied to the initial point, the finite difference equation did generate a numerical solution. See item 6.

4. Set of nonlinear differential equations. Since a set of nonlinear ordinary differential equations leads to a set of nonlinear algebraic or transcendental equations, we must deal with nonlinear problems. Because no general method is available to solve finite sets of nonlinear equations, the nonlinear algebraic or transcendental equations are commonly solved by some kind of a linearizing technique coupled with an iterative scheme. Two techniques are frequently used: the Newton-Raphson method, and direct linearization of nonlinear terms. The Newton-Raphson method is discussed in Section 6.2, Chapter 6, where an illustration of its application to a set of nonlinear equations is given. The direct linearization of the nonlinear terms is carried out by replacing nonlinear terms with certain linear approximations. For example, y_i^2 may be replaced by $y_i \bar{y}_i$, y_i^3 by $y_i \bar{y}_i^2$, $y_i v_{j+1}$ by $\frac{1}{2}(y_i \bar{v}_{j+1} + \bar{y}_i v_{j+1})$, where $\bar{y}_i$ and $\bar{v}_{j+1}$ are nominal or trial values. Since there are many linearization possibilities, how the terms are linearized has an important bearing on the problem. Once the linearized equations have been solved, it is necessary to check whether $y_i^{(k)} = \bar{y}_i^{(k)}$, $i = 1,2,\ldots,N$, for the kth

iteration. If $y_i^{(k)}$ does not equal $\bar{y}_i^{(k)}$ within a small tolerance, then a new choice of trial solution, namely, $\bar{y}_i^{(k+1)}$ must be made and the linearized equations solved again. The manner in which $\bar{y}_i^{(k+1)}$ is formed may be critical in numerically sensitive problems. Holt [13] recommends forming $\bar{y}_i^{(k+1)} = (\bar{y}_i^{(k)} + y_i^{(k)})/2$. While this is one of the simplest possibilities, an entire class of schemes suggests itself, such as

$$\bar{y}_i^{(k+1)} = \sum_{j=m}^{k} w_j \bar{y}_i^{(j)} + w_{k+1} y_i^{(k)}, \qquad i = 1, 2, \ldots, N, \tag{8.4.1}$$

where w_j, $j = m, \ldots, k$ $(m < k)$ are weights assigned to the mth, $(m+1)$st, etc., profiles. If $m = k$ and $w_k = 0$, $w_{k+1} = 1$, we have $\bar{y}_i^{(k+1)} = y_i^{(k)}$. If $m = k$, $w_k = \frac{1}{2}$, $w_{k+1} = \frac{1}{2}$, then $\bar{y}_i^{(k+1)} = (\bar{y}_i^{(k)} + y_i^{(k)})/2$.

The values for the weights are assigned by the investigator. The adequacy of any scheme for forming new trial values is a function of both the type of equations and the closeness of the trial solution to the true solution of the set of nonlinear algebraic equations.

There are two distinct paths to the numerical solution of nonlinear two-point boundary value problems. We may linearize the nonlinear differential equations, and then solve the resulting sequence of linear algebraic equations. Or we may replace the original nonlinear differential equations by nonlinear finite difference equations, and then solve the resulting set of nonlinear algebraic or transcendental equations by replacing them with a sequence of linear algebraic equations. The question whether these two paths are equivalent has been studied quite generally by Ortega and Rheinboldt [14]. Roughly speaking, their conclusion is that for most problems met in practice, the order of linearization and discretization is immaterial.

5. Initial trial solution. For the linearized set of algebraic equations, whether obtained by the Newton-Raphson method or any other linearization technique, an initial trial solution $\mathbf{y}^{(0)} = \{y_1^{(0)}, y_2^{(0)}, \ldots, y_N^{(0)}\}$ must be chosen. The choice of $\mathbf{y}^{(0)}$ is critical to the success of the method. The problem of finding a $\mathbf{y}^{(0)}$ which will lead to a solution is an open one. There are a number of different schemes for obtaining the initial trial solution, but none is guaranteed to be successful. One possibility is to guess the set of missing initial conditions and integrate the nonlinear differential equations forward. To carry this thought one step further, we may try one of the shooting methods. Another possibility is to replace the original nonlinear differential equations with a simpler set of equations and try for an analytical solution or employ a shooting method.

6. Example. To illustrate the setting up of the finite difference equations

for a set of nonlinear differential equations, consider the following example, due to Holt [13]:

$$\Psi''' + \frac{3-n}{2}\Psi\Psi'' + n(\Psi')^2 - 1 + G^2 - s\Psi' = 0, \tag{8.4.2}$$

$$G'' + \frac{3-n}{2}\Psi G' + (n-1)\Psi' G - s(G-1) = 0, \tag{8.4.3}$$

with the boundary conditions

$$\Psi(0) = \Psi'(0) = G(0) = 0, \qquad \Psi'(\infty) = 0, \qquad G(\infty) = 1, \tag{8.4.4}$$

where n and s are specified constants. Holt gives as the linearized finite difference equations

$$\left[-4\left(1+\frac{sh^2}{3}\right)\right]\Psi_1 + \left[\left(1+\frac{sh^2}{12}\right)\right]\Psi_2 = \left[\frac{2}{3}h^3\right], \tag{8.4.5}$$

$$\begin{aligned}
&[-1]\Psi_{i-2} + [2+(3-n)h\bar{\Psi}_i - nh(\bar{\Psi}_{i+1}-\bar{\Psi}_{i-1}) + sh^2]\Psi_{i-1} \\
&\quad + [(3-n)h(\bar{\Psi}_{i+1}-2\bar{\Psi}_i+\bar{\Psi}_{i-1}) - 2(3-n)h\bar{\Psi}_i]\Psi_i \\
&\quad + [4h^3\bar{G}_i]G_i + [-2+(3-n)h\bar{\Psi}_i + nh(\bar{\Psi}_{i+1}-\bar{\Psi}_{i-1}) - sh^2]\Psi_{i+1} \\
&\quad + [1]\Psi_{i+2} = [(3-n)h\bar{\Psi}_i(\bar{\Psi}_{i+1}-2\bar{\Psi}_i+\bar{\Psi}_{i-1}) \\
&\qquad\qquad + \frac{nh}{2}(\bar{\Psi}_{i+1}-\bar{\Psi}_{i-1})^2 + 2h^3(1+\bar{G}_i^{\,2})],
\end{aligned} \tag{8.4.6}$$

$$\begin{aligned}
&\left[-\frac{n-1}{2}h\bar{G}_i\right]\Psi_{i-1} + \left[1 - \frac{(3-n)}{4}h\bar{\Psi}_i\right]G_{i-1} \\
&\quad + \left[\frac{3-n}{4}h(\bar{G}_{i+1}-\bar{G}_{i-1})\right]\Psi_i \\
&\quad + \left[-2 + \frac{n-1}{2}h(\bar{\Psi}_{i+1}-\bar{\Psi}_{i-1}) - sh^2\right]G_i \\
&\quad + \left[\frac{n-1}{2}h\bar{G}_i\right]\Psi_{i+1} + \left[1 + \frac{3-n}{4}h\bar{\Psi}_i\right]G_{i+1} \\
&\quad = \left[\frac{3-n}{4}h\bar{\Psi}_i(\bar{G}_{i+1}-\bar{G}_{i-1}) + \frac{n-1}{2}h\bar{G}_i(\bar{\Psi}_{i+1}-\bar{\Psi}_{i-1}) - sh^2\right],
\end{aligned} \tag{8.4.7}$$

with the boundary conditions

$$\Psi_0 = G_0 = 0, \quad G_N = 1, \quad \Psi_{-1} = \Psi_1 - \frac{h^3}{3}, \quad \Psi_{N+1} = \Psi_{N-1}. \tag{8.4.8}$$

In these equations $\bar{\Psi}_i$ and $\bar{G}_i$ are the trial variables. Equations (8.4.6) and (8.4.7) for $i = 1, 2, \ldots, N-1$ constitute a set of $2N-2$ equations in the $2N-1$ variables $\Psi_1, \Psi_2, \ldots, \Psi_N$ and $G_1, G_2, \ldots, G_{N-1}$. The additional equation required to produce a system of $(2N-1)$ equations in $2N-1$ variables was obtained by differentiating (8.4.2) and evaluating it at $t = 0$. Equation (8.4.5) is the finite difference approximation to the additional equation. Holt reports that the additional equation was originally evaluated at $t = \infty$, but this gave rise to oscillations in the solution of the linearized equations.

If (8.4.5), (8.4.6), and (8.4.7) are written with the coefficients indicated by parentheses, we can easily determine the structure of the equations:

$$(\ )\Psi_1 + (\ )\Psi_2 = R_1, \tag{8.4.5a}$$

$$(\ )\Psi_{i-2} + (\ )\Psi_{i-1} + (\ )\Psi_i + (\ )G_i + (\ )\Psi_{i+1} + (\ )\Psi_{i+2} = R_i, \tag{8.4.6a}$$

$$(\ )\Psi_{i-1} + (\ )G_{i-1} + (\ )\Psi_i + (\ )G_i + (\ )\Psi_{i+1} + (\ )G_{i+1} = r_i, \tag{8.4.7a}$$

where R_1, R_i, r_i are the right-hand sides of (8.4.5), (8.4.6), (8.4.7), respectively. Table 8.1 exhibits the banded structure of the matrix of equations (8.4.5), (8.4.6), (8.4.7). In contrast to our second-order example, it is not tridiagonal.

For boundary value problems, extra care is needed in setting up the equations at the initial and terminal conditions. For example, in (8.4.6), for $i = 1$, the variable Ψ_{-1} is replaced by $\Psi_1 - h^3/3$, which comes from (8.4.8). On collecting terms in Ψ_1 in (8.4.6), a 1 is subtracted from the coefficient of Ψ_1 and $-(h^3/3)$ is added to the right-hand side of (8.4.6). For $i = N-1$ in the $(2N-2)$nd equation for (8.4.6), a 1 is added to the coefficient of ψ_{N-1}, again by virtue of (8.4.8). In addition, for $i = N-1$ for (8.4.7), G_N is brought over to the right-hand side by virtue of the boundary conditions (8.4.8).

The initial trial solutions were obtained by taking

$$\bar{G}_i = 1 - e_i^{-t_i}, \qquad \bar{\Psi}_i = a\bar{G}_i$$

for $a = -0.25$, $n = s = 0.5$, and also for $a = -4$, $n = 1$, $s = 10$.

The solution of this problem has been presented by Holt over a range of n and s for $t(\infty)$ ranging from 12 to 200. The problem is stable for only certain combinations of n and s. In addition, as $n \rightarrow 1$ and $s \rightarrow 0$, the solution

Table 8.1

Index i	*Eq. no.*	Ψ_1	Ψ_2	G_1	Ψ_3	G_2	Ψ_4	G_3	Ψ_5	G_4	Ψ_6	G_5	...	Ψ_{N-3}	G_{N-2}	Ψ_{N-2}	G_{N-1}	Ψ_{N-1}	Ψ_N
1	(8.4.5)	×	×																
	(8.4.6)	×	×	×	×														
	(8.4.7)	×	×	×		×													
2	(8.4.6)	×	×		×	×	×												
	(8.4.7)	×	×	×	×	×		×											
3	(8.4.6)	×	×		×		×	×	×										
	(8.4.7)		×		×	×	×	×		×									
4	(8.4.6)		×		×		×		×	×	×								
	(8.4.7)				×		×	×	×	×		×							
$N-1$	(8.4.6)													×		×	×	×	×
	(8.4.7)														×	×	×	×	×

exhibits extreme changes in its oscillatory behavior. In Sections 7.4 and 7.9, Chapter 7, this problem has been solved by shooting methods despite its numerical sensitivity to small changes in the initial conditions. For $n = -0.1$ and $s = 0.2$, the shooting method with continuation did not produce a solution beyond $t(\infty) = 11.3$, while quasilinearization with continuation did not produce a solution beyond $t(\infty) = 13.3$. In contrast, Holt, by solving the finite difference approximation equations to the quasilinear equations, obtained a solution for $t(\infty) = 20.0$.

8.5. SOME MATRIX THEORY

Once the set of finite difference equations is put into matrix form (assuming that the equations are linear to begin with, or have been linearized), the existence of a solution to the linear algebraic equations, and the convergence properties of the method of solution, if an iterative method is employed, can be determined from the properties of the matrix of coefficients. The area of matrix theory applicable to this kind of problem is by now highly developed. The reader is referred to Varga [10] as the standard text in the field. In this section we state only two of the most important theorems to show how existence and convergence can be inferred from the matrix of coefficients. In Section 8.9, in connection with an interesting method of solution of initial value problems by boundary value techniques (in contrast to the main emphasis of this book, which is the solution of boundary value problems by initial value techniques) we give the results of an application of matrix theory to a specific problem.

Two concepts which turn out to be of great importance are "diagonally dominant" and "irreducible"; their definitions follow.

An $n \times n$ matrix $\mathbf{A}$ is said to be *reducible* if there exists an $n \times n$ permutation matrix $\mathbf{P}$ such that

$$\mathbf{PAP}^T = \begin{pmatrix} \mathbf{A}_{11} & \mathbf{A}_{12} \\ \mathbf{O} & \mathbf{A}_{22} \end{pmatrix},$$

where $\mathbf{A}_{11}$ is an $r \times r$ matrix, $\mathbf{A}_{22}$ is an $(n-r)\times(n-r)$ matrix, $\mathbf{A}_{12}$ is an $r\times(n-r)$ matrix, and $\mathbf{O}$ is an $(n-r)\times r$ matrix, all of whose elements are zero. (Recall that a permutation matrix has in each row and column exactly one element equal to unity, the remaining elements being zero. Premultiplication by a permutation matrix effects row interchanges, postmultiplication effects column interchanges.)

A matrix which is not reducible is said to be *irreducible*.

It might be pointed out that in practice the decision whether or not a given matrix is reducible is often not made by determining **P**, but by a study of the *graph* of the matrix.

An $n \times n$ matrix **A** is said to be *diagonally dominant* if

$$|a_{i,i}| \geqq \sum_{j=1, j \neq i}^{n} |a_{i,j}|, \quad i = 1, 2, \ldots, n. \tag{8.5.1}$$

If strict inequality holds in (8.5.1), then **A** is said to be *strictly diagonally dominant.* If **A** is irreducible and if (8.5.1) holds with strict inequality for at least one value of i, **A** is said to be *irreducibly diagonally dominant.*

The principal result with regard to the existence of the inverse of **A** is the following.

THEOREM 8.1. *Let* **A** *be an* $n \times n$ *matrix which is either strictly diagonally dominant or irreducibly diagonally dominant. Then* **A** *is nonsingular, that is, its inverse* $\mathbf{A}^{-1}$ *exists.*

As an application of this theorem, consider the matrix in (8.2.8). Here $a_{ii} = 2 + h^2$ and, except for $i = 1$ and $i = N$, $a_{i,i-1} = a_{i,i+1} = -1$, all other a_{ij} are zero. Of course, $a_{12} = -1$ while $a_{1j} = 0, j > 2$, and $a_{N,N-1} = -1$ while $a_{N,j} = 0$, $j < N-1$. Thus, for $i \neq 1$ or N, $\sum_{j=1,j\neq i}^{n} |a_{ij}| = 2$ while, for $i = 1$ or N, $\sum_{j=1,j\neq i}^{n} |a_{ij}| = 1$. Therefore, for any $h > 0$ and all i,

$$2 + h^2 = |a_{ii}| > \sum_{j=1, j\neq i}^{n} |a_{ij}|.$$

It follows that the matrix in (8.2.8) is strictly diagonally dominant, and therefore has an inverse (that is, Eqs. (8.2.8) have a solution) for any $h > 0$ and any N.

For N relatively small, Eq. (8.2.8) will in practice usually be solved by a variant of Gaussian elimination while, for N relatively large, (8.2.8) will usually be solved by an iterative method, often either the Jacobi or the Gauss-Seidel method. The two methods for the solution of the linear equations $\mathbf{Ay} = \mathbf{b}$ have in common that **A** is decomposed into the sum of matrices:

$$\mathbf{A} = \mathbf{D} - \mathbf{L} - \mathbf{U},$$

where **D** is the diagonal matrix whose elements along the main diagonal are a_{ii}, and **U** and **L** are upper and lower triangular matrices, respectively, whose nonzero elements are the negatives of the corresponding elements of **A**.

Then the Jacobi iterative method for the solution of $\mathbf{Ay}=\mathbf{b}$ can be expressed in the form

$$\mathbf{y}^{(m+1)} = \mathbf{D}^{-1}(\mathbf{L}+\mathbf{U})\mathbf{y}^{(m)}+\mathbf{D}^{-1}\mathbf{b},$$

and the Gauss-Seidel iterative method can be expressed as

$$\mathbf{y}^{(m+1)} = (\mathbf{D}-\mathbf{L})^{-1}\mathbf{U}\mathbf{y}^{(m)}+(\mathbf{D}-\mathbf{L})^{-1}\mathbf{b}.$$

In the Jacobi method all the elements of $\mathbf{y}^{(m)}$ are computed on the mth iteration before the elements of $\mathbf{y}^{(m+1)}$ are computed. In the Gauss-Seidel method, on the other hand, when $y_1^{(m)}$ has been found, it is used to compute $y_2^{(m)}$; $y_1^{(m)}, y_2^{(m)}$ are used to compute $y_3^{(m)}$; and so on. In each method an initial vector $\mathbf{y}^{(0)}$ must be available to start the iteration.

The principal question is then whether these iterative methods converge to the solution $\mathbf{y}$ of $\mathbf{Ay} = \mathbf{b}$, given some initial vector $\mathbf{y}^{(0)}$. Again diagonal dominance is the key, as indicated by the following.

THEOREM 8.2. *Let* $\mathbf{A}$ *be strictly or irreducibly diagonally dominant. Then both the Jacobi and the Gauss-Seidel methods are convergent for any initial vector* $\mathbf{y}^{(0)}$.

Thus, application of either of the two methods to the solution of (8.2.8) would eventually lead to a solution of the set of equations for any initial guess.

Further developments of the subject are concerned with the *rate* of convergence of the various methods. A proper discussion of this topic can become rather complicated, but as a rule of thumb for the practical computer it can be said that the Gauss-Seidel method is often "faster" than the Jacobi method.

8.6. NUMERICAL EXAMPLES

To illustrate finite difference methods, and to compare them with previous methods, two numerical examples are given. In the first example we compare finite difference methods with shooting methods and the exact solution. Good agreement is obtained here.

In Example 2, the linear two-point boundary value problem is characterized by a boundary condition at infinity. For a short range [0, 5] the shooting method gave results comparable to the finite difference methods, while beyond that the shooting method failed. This failure exemplifies the need to supplement shooting methods with other schemes.

Example 1 (The two-point boundary value problem (8.2.1), (8.2.2) of Section 8.2). This problem has the exact answer

$$y(t) = \frac{2 \sinh t}{\sinh 1} - t.$$

In Tables 8.2A and 8.2B we tabulate the exact answer, the solution of (8.2.8), and the solution by the method of adjoints for $h = 0.25$ and 0.10. In Table 8.2C the solutions, by the method of adjoints for $h = 0.05$, 0.02, and 0.01 are also given. Relatively good agreement is obtained among all the methods. Both Varga [10] and Fox [3] discuss this example.

Table 8.2A [a]

t	$y(t)$ (*exact*)	$y(t)$ (*finite diff.*) $h = 0.25$	$y(t)$ (*adjoints*) $h = 0.25$
0.00	0.0	0.0	0.0
0.25	1.7990479 (10^{-1})	1.8022950 (10^{-1})	1.7991958 (10^{-1})
0.50	3.8681887 (10^{-1})	3.8734835 (10^{-1})	3.8684909 (10^{-1})
0.75	6.4944842 (10^{-1})	6.4992647 (10^{-1})	6.4949388 (10^{-1})
1.00	1.0000000	1.0000000	1.0000000

Table 8.2B [a]

t	$y(t)$ (*exact*)	$y(t)$ (*finite diff.*) $h = 0.10$	$y(t)$ (*adjoints*) $h = 0.10$
0.00	0.0	0.0	0.0
0.10	7.0467410 (10^{-2})	7.0489380 (10^{-2})	7.0467564 (10^{-2})
0.20	1.4264090 (10^{-1})	1.4268364 (10^{-1})	1.4264122 (10^{-1})
0.30	2.1824367 (10^{-1})	2.1830475 (10^{-1})	2.1824415 (10^{-1})
0.40	2.9903319 (10^{-1})	2.9910891 (10^{-1})	2.9903383 (10^{-1})
0.50	3.8681887 (10^{-1})	3.8690415 (10^{-1})	3.8681967 (10^{-1})
0.60	4.8348014 (10^{-1})	4.8356844 (10^{-1})	4.8348108 (10^{-1})
0.70	5.9098524 (10^{-1})	5.9106841 (10^{-1})	5.9098632 (10^{-1})
0.80	7.1141095 (10^{-1})	7.1147906 (10^{-1})	7.1141217 (10^{-1})
0.90	8.4696337 (10^{-1})	8.4700451 (10^{-1})	8.4696472 (10^{-1})
1.00	1.0000000	1.0000000	1.0000014

[a] Exact and finite difference solutions taken from Varga [10, pp. 299–301].

Table 8.2C [a]

t	$y(t)$ (*adjoints*) $h = 0.05$	$y(t)$ (*adjoints*) $h = 0.02$	$y(t)$ (*adjoints*) $h = 0.01$
0.00	0.0	0.0	0.0
0.10	7.0467416(10⁻²)	7.0467407(10⁻²)	7.0467406(10⁻²)
0.20	1.4264092(10⁻¹)	1.4264090(10⁻¹)	1.4264090(10⁻¹)
0.30	2.1824370(10⁻¹)	2.1824367(10⁻¹)	2.1824367(10⁻¹)
0.40	2.9903324(10⁻¹)	2.9903320(10⁻¹)	2.9903320(10⁻¹)
0.50	3.8681893(10⁻¹)	3.8681888(10⁻¹)	3.8681888(10⁻¹)
0.60	4.8348020(10⁻¹)	4.8348015(10⁻¹)	4.8348014(10⁻¹)
0.70	5.9098531(10⁻¹)	5.9098524(10⁻¹)	5.9098524(10⁻¹)
0.80	7.1141103(10⁻¹)	7.1141096(10⁻¹)	7.1141096(10⁻¹)
0.90	8.4696346(10⁻¹)	8.4696338(10⁻¹)	8.4696338(10⁻¹)
1.00	1.0000000	1.0000000	1.0000000

[a] Exact and finite difference solutions taken from Varga [10, pp. 299–301].

Example 2. $\ddot{y}+2t\dot{y}-2ny = 0, \quad y(0) = \beta, \quad y(\infty) = 0,$

where n and β are specified constants. As a scaling device we make the change of variables

$$y = ve^{-t^2/2}$$

and, on eliminating y, obtain

$$\ddot{v}-(2n+1+t^2)v = 0, \qquad v(0) = \beta, \qquad v(\infty) = 0.$$

While this problem is a linear problem it is interesting because of the boundary condition at infinity. A finite difference approximation is given by

$$[1-\tfrac{1}{12}h^2(2n+1+t_{i-1}^2)]v_{i-1}-[2+\tfrac{5}{6}h^2(2n+1+t_i^2)]v_i$$
$$+[1-\tfrac{1}{12}h^2(2n+1+t_{i+1}^2)]v_{i+1} = 0, \qquad i = 1,2,\ldots,N-1,$$
$$v_0 = \beta, \qquad v_N = 0.$$

To solve this problem we must approximate the infinite interval by specifying a final value for t. This set of equations is a tridiagonal diagonally dominant matrix, so we would anticipate a solution. In Tables 8.3A, 8.3B, and 8.3C are listed the results of Holt [13] for several combinations of β, n, N. As a comparison, solutions by the method of complementary functions are also given. While Holt claimed that his use of shooting methods gave poor answers beyond $t = 3.5$, our results indicate that the method of complementary functions give satisfactory results over a little wider range.

Table 8.3A [a]

t	$y(t)$ (*complementary functions*) $t(\infty) = 8.0$ $\beta = 1/\sqrt{\pi}$ $n = 1.0$ $h = 0.005$	$y(t)$ (*finite diff.*) $t(\infty) = 10.0$ $\beta = 1/\sqrt{\pi}$ $n = 1.0$ $h = 0.005$
0.00	$5.6418960(10^{-1})$	$5.642(10^{-1})$
0.01	$5.5424601(10^{-1})$	$5.542(10^{-1})$
0.10	$4.6982210(10^{-1})$	$4.698(10^{-1})$
0.50	$1.9964123(10^{-1})$	$1.996(10^{-1})$
1.00	$5.0254543(10^{-2})$	$5.026(10^{-2})$
1.50	$8.6228645(10^{-3})$	$8.624(10^{-3})$
2.00	$9.7802274(10^{-4})$	$9.782(10^{-4})$
2.50	$7.1762073(10^{-5})$	$7.177(10^{-5})$
3.00	$3.3550350(10^{-6})$	$3.356(10^{-6})$
3.50	$9.8869071(10^{-8})$	—
4.00	$1.8221222(10^{-9})$	$1.823(10^{-9})$
4.50	$2.0858736(10^{-11})$	—
5.00	$1.2367523(10^{-13})$	$1.482(10^{-13})$
5.50	$-2.6256552(10^{-14})$	—
6.00	$-2.9349128(10^{-14})$	$1.747(10^{-18})$
7.00	$-3.4242684(10^{-14})$	$2.931(10^{-24})$
8.00	$-3.9134491(10^{-14})$	$6.912(10^{-31})$

[a] Finite difference solutions taken from Holt [13].

For large values of t, the method of complementary functions gives absurd answers. Clearly, here is an example where finite difference approximation is the superior method.

8.7. PARALLEL OR MULTIPLE SHOOTING

We have seen that finite difference methods may be successful in sensitive or unstable problems where shooting methods fail. An interesting method proposed by Morrison *et al.* [15] and studied more thoroughly by Keller [16] is *multiple* or *parallel shooting* which has proved to be effective in the numerical solution of sensitive problems. Multiple shooting may be thought of as a compromise between finite difference methods and shooting, in which the interval $[a, b]$ over which the solution to the two-point boundary value problem is sought, is divided into a number of subintervals $[t_j, t_{j+1}]$, as in

Table 8.3B [a]

t	$y(t)$ (complementary functions) $t(\infty) = 10.0$ $\beta = 0.25$ $n = 2.0$ $h = 0.005$	$y(t)$ (finite diff.) $t(\infty) = 10.0$ $\beta = 0.25$ $n = 2.0$ $h = 0.005$
0.00	2.5000000 (10^{-1})	2.500 (10^{-1})
0.01	2.4440791 (10^{-1})	2.444 (10^{-1})
0.10	1.9839316 (10^{-1})	1.984 (10^{-1})
0.50	6.9964723 (10^{-2})	6.997 (10^{-2})
1.00	1.4197530 (10^{-2})	1.420 (10^{-2})
1.50	2.0065651 (10^{-3})	2.007 (10^{-3})
2.00	1.9141103 (10^{-4})	1.914 (10^{-4})
2.50	1.2035414 (10^{-5})	1.204 (10^{-5})
3.00	4.9007176 (10^{-7})	4.901 (10^{-7})
4.00	2.0999802 (10^{-9})	2.101 (10^{-9})
5.00	−3.6865462 (10^{-14})	1.403 (10^{-14})
6.00	−7.2849101 (10^{-14})	1.400 (10^{-19})
7.00	−9.8795539 (10^{-14})	2.034 (10^{-25})
8.00	−1.2873356 (10^{-13})	4.224 (10^{-32})

[a] Finite difference solution taken from Holt [13].

finite difference methods, over each of which a shooting method is applied. However, in contrast to pure finite difference methods, the number of subintervals is usually relatively small. In our description of multiple shooting we follow Keller [16].

Consider the nonlinear two-point boundary value problem on the interval $[t_0, t_f]$:

$$\dot{\mathbf{y}} = \mathbf{g}(\mathbf{y}, t), \qquad \mathbf{B}_0\mathbf{y}(t_0)+\mathbf{B}_f\mathbf{y}(t_f) = \mathbf{c}, \tag{8.7.1}$$

where $\mathbf{y}$, $\mathbf{g}$, and $\mathbf{c}$ are $n\times 1$ vectors, and $\mathbf{B}_0$ and $\mathbf{B}_f$ are $n\times n$ matrices. Divide $[t_0, t_f]$ into N subintervals by the points t_j, where $t_{N+1} = t_f$, and on each subinterval introduce the new independent variable $\tau = (t-t_{j-1})/(t_j-t_{j-1})$, the new dependent variable n-vector $\mathbf{y}_j(\tau) = \mathbf{y}(t_{j-1}+(t_j-t_{j-1})\tau)$, and the new n-vector $\mathbf{g}_j(\mathbf{y}_j, \tau) = (t_j-t_{j-1})\mathbf{g}(\mathbf{y}_j, t_{j-1}+(t_j-t_{j-1})\tau)$. Since $d/d\tau = (t_j-t_{j-1})\, d/dt$, the problem stated in (8.7.1) may be written

$$\begin{aligned} \frac{d\mathbf{y}_j}{d\tau} &= \mathbf{g}_j(\mathbf{y}_j(\tau), \tau), \qquad j = 1, \ldots, N, \\ \mathbf{B}_0\mathbf{y}_1(0)+\mathbf{B}_f\mathbf{y}_N(1) &= \mathbf{c}, \end{aligned} \tag{8.7.2}$$

Table 8.3C [a]

t	$y(t)$ (complementary functions) $t(\infty) = 10.0$ $\beta = 1.0$ $n = 0$ $h = 0.005$	$y(t)$ (finite diff.) $t(\infty) = 14.0$ $\beta = 1.0$ $n = 0$ $h = 0.02$
0.00	1.0000000	1.00000
0.02	$9.7743542(10^{-1})$	$9.77435(10^{-1})$
0.10	$8.8753708(10^{-1})$	$8.87538(10^{-1})$
0.50	$4.7950012(10^{-1})$	$4.79501(10^{-1})$
1.00	$1.5729920(10^{-1})$	$1.57300(10^{-1})$
1.50	$3.3894853(10^{-2})$	$3.38951(10^{-2})$
2.00	$4.6777349(10^{-3})$	$4.67778(10^{-3})$
2.50	$4.0695201(10^{-4})$	$4.06957(10^{-4})$
3.00	$2.2090497(10^{-5})$	$2.20908(10^{-5})$
3.50	$7.4309837(10^{-7})$	$7.43109(10^{-7})$
4.00	$1.5417257(10^{-8})$	$1.54175(10^{-8})$
4.50	$1.9661529(10^{-10})$	$1.96619(10^{-10})$
5.00	$1.5366706(10^{-12})$	$1.53749(10^{-12})$
5.50	$6.6046927(10^{-15})$	$7.35798(10^{-15})$
6.00	$-7.3163560(10^{-16})$	$2.15201(10^{-17})$
6.50	$-7.5311690(10^{-16})$	$3.84222(10^{-20})$
7.00	$-7.5311525(10^{-16})$	$4.18390(10^{-23})$
8.00	$-7.5315520(10^{-16})$	$1.12244(10^{-29})$

[a] Finite difference solutions taken from Holt [13].

where $0 \leq \tau \leq 1$. But notice that, where originally n functions $y_i(t)$, the components of $\mathbf{y}(t)$, were to be found, now Nn functions $y_{ji}(\tau)$, $j = 1, \ldots, N$, $i = 1, 2, \ldots, n$, the components of $\mathbf{y}_j(\tau)$, are to be computed, so the dimensionality of the problem has been multiplied by N. Moreover, since the boundary conditions in (8.7.2) impose only n conditions on the $y_{ji}(\tau)$, $Nn - n = (N-1)n$ additional conditions are needed, and they are furnished by the continuity conditions $\mathbf{y}_{j+1}(0) = \mathbf{y}_j(1)$, or

$$\mathbf{y}_{j+1}(0) - \mathbf{y}_j(1) = 0, \qquad j = 1, 2, \ldots, N-1. \tag{8.7.2'}$$

Equation (8.7.2′) merely expresses the requirement that the representation of the solution $\mathbf{y}(t)$ on the subintervals $[t_{j-1}, t_j]$ and $[t_j, t_{j+1}]$ agree at their common point, t_j.

In summary, if Nn vectors $\mathbf{Y} = (\mathbf{y}_j)$, $\mathbf{G}(\mathbf{Y}, \tau) = (\mathbf{g}_j(\mathbf{y}_j, \tau))$, and $\boldsymbol{\gamma} = (\mathbf{c}, 0, \ldots, 0)^T$ are introduced, $j = 1, 2, \ldots N$, along with the two $Nn \times Nn$ matrices

$$\mathbf{U} = \begin{bmatrix} \mathbf{B}_0 & \mathbf{O} & \mathbf{O} & \ldots & \mathbf{O} \\ \mathbf{O} & \mathbf{I} & \mathbf{O} & \ldots & \mathbf{O} \\ \mathbf{O} & \mathbf{O} & \mathbf{I} & \ldots & \mathbf{O} \\ \cdots & \cdots & \cdots & \cdots & \cdots \\ \mathbf{O} & \mathbf{O} & \mathbf{O} & \ldots & \mathbf{I} \end{bmatrix}, \qquad \mathbf{V} = \begin{bmatrix} \mathbf{O} & \mathbf{O} & \mathbf{O} & \ldots & \mathbf{B}_f \\ -\mathbf{I} & \mathbf{O} & \mathbf{O} & \ldots & \mathbf{O} \\ \mathbf{O} & -\mathbf{I} & \mathbf{O} & \ldots & \mathbf{O} \\ \cdots & \cdots & \cdots & \cdots & \cdots \\ \mathbf{O} & \mathbf{O} & \mathbf{O} & -\mathbf{I} & \mathbf{O} \end{bmatrix},$$

where all the displayed entries in $\mathbf{U}$ and $\mathbf{V}$ are themselves $n \times n$ matrices, then problem (8.7.1) has been transformed into the two-point boundary value problem

$$\begin{aligned} \frac{d\mathbf{Y}}{d\tau} &= \mathbf{G}(\mathbf{Y}, \tau), \qquad 0 \le \tau \le 1, \\ \mathbf{U}\mathbf{Y}(0) + \mathbf{V}\mathbf{Y}(1) &= \boldsymbol{\gamma}. \end{aligned} \tag{8.7.3}$$

Problem (8.7.3) can be solved by one of the shooting methods described previously (Chapters 3 and 4), and it can be shown that, under reasonable conditions on $\mathbf{g}(t, \mathbf{y})$, problems (8.7.1) and (8.7.3) are equivalent.

It should be pointed out that, because of the special form of the vectors $\mathbf{Y}(\tau)$ and $\mathbf{G}(\mathbf{Y}, \tau)$, the system of Nn initial value problems corresponding to (8.7.3), as well as the associated system of Nn variational equations

$$\frac{d\mathbf{X}}{d\tau} = \frac{\partial \mathbf{G}(\mathbf{Y}, \tau)}{\partial \mathbf{Y}} \mathbf{X},$$

which must be solved in applying one of the shooting methods, actually consists of N *independent* systems of n differential equations. The only coupling of the N systems comes about in solving the linear algebraic equations for the missing initial conditions in (8.7.3). Thus a code for solving initial value problems for systems of n differential equations as in (8.7.1) could be used serially N times in the solution of (8.7.3). Keller [16] points out that the set of N systems of initial value problems are "ideally suited for computation on *parallel computers* (which are at present being designed)".

The power of multiple shooting in dealing with sensitive problems is explained by Keller as follows. In solving the initial value problems associated with (8.7.1) and (8.7.3), respectively, by the same integration scheme of order p and step size h, the errors in the numerical solution can be bounded by

$$|\mathbf{y}(t_i)-\mathbf{y}_i| \leqq h^p M_1 \exp[K_1|t_0-t_f|],$$

$$|\mathbf{Y}(\tau_i)-\mathbf{Y}_i| \leqq h^p M_2 \exp[K_2|\tau_i|],$$

where $K_2 \approx K_1|t_f-t_0|/N$. Therefore at $\tau_i = 1$, which corresponds to $t_i = t_f$, the bound on the error in multiple shooting is proportional to (exp $(K_1|t_f-t_0|)^{1/N}$) rather than to $\exp(K_1|t_f-t_0|)$. Thus the growth of solutions for values of the missing initial conditions slightly different from their true values, so characteristic of sensitive problems, can be drastically curtailed if N is taken large enough.

Keller also indicates variations of multiple shooting which are possible by shooting forward on certain subintervals $[t_j, t_{j+1}]$ and backward on other subintervals $[t_k, t_{k+1}]$. This can be done to reduce the effective order of the system (8.7.3) or to take advantage of knowledge of the growth or decay of the solutions. In particular, when $N = 2$ and shooting is forward from the left end point and backward from the right end point, then multiple shooting reduces to an old procedure known as "matching in the middle".

M.R. Osborne [17] has employed a combination of parallel shooting and continuation (Chapter 7) to solve numerically sensitive two-point boundary value problems.

8.8. QUASILINEARIZATION AND FINITE DIFFERENCE EQUATIONS

In Section 8.7 we discussed multiple shooting, which combines the features of shooting and finite difference methods. Another method which lies between shooting and finite difference has been suggested by Sylvester and Meyer [18] in connection with quasilinearization. Recall that in quasilinearization (discussed in detail in Chapter 5) the original nonlinear two-point boundary value is replaced by a sequence of linear two-point boundary value problems, each of which may be solved by the method of adjoints or by the method of complementary functions, that is, by a shooting method.

Sylvester and Meyer have suggested solving by finite difference methods the error equations associated with the quasilinear process. Suppose we are given the two-point boundary value problem

$$\dot{y}_i = g_i(y_1, y_2, \ldots, y_n, t), \qquad i = 1, 2, \ldots, n, \tag{8.8.1}$$

with the linear boundary conditions

$$\mathbf{B}_0\, \mathbf{y}(t_0) + \mathbf{B}_f\, \mathbf{y}(t_f) = \mathbf{c}, \tag{8.8.2}$$

where $\mathbf{B}_0, \mathbf{B}_f$ are $n \times n$ constant coefficient matrices associated with the initial and terminal conditions, and $\mathbf{c}$ is an $n \times 1$ vector of constants.

The quasilinear equations may be written

$$\dot{y}_i^{(k+1)} = g_i(y_1^{(k)}, y_2^{(k)}, \ldots, y_n^{(k)}, t) + \sum_{j=1}^{n} \frac{\partial g_i}{\partial y_j}(y_j^{(k+1)} - y_j^{(k)}),$$

$$i = 1, 2, \ldots, n, \qquad (8.8.3)$$

where $\partial g_i/\partial y_j$ are evaluated at $(y_1^{(k)}, y_2^{(k)}, \ldots, y_n^{(k)}, t)$. If we subtract $\dot{y}_i^{(k)}$ from both sides of (8.8.3), we obtain the error expression

$$\dot{\varepsilon}_i^{(k)} - \sum_{j=1}^{n} \frac{\partial g_i}{\partial y_j} \varepsilon_j^{(k)} = g_i(y_1^{(k)}, y_2^{(k)}, \ldots, y_n^{(k)}, t) - \dot{y}_i^{(k)}, \quad i = 1, 2, \ldots, n, \quad (8.8.4)$$

where

$$\varepsilon_i^{(k)}(t) = y_i^{(k+1)}(t) - y_i^{(k)}(t), \qquad i = 1, 2, \ldots, n, \qquad (8.8.5a)$$

$$\dot{\varepsilon}_i^{(k)}(t) = \dot{y}_i^{(k+1)}(t) - \dot{y}_i^{(k)}(t), \qquad i = 1, 2, \ldots, n. \qquad (8.8.5b)$$

In matrix form, (8.8.4) is written

$$\dot{\boldsymbol{\varepsilon}} - \mathbf{A}(\mathbf{y}, t)\boldsymbol{\varepsilon} = \mathbf{g}(\mathbf{y}, t) - \dot{\mathbf{y}}, \qquad (8.8.6)$$

where $\mathbf{A}$ is an $n \times n$ matrix whose i,j element is $\partial g_i/\partial y_j$. The boundary conditions are

$$\mathbf{B}_0\boldsymbol{\varepsilon}(t_0) + \mathbf{B}_f\boldsymbol{\varepsilon}(t_f) = \mathbf{0}. \qquad (8.8.7)$$

The superscript for the iteration count has been suppressed because it appears in every term and is identical in all terms.

Suppose that the interval $[t_0, t_f]$ is divided into m subintervals. Then the matrix of equations must be written for each discrete point in the interval $[t_0, t_f]$, and a central difference finite approximation to (8.8.6) may be written

$$\frac{\boldsymbol{\varepsilon}_{i+1} - \boldsymbol{\varepsilon}_i}{h_i} - \mathbf{A}(\mathbf{w}_i, \bar{t}_i)\frac{\boldsymbol{\varepsilon}_i + \boldsymbol{\varepsilon}_{i+1}}{2} = -\left[\frac{\mathbf{y}_{i+1} - \mathbf{y}_i}{h_i} - \mathbf{g}(\mathbf{w}_i, \bar{t}_i)\right],$$

$$i = 1, 2, \ldots, m+1, \qquad (8.8.8)$$

where

$\boldsymbol{\varepsilon}_i$ = $n \times 1$ vector with components $\varepsilon_1(t_i), \varepsilon_2(t_i), \ldots, \varepsilon_n(t_i)$,
$\mathbf{y}_i$ = $n \times 1$ vector with components $y_1(t_i), y_2(t_i), \ldots, y_n(t_i)$,

$$\begin{aligned} \mathbf{w}_i &= n\times 1 \text{ vector} = \tfrac{1}{2}(\mathbf{y}_i+\mathbf{y}_{i+1}), \\ \bar{t}_i &= \tfrac{1}{2}(t_i+t_{i+1}), \\ h_i &= t_{i+1}-t_i. \end{aligned}$$

The boundary conditions appear as

$$\mathbf{B}_0\boldsymbol{\varepsilon}_0+\mathbf{B}_f\boldsymbol{\varepsilon}_{m+1} = \mathbf{0}, \tag{8.8.9}$$

where $\boldsymbol{\varepsilon}_0$ and $\boldsymbol{\varepsilon}_{m+1}$ are the initial and terminal vectors. Equations (8.8.8) and (8.8.9) consist of $n(m+1)$ equations in the $n(m+1)$ components of $\boldsymbol{\varepsilon}_i$, $i = 1,2,\ldots,m+1$.

Equation (8.8.8) may be expressed more compactly as

$$\boldsymbol{\varepsilon}_i+\mathbf{D}_i\boldsymbol{\varepsilon}_{i+1} = \mathbf{s}_i, \qquad i = 1,2,\ldots,m; \tag{8.8.10}$$

where

$$\begin{aligned} \mathbf{D}_i &= n\times n \text{ matrix} = [\mathbf{I}+\tfrac{1}{2}h_i\mathbf{A}_i]^{-1}[-\mathbf{I}+\tfrac{1}{2}h_i\mathbf{A}_i] = \mathbf{I}-2(\mathbf{I}+\tfrac{1}{2}h_i\mathbf{A}_i)^{-1}, \\ \mathbf{s}_i &= n\times 1 \text{ vector} = [\mathbf{I}+\tfrac{1}{2}h_i\mathbf{A}_i]^{-1}[\mathbf{y}_{i+1}-\mathbf{y}_i-h_i\mathbf{g}_i], \\ \mathbf{A}_i &= n\times n \text{ matrix} = \mathbf{A}(\mathbf{w}_i,\bar{t}_i), \\ \mathbf{g}_i &= n\times 1 \text{ vector} = \mathbf{g}(\mathbf{w}_i,\bar{t}_i); \\ \mathbf{I} &= n\times n \text{ identity matrix}. \end{aligned} \tag{8.8.11}$$

The boundary value problem may be written in partitioned matrix form, of order $n(m+1)$, as

$$\begin{bmatrix} \mathbf{I} & \mathbf{D}_1 & & & \\ & \mathbf{I} & \mathbf{D}_2 & & \\ & & \ddots & \ddots & \\ & & & \mathbf{I} & \mathbf{D}_m \\ \mathbf{B}_0 & & & & \mathbf{B}_f \end{bmatrix} \begin{bmatrix} \boldsymbol{\varepsilon}_1 \\ \boldsymbol{\varepsilon}_2 \\ \vdots \\ \boldsymbol{\varepsilon}_m \\ \boldsymbol{\varepsilon}_{m+1} \end{bmatrix} = \begin{bmatrix} \mathbf{s}_1 \\ \mathbf{s}_2 \\ \vdots \\ \mathbf{s}_m \\ \mathbf{0} \end{bmatrix}. \tag{8.8.12}$$

Provided the inverse exists, the solution vector $\boldsymbol{\varepsilon}_1,\boldsymbol{\varepsilon}_2,\ldots,\boldsymbol{\varepsilon}_{m+1}$ gives new estimates of the solution vector $\mathbf{y}$ at the discrete points. New trial values of $\mathbf{y}_i$ are formed by (8.8.5); that is,

$$\mathbf{y}_i^{(k+1)} = \mathbf{y}_i^{(k)}+\boldsymbol{\varepsilon}_i^{(k)}, \qquad i = 1,2,\ldots,m+1, \tag{8.8.13}$$

where the superscripts are introduced to indicate the iteration number. With the new trial values of $\mathbf{y}_i$, Eq. (8.8.12) is solved again. The process is repeated until the error solution vector from (8.8.12) is sufficiently small.

Rather than take the inverse of the matrix in (8.8.12), Sylvester and Meyer have proposed a simpler procedure. If each side of (8.8.12) is premultiplied by the square partitioned matrix of order $n(m+1)$,

$$\begin{bmatrix} \mathbf{I} & & & & \\ & \mathbf{I} & & & \\ & & \mathbf{I} & & \\ & & & \ddots & \\ \mathbf{F}_1 & \mathbf{F}_2 & \cdots & \mathbf{F}_m & \mathbf{I} \end{bmatrix}$$

where

$$\mathbf{F}_1 = -\mathbf{B}_0,$$

$$\mathbf{F}_i = (-1)^i \mathbf{B}_0 \prod_{j=1}^{i-1} \mathbf{D}_j, \qquad i = 2, 3, \ldots, m,$$

the following partitioned matrix system is obtained:

$$\begin{bmatrix} \mathbf{I} & \mathbf{D}_1 & & & \\ & \mathbf{I} & \mathbf{D}_2 & & \\ & & \ddots & \ddots & \\ & & & \mathbf{I} & \mathbf{D}_m \\ & & & & \mathbf{G} \end{bmatrix} \begin{bmatrix} \boldsymbol{\varepsilon}_1 \\ \boldsymbol{\varepsilon}_2 \\ \vdots \\ \boldsymbol{\varepsilon}_m \\ \boldsymbol{\varepsilon}_{m+1} \end{bmatrix} = \begin{bmatrix} \mathbf{s}_1 \\ \mathbf{s}_2 \\ \vdots \\ \mathbf{s}_m \\ -\mathbf{B}_0\mathbf{p} \end{bmatrix} \qquad (8.8.14)$$

where

$$\mathbf{G} = n \times n \text{ matrix} = (-1)^m \mathbf{B}_0 \left(\prod_{i=1}^{m} \mathbf{D}_i \right) + \mathbf{B}_f,$$

$$\mathbf{p} = \mathbf{s}_1 + \sum_{i=2}^{m} (-1)^{i-1} \left(\prod_{j=1}^{i-1} \mathbf{D}_j \right) \mathbf{s}_i. \qquad (8.8.15)$$

If $\mathbf{G}$ is not singular, (8.8.14) may be solved backward starting with

$$\begin{aligned} \boldsymbol{\varepsilon}_{m+1} &= -\mathbf{G}^{-1}(\mathbf{B}_0\mathbf{p}), \\ \boldsymbol{\varepsilon}_{m+1-i} &= \mathbf{s}_{m+1-i} - \mathbf{D}_{m+1-i}\boldsymbol{\varepsilon}_{m+2-i}, \qquad i = 1, 2, \ldots, m. \end{aligned} \qquad (8.8.16)$$

The success of this method depends crucially on whether $\mathbf{G}$ is well conditioned. If $\mathbf{G}$ is ill-conditioned, the error in $\boldsymbol{\varepsilon}_{m+1}$ will be propagated by the recursion relations (8.8.16) to give meaningless results.

We can establish qualitatively the types of problems which lead to $\mathbf{G}$ being well- or ill-conditioned. From the definition of $\mathbf{D}_i$ it follows that, when the elements of $h_i\mathbf{A}_i$ and h_i are small relative to 1,

$$\prod_{i=1}^{m} \mathbf{D}_i = (-1)^m \left(\mathbf{I} - \sum_{i=1}^{m} h_i\mathbf{A}_i + \mathrm{O}(h^2)\right), \tag{8.8.17}$$

$$\mathbf{G} = \mathbf{B}_0 + \mathbf{B}_f - \sum_{i=1}^{m} h_i\mathbf{B}_0\mathbf{A}_i + \mathrm{O}(h^2), \tag{8.8.18}$$

where $h = \max h_i$. For small h,

$$\det \mathbf{G} \approx \det(\mathbf{B}_0 + \mathbf{B}_f). \tag{8.8.19}$$

It appears therefore that the determinant of the matrix of the boundary conditions is the clue to the nature of the condition of the matrix $\mathbf{G}$. For problems where the $\det(\mathbf{B}_0+\mathbf{B}_f) = 0, h$ must be chosen large enough that the det $\mathbf{G}$ is not small and yet h must be small enough that the finite difference equations are "reasonable" approximations to the differential equations. For problems where $\det(\mathbf{B}_0+\mathbf{B}_f) = 0$ and the elements of $h_i\mathbf{A}_i$ are not small, then the conditioning of $\mathbf{G}$ depends in some complicated way on h_i and the iteration vector $\mathbf{y}^{(k)}$. For problems where $\det(\mathbf{B}_0+\mathbf{B}_f)$ is large relative to zero, $\mathbf{G}$ may or may not be ill conditioned. In this case the influence of the boundary conditions is diminished. Since $\mathbf{G}$ is a product of the $\mathbf{D}_i$ matrices, round-off error in forming $\prod_{i=1}^{m} \mathbf{D}_i$ may create an ill-conditioned matrix $\mathbf{G}$.

Treatment of the quasilinear process equations as finite difference equations has several advantages. As discussed in this chapter, for two-point boundary value problems which are very sensitive to modest changes in the initial conditions, finite difference methods may succeed where shooting methods fail. The error equations of the quasilinear process, when expressed as finite difference equations, also possess this characteristic. The stylized format of (8.8.14), (8.8.15), and (8.8.16) permits a general computer code to be written for all nonlinear problems treated as a quasilinear process problem. Only the partial derivatives $\partial g_i/\partial y_j$ in the matrix of (8.8.16) and $\mathbf{g}_i$ need to be determined for each problem and introduced into the code. The method lends itself to an efficient computer program because only the $\mathbf{D}_i$ matrices and the $\mathbf{y}_i$ and $\mathbf{s}_i$ vectors need to to be stored. If the process converges, it does so at a quadratic rate, which is characteristic of Newton-Raphson-type methods, as discussed in Chapter 6. One important extension of the method is its applicability to the linear multipoint boundary value problem. If, for example, the linear boundary conditions are

$$\mathbf{B}_0\mathbf{y}(t_0)+\mathbf{B}_j\mathbf{y}(t_j)+\mathbf{B}_f\mathbf{y}(t_f) = \mathbf{c},$$

then the matrix $\mathbf{B}_j$ is entered into the last row of the matrix (8.8.12) in the appropriate column. The definitions of $\mathbf{F}_i$ must be altered to accommodate the presence of $\mathbf{B}_j$.

8.9. BOUNDARY VALUE VERSION OF INITIAL VALUE PROBLEM

In this section we digress temporarily to present an interesting application of boundary value techniques to initial value problems. This is of course just the reverse of our main theme, shooting methods, which are applications of initial value techniques to boundary value problems.

In general, boundary value problems are more difficult to solve numerically than initial value problems, and instabilities are harder to overcome. However, there are occasions when it is advantageous to convert an initial value problem into a boundary value problem. On the basis of practical computational experience, Fox and Mitchell [19] concluded that conversion of an initial value problem into a boundary value problem and subsequent solution of the boundary value problem by finite difference methods may give good results for proper choice of the step size h when the original problem is unstable, but may give poor results when the original problem is stable.

In order to put this technique on a firmer theoretical foundation, we present on the basis of the work of Usmani [20], the finite difference approximation formulas to a two-point boundary value problem (derived from an initial value problem) and the conditions under which a solution to the two-point boundary value problem exists.

Consider the first-order system

$$\dot{y} = f(t)y+g(t) \tag{8.9.1}$$

and the initial condition

$$y(t_0) = y_0. \tag{8.9.2}$$

The initial value problem can be converted into a two-point boundary value problem by differentiating (8.9.1) and eliminating the first derivatives:

$$\ddot{y} = (f^2+\dot{f})y+(fg+\dot{g}) = F(t)y+G(t), \tag{8.9.3}$$

where

$$y(t_0) = y_0,$$
$$\dot{y}(t_N) = f(t_N)y_N + g(t_N). \tag{8.9.4}$$

The second boundary condition comes from the fact that (8.9.1) holds for every t.

Suppose (8.9.3) is approximated by the finite difference equation

$$y_n - 2y_{n+1} + y_{n+2} = \frac{h^2}{12}(\ddot{y}_n + 10\ddot{y}_{n+1} + \ddot{y}_{n+2}), \tag{8.9.5}$$

which has a truncation error $-(h^6/240)y^{(6)}$ and $h = (t_N - t_0)/N$. On replacing the second-order derivatives in (8.9.5) by (8.9.3) at the points t_n, t_{n+1}, t_{n+2} we have

$$\left(-1 + \frac{h^2}{12}F_n\right)y_n + \left(2 + \frac{10h^2}{12}F_{n+1}\right)y_{n+1} + \left(-1 + \frac{h^2}{12}F_{n+2}\right)y_{n+2}$$
$$= -\frac{h^2}{12}(G_n + 10G_{n+1} + G_{n+2}), \qquad n = 0, 1, 2, \ldots, N-2. \tag{8.9.6}$$

Equation (8.9.6) represents a set of $(N-1)$ equations in N unknowns, y_i, $i = 1, 2, \ldots, N$. Another equation is needed which does not introduce more variables, and which has a truncation error comparable to (8.9.5).

One class of equations which meets these criteria is

$$\sum_{i=0}^{K} a_i y_{n+i} = h \sum_{i=0}^{K} b_i \dot{y}_{n+i} + h^2 \sum_{i=0}^{K} c_i \ddot{y}_{n+i} + \ldots, \tag{8.9.7}$$

where a_i, b_i, c_i, and K are constants.

A specific equation from this class which will be used here is

$$-y_{N-1} + y_N = \frac{h}{2}(\dot{y}_{N-1} + \dot{y}_N) + \frac{h^2}{10}(\ddot{y}_{N-1} - \ddot{y}_N) + \frac{h^3}{120}(\dddot{y}_{N-1} + \dddot{y}_N) \tag{8.9.8}$$

with a truncation error of $-(h^7 y^{(7)}/100,800)$. This equation introduces only y_N and y_{N-1}, either directly or indirectly through its derivatives. To eliminate the derivatives in (8.9.8) we replace the first derivatives by (8.9.1), the second derivatives by (8.9.3), and the third derivatives by

$$\dddot{y} = R(t)y + H(t), \tag{8.9.9}$$

where

$$R(t) = Ff + \dot{F}, \tag{8.9.10}$$

$$H(t) = Fg + \dot{G}. \tag{8.9.11}$$

The resulting expression is

$$-\left(1+\frac{h}{2}f_{N-1}+\frac{h^2}{10}F_{N-1}+\frac{h^3}{120}R_{N-1}\right)y_{N-1}$$
$$+\left(1-\frac{h}{2}f_N+\frac{h^2}{10}F_N-\frac{h^3}{120}R_N\right)y_N$$
$$=\frac{h}{2}(g_{N-1}+g_N)+\frac{h^2}{10}(G_{N-1}-G_N)+\frac{h^3}{120}(H_{N-1}+H_N). \qquad (8.9.12)$$

Thus the set of $(N-1)$ equations given by (8.9.6) and the one equation given by (8.9.12) constitute a set of N equations in N unknowns.

The $(N-1)$ Eqs.(8.9.6) plus the single Eq. (8.9.12) can be written in matrix form as

$$\mathbf{Dy}+\mathbf{b}=0, \qquad (8.9.13)$$

where

$$\mathbf{D}=\begin{bmatrix} A_1 & B_1 & & & \\ B_1 & A_2 & B_3 & & \\ \cdot & \cdot & \cdot & & \\ & \cdot & \cdot & \cdot & \\ & \cdot & \cdot & \cdot & \\ & B_{N-1} & A_{N-1} & B_N & \\ & & B^*_{N+1} & A_N & \end{bmatrix}, \quad \mathbf{y}=\begin{bmatrix} y_1 \\ y_2 \\ \cdot \\ \cdot \\ \cdot \\ \cdot \\ y_N \end{bmatrix}, \quad \mathbf{b}=\begin{bmatrix} c_1 \\ c_2 \\ \cdot \\ \cdot \\ \cdot \\ \cdot \\ c_N \end{bmatrix}, \qquad (8.9.14)$$

$$A_i = 2+\frac{10}{12}h^2F_i, \qquad i=1,2,\ldots,N-1,$$

$$A_N = 1-\frac{h}{2}f_N+\frac{h^2}{10}F_N-\frac{h^3}{120}R_N,$$

$$B_i = -1+\frac{h^2}{12}F_i, \qquad i=0,1,2,\ldots,N,$$

$$B^*_{N+1} = -1-\frac{h}{2}f_{N-1}-\frac{h^2}{10}F_{N-1}-\frac{h^3}{120}R_{N-1},$$

$$c_i = \frac{h^2}{12}(G_{i-1}+10G_i+G_{i+1}), \qquad i = 2, \ldots, N-1,$$

$$c_1 = \frac{h^2}{12}(G_0+10G_1+G_2)+B_0 y_0,$$

$$c_N = -\frac{h}{2}(g_{N-1}+g_N)-\frac{h^2}{10}(G_{N-1}-G_N)-\frac{h^3}{120}(H_{N-1}+H_N).$$

Under the conditions that $F(t) \geq 0$ and $f(t) \leq 0$ over $[t_0, t_N]$ and for h sufficiently small, Usmani shows that the matrix $\mathbf{D}$ has an inverse, $\mathbf{D}^{-1}$ is a positive matrix, and $\mathbf{D}$ is a monotone matrix, all of which are used to prove the following theorem. (See Varga [10] for a discussion of positive and monotone matrices.)

THEOREM 8.3. *If the two-point boundary value problem*

$$\ddot{y} = (f^2+\dot{f})y+(fg+\dot{g}) = F(t)y+G(t),$$
$$y(t_0) = y_0,$$
$$\dot{y}(t_N) = f(t_N)y(t_N)+g(t_N)$$

(derived from the initial value problem $\dot{y} = f(t)y+g(t)$, $y(t_0) = y_0$) is replaced by the finite difference equations

$$y_n-2y_{n+1}+y_{n+2} = \frac{h^2}{12}(\ddot{y}_n+10\ddot{y}_{n+1}+\ddot{y}_{n+2}), \qquad n = 0, 1, \ldots, N-2,$$

$$-y_{N-1}+y_N = \frac{h}{2}(\dot{y}_{N-1}+\dot{y}_N)+\frac{h^2}{10}(\ddot{y}_{N-1}-\ddot{y}_N)+\frac{h^3}{120}(\dddot{y}_{N-1}+\dddot{y}_N),$$

then, if

(i) $f(t) \leq 0$, $F(t) \geq 0$ over $[t_0, t_N]$,

(ii) $y(t) \in C^6$, $f(t), g(t) \in C^3$,

the maximum error between the true solution of the boundary value problem, $y(t_n)$, and the computed solution from the finite difference equations, y_n, is bounded by

$$\max_n |y(t_n)-y_n| \leq \frac{h^4 M_6}{480}[(t_N-t_0)^2+h(t_N-t_0)], \qquad n = 0, 1, \ldots, N,$$

where $\max \left| \dfrac{d^6 y}{dt^6} \right| \leq M_6$.

REFERENCES

1. L. Collatz, *The Numerical Treatment of Differential Equations*, 3rd ed., Springer-Verlag, Berlin, 1960.
2. G. E. Forsythe and W. R. Wasow, *Finite Difference Methods for Partial Differential Equations*, Wiley, New York, 1960.
3. L. Fox, *Numerical Solution of Two-Point Boundary Problems in Ordinary Differential Equations*, Oxford, London, 1957.
4. P. Henrici, *Discrete Variable Methods in Ordinary Differential Equations*, Wiley, New York, 1962.
5. P. Henrici, *Error Propagation for Difference Methods*, Wiley, New York, 1963.
6. F. B. Hildebrand, *Methods of Applied Mathematics*, Prentice-Hall, Englewood Cliffs, N.J., 1952.
7. F. B. Hildebrand, *Finite-Difference Equations and Simulation*, Prentice-Hall, Englewood Cliffs, N.J., 1968.
8. E. Issacson and H. B. Keller, *Analysis of Numerical Methods*, Wiley, New York, 1966.
9. B. Noble, *Numerical Methods*, Vol. 2, Oliver and Boyd, London, or Interscience, New York, 1964.
10. R. S. Varga, *Matrix Iterative Analysis*, Prentice-Hall, Englewood Cliffs, N.J., 1962.
11. V. Pereyra, On Improving an Approximate Solution of a Functional Equation by Deferred Corrections, *Numer. Math.*, **8** (1966), 376–391.
12. V. Pereyra, Iterated Deferred Corrections for Nonlinear Operator Equations, *Numer. Math.*, **10** (1967), 316–323.
13. J. F. Holt, Numerical Solution of Nonlinear Two-Point Boundary Value Problems by Finite Difference Methods, *Comm. ACM*, (6) **7** (1964), 363–373.
14. J. M. Ortega and W. C. Rheinboldt, On Discretization and Differentiation of Operators with Application to Newton's Method, *SIAM J. Numer. Anal.* **3** (1966), 143–156.
15. D. D. Morrison, J. D. Riley, and J. F. Zancanaro, Multiple Shooting Method for Two-Point Boundary Value Problems, *Comm. ACM*, (12) **5** (Dec. 1962), 613–614.
16. H. B. Keller, *Numerical Methods for Two-Point Boundary Value Problems*, Blaisdell, Waltham, Mass., 1968.
17. M. R. Osborne, On Shooting Methods for Boundary Value Problems, *J. Math. Anal. Appl.*, (2) **27** (Aug. 1969), 417–433.
18. R. Sylvester and F. Meyer, Two-Point Boundary Value Problems by Quasilinearization, *SIAM J. Appl. Math.*, (2) **13** (June 1965), 586–602.
19. L. Fox, and A. R. Mitchell, Boundary Value Techniques for the Numerical Solution of Initial-Value Problems in Ordinary Differential Equations, *Quart. J. Mech. Appl. Math.*, **10** (1957), Pt. 2, 232–243.
20. R. A. Usmani, Boundary Value Techniques for the Numerical Solution of Certain Initial Value Problems in Ordinary Differential Equations, *J. Assoc. Comput. Mach.*, (2) **13** (April 1966), 287–295.

General References

P. G. Ciarlet, M. H. Schultz, and R. S. Varga, Numerical Methods of High-Order Accuracy for Nonlinear Boundary Value Problems. I. One Dimensional Problems, *Numer. Math.*, **9** (1967), 394–430.

P. G. Ciarlet, M. H. Schultz, and R. S. Varga, Numerical Methods of High-Order Accuracy for Nonlinear Boundary Value Problems. II. Nonlinear Boundary Conditions, *Numer. Math.*, **11** (1968), 331–345.

P. G. Ciarlet, M. H. Schultz, and R. S. Varga, Numerical Methods of High-Order Accuracy for Nonlinear Boundary Value Problems. III. Eigenvalue Problems, *Numer. Math.*, **12** (1968), 120–133.

P. G. Ciarlet, M. H. Schultz, and R. S. Varga, Numerical Methods of High-Order Accuracy for Nonlinear Boundary Value Problems. IV. Periodic Boundary Conditions, *Numer. Math.* **12** (1968), 266–279.

M. Lees, Discrete Methods for Nonlinear Two-Point Boundary Value Problems, in J. H. Bramble, ed., *Numerical Solution of Partial Differential Equations*, Academic Press, New York, 1966.

J. Todd, *Survey of Numerical Analysis*, McGraw-Hill, New York, 1962.

Chapter 9

SECOND ORDER NEWTON-RAPHSON METHODS

9.1. INTRODUCTION

In this chapter we discuss second-order Newton-Raphson methods. These are methods in which the solution of the nonlinear operator equation $P(x) = 0$ is effected by approximating the nonlinear operator P by the first- and second-order terms of its Taylor's series expansions. In principle, arbitrarily high-order schemes are possible, and one can find, in the Russian literature in particular, discussions of abstract nth-order Newton-Raphson methods (see, for example, Moore [1] and Grebenjuk [2]). However, the difficulties of implementing schemes of higher order than the second are such that they are rarely used in practice. In fact, it is rather difficult to apply even a second-order Newton-Raphson method to the solution of two-point boundary value problems for systems of n equations because, in addition to the n^2 first-order partial derivatives that would also be required for the first-order Newton-Raphson method, $n^2(n+1)/2$ *additional* second-order partial derivatives are required. Another difficulty in the application of certain second-order Newton-Raphson methods is that a better initial approximation is necessary for convergence than that required by first-order methods (regular or modified), contrary to naive expectation.

Because of the two difficulties cited, second-order Newton-Raphson methods have not found much application in the routine solution of two-point boundary value problems. However, because problems may arise in which second-order methods would be useful, and in order to show the implications of the abstract treatment of second-order Newton-Raphson methods in such references as Moore [1], we develop these methods and present some numerical experience with them.

Following our procedure in Chapter 6, first we discuss second-order Newton-Raphson methods as applied to a single nonlinear equation, then to a finite set (or system) of nonlinear equations, and finally to the nonlinear operator equation $P(x) = 0$ on the Banach space X. We then specialize the abstract treatment to the solution of two-point boundary value problems. Since, as we have indicated, the calculation of the required partial derivatives is an important part of the computational process, we pay special attention to this aspect of the method. Finally we present some computational experience with the second-order method applied to the problem of McGill and Kenneth [3], which we also used for an illustrative example in Chapter 6, Section 6.8.

9.2. SECOND-ORDER NEWTON-RAPHSON METHODS

In Chapter 6, Section 6.2, we discussed the first-order Newton-Raphson method for a single algebraic equation. We may extend this treatment to second-order terms.

Suppose we want to solve the scalar equation

$$\varphi(x) = 0 \tag{9.2.1}$$

by the second-order Newton-Raphson method. We expand $\varphi(x)$ in a Taylor's series expansion around a nominal trial solution solution $x^{(0)}$:

$$\varphi(x) = \varphi(x^{(0)})+\varphi'(x^{(0)})(x-x^{(0)})+\tfrac{1}{2}\varphi''(x^{(0)})(x-x^{(0)})(x-x^{(0)}) + \text{higher-order terms.} \tag{9.2.2}$$

Neglecting third-order and higher-order terms and setting $\varphi(x) = 0$, we write (9.2.2) as

$$0 = \varphi(x^{(0)})+\varphi'(x^{(0)})(x-x^{(0)})+\tfrac{1}{2}\varphi''(x^{(0)})(x-x^{(0)})(x-x^{(0)}). \tag{9.2.3}$$

This is a quadratic in $(x-x^{(0)})$. Since (9.2.3) is a scalar equation, it may be solved by the quadratic formula. A more general method which may also be applied to sets of algebraic equations and to operator equations, however, is the following technique. We solve (9.2.3), itself an approximation, by approximate means. First the equation

$$\varphi(x^{(0)})+\varphi'(x^{(0)})(x-x^{(0)}) = 0 \tag{9.2.4}$$

is solved for $\tilde{x}$, an approximation to x. The $\tilde{x}$ may then be introduced into the quadratic member in (9.2.3) in two ways. The first way is to write (9.2.3) as

$$\varphi(x^{(0)})+\varphi'(x^{(0)})(x-x^{(0)})+\tfrac{1}{2}\varphi''(x^{(0)})(\tilde{x}-x^{(0)})(\tilde{x}-x^{(0)}) = 0, \tag{9.2.5}$$

where $\tilde{x}$ appears in both terms of the quadratic. The solution to (9.2.3) is then written

$$x = x^{(0)} - [\varphi'(x^{(0)})^{-1}\varphi(x^{(0)}) + \tfrac{1}{2}\varphi'(x^{(0)})^{-1}\varphi''(x^{(0)})(\tilde{x}-x^{(0)})(\tilde{x}-x^{(0)})]. \tag{9.2.6}$$

The general formula for this process, which is known as the method of "tangent parabolas" in the Russian literature, is

$$x^{(k+1)} = x^{(k)} - [\varphi'(x^{(k)})^{-1}\varphi(x^{(k)}) + \tfrac{1}{2}\varphi'(x^{(k)})^{-1}\varphi''(x^{(k)})(\tilde{x}^{(k+1)}-x^{(k)})(\tilde{x}^{(k+1)}-x^{(k)})] \tag{9.2.7}$$

or

$$x^{(k+1)} = x^{(k)} - \varphi'(x^{(k)})^{-1}[\varphi(x^{(k)}) + \tfrac{1}{2}\varphi''(x^{(k)})(\tilde{x}^{(k+1)}-x^{(k)})^2], \tag{9.2.8}$$

where

$$\tilde{x}^{(k+1)} = x^{(k)} - \varphi'(x^{(k)})^{-1}\varphi(x^{(k)}). \tag{9.2.9}$$

The second way to write (9.2.3) is

$$\varphi(x^{(0)}) + \varphi'(x^{(0)})(x-x^{(0)}) + \tfrac{1}{2}\varphi''(x^{(0)})(x-x^{(0)})(\tilde{x}-x^{(0)}) = 0, \tag{9.2.10}$$

where $\tilde{x}$ appears in only one term of the quadratic. Rearranging (9.2.10), we have

$$\varphi(x^{(0)}) + [\varphi'(x^{(0)}) + \tfrac{1}{2}\varphi''(x^{(0)})(\tilde{x}-x^{(0)})](x-x^{(0)}) = 0 \tag{9.2.11}$$

or

$$x = x^{(0)} - [\varphi'(x^{(0)}) + \tfrac{1}{2}\varphi''(x^{(0)})(\tilde{x}-x^{(0)})]^{-1}\varphi(x^{(0)}). \tag{9.2.12}$$

The general formula for the second process, known as the method of "tangent hyperbolas", is

$$x^{(k+1)} = x^{(k)} - [\varphi'(x^{(k)}) + \tfrac{1}{2}\varphi''(x^{(k)})(\tilde{x}^{(k+1)}-x^{(k)})]^{-1}\varphi(x^{(k)}), \tag{9.2.13}$$

where

$$\tilde{x}^{(k+1)} = x^{(k)} - \varphi'(x^{(k)})^{-1}\varphi(x^{(k)}). \tag{9.2.14}$$

9.3. SETS OF EQUATIONS

Now suppose we want to solve two nonlinear algebraic equations,

$$\varphi_1(x_1, x_2) = 0, \tag{9.3.1}$$

$$\varphi_2(x_1, x_2) = 0. \tag{9.3.2}$$

We again write a Taylor's series expansion through second-order terms around a nominal solution $x_1^{(k)}, x_2^{(k)}$:

$$\begin{aligned}\varphi_1(x_1^{(k+1)}, x_2^{(k+1)}) &= \varphi_1(x_1^{(k)}, x_2^{(k)}) + \frac{\partial \varphi_1}{\partial x_1}(x_1^{(k+1)} - x_1^{(k)}) \\ &+ \frac{\partial \varphi_1}{\partial x_2}(x_2^{(k+1)} - x_2^{(k)}) + \frac{1}{2}\left[\frac{\partial^2 \varphi_1}{\partial x_1^2}(x_1^{(k+1)} - x_1^{(k)})^2\right. \\ &\left. + 2\frac{\partial^2 \varphi_1}{\partial x_1 \partial x_2}(x_1^{(k+1)} - x_1^{(k)})(x_2^{(k+1)} - x_2^{(k)}) + \frac{\partial^2 \varphi_1}{\partial x_2^2}(x_2^{(k+1)} - x_2^{(k)})^2\right],\end{aligned} \tag{9.3.3}$$

$$\begin{aligned}\varphi_2(x_1^{(k+1)}, x_2^{(k+1)}) &= \varphi_2(x_1^{(k)}, x_2^{(k)}) + \frac{\partial \varphi_2}{\partial x_1}(x_1^{(k+1)} - x_1^{(k)}) \\ &+ \frac{\partial \varphi_2}{\partial x_2}(x_2^{(k+1)} - x_2^{(k)}) + \frac{1}{2}\left[\frac{\partial^2 \varphi_2}{\partial x_1^2}(x_1^{(k+1)} - x_1^{(k)})^2\right. \\ &\left. + 2\frac{\partial^2 \varphi_2}{\partial x_1 \partial x_2}(x_1^{(k+1)} - x_1^{(k)})(x_2^{(k+1)} - x_2^{(k)}) + \frac{\partial^2 \varphi_2}{\partial x_2^2}(x_2^{(k+1)} - x_2^{(k)})^2\right].\end{aligned} \tag{9.3.4}$$

Setting $\varphi_1(x_1^{(k+1)}, x_2^{(k+1)}) = 0$ and $\varphi_2(x_1^{(k+1)}, x_2^{(k+1)}) = 0$, we express (9.3.3) and (9.3.4) as

$$\begin{bmatrix} \varphi_1(x_1^{(k)}, x_2^{(k)}) \\ \varphi_2(x_1^{(k)}, x_2^{(k)}) \end{bmatrix} + \begin{bmatrix} \dfrac{\partial \varphi_1}{\partial x_1} & \dfrac{\partial \varphi_1}{\partial x_2} \\ \dfrac{\partial \varphi_2}{\partial x_1} & \dfrac{\partial \varphi_2}{\partial x_2} \end{bmatrix} \begin{bmatrix} x_1^{(k+1)} - x_1^{(k)} \\ x_2^{(k+1)} - x_2^{(k)} \end{bmatrix}$$

$$+\frac{1}{2}\left\{ \begin{bmatrix} \dfrac{\partial^2 \varphi_1}{\partial x_1^2} & \dfrac{\partial^2 \varphi_1}{\partial x_2^2} \\ \dfrac{\partial^2 \varphi_2}{\partial x_1^2} & \dfrac{\partial^2 \varphi_2}{\partial x_2^2} \end{bmatrix} \begin{bmatrix} (x_1^{(k+1)} - x_1^{(k)})^2 \\ (x_2^{(k+1)} - x_2^{(k)})^2 \end{bmatrix} \right.$$

$$\left. +2 \begin{bmatrix} \dfrac{\partial^2 \varphi_1}{\partial x_1 \partial x_2} (x_1^{(k+1)} - x_1^{(k)})(x_2^{(k+1)} - x_2^{(k)}) \\ \dfrac{\partial^2 \varphi_2}{\partial x_1 \partial x_2} (x_1^{(k+1)} - x_1^{(k)})(x_2^{(k+1)} - x_2^{(k)}) \end{bmatrix} \right\} = \begin{bmatrix} 0 \\ 0 \end{bmatrix}. \tag{9.3.5}$$

If $\tilde{x}_1^{(k+1)}, \tilde{x}_2^{(k+1)}$ is the solution of

$$\begin{bmatrix} \varphi_1(x_1^{(k)}, x_2^{(k)}) \\ \varphi_2(x_1^{(k)}, x_2^{(k)}) \end{bmatrix} + \begin{bmatrix} \dfrac{\partial \varphi_1}{\partial x_1} & \dfrac{\partial \varphi_1}{\partial x_2} \\ \dfrac{\partial \varphi_2}{\partial x_1} & \dfrac{\partial \varphi_2}{\partial x_2} \end{bmatrix} \begin{bmatrix} x_1^{(k+1)} - x_1^{(k)} \\ x_2^{(k+1)} - x_2^{(k)} \end{bmatrix} = \begin{bmatrix} 0 \\ 0 \end{bmatrix}, \tag{9.3.6}$$

then the solution of the second-order Newton-Raphson method may be obtained again in two ways corresponding to the two ways in Section 9.2. Using $\tilde{x}_1^{(k+1)}$ and $\tilde{x}_2^{(k+1)}$ in both terms of the quadratic term, we have for the method of tangent parabolas

$$
\begin{bmatrix} x_1^{(k+1)} \\ \\ x_2^{(k+1)} \end{bmatrix} = \begin{bmatrix} x_1^{(k)} \\ \\ x_2^{(k)} \end{bmatrix} - \left\{ \begin{bmatrix} \dfrac{\partial \varphi_1}{\partial x_1} & \dfrac{\partial \varphi_1}{\partial x_2} \\ \\ \dfrac{\partial \varphi_2}{\partial x_1} & \dfrac{\partial \varphi_2}{\partial x_2} \end{bmatrix}^{-1} \begin{bmatrix} \varphi_1(x_1^{(k)}, x_2^{(k)}) \\ \\ \varphi_2(x_1^{(k)}, x_2^{(k)}) \end{bmatrix} + \frac{1}{2} \begin{bmatrix} \dfrac{\partial \varphi_1}{\partial x_1} & \dfrac{\partial \varphi_1}{\partial x_2} \\ \\ \dfrac{\partial \varphi_2}{\partial x_1} & \dfrac{\partial \varphi_2}{\partial x_2} \end{bmatrix}^{-1} \right.
$$

$$
\left. \times \begin{bmatrix} \dfrac{\partial^2 \varphi_1}{\partial x_1^2} (\tilde{x}_1^{(k+1)} - x_1^{(k)})^2 + \dfrac{\partial^2 \varphi_1}{\partial x_2^2} (\tilde{x}_2^{(k+1)} - x_2^{(k)})^2 + 2 \dfrac{\partial^2 \varphi_1}{\partial x_1 \partial x_2} (\tilde{x}_1^{(k+1)} - x_1^{(k)})(\tilde{x}_2^{(k+1)} - x_2^{(k)}) \\ \\ \dfrac{\partial^2 \varphi_2}{\partial x_1^2} (\tilde{x}_1^{(k+1)} - x_1^{(k)})^2 + \dfrac{\partial^2 \varphi_2}{\partial x_2^2} (\tilde{x}_2^{(k+1)} - x_2^{(k)})^2 + 2 \dfrac{\partial^2 \varphi_2}{\partial x_1 \partial x_2} (\tilde{x}_1^{(k+1)} - x_1^{(k)})(\tilde{x}_2^{(k+1)} - x_2^{(k)}) \end{bmatrix} \right\} . \quad (9.3.7)
$$

The alternative version, the method of tangent hyperbolas, where $\tilde{x}_1^{(k+1)}$ and $\tilde{x}_2^{(k+1)}$ appear to the first power only in the quadratic member, follows. Equations (9.3.3) and (9.3.4) may be written as

$$
\begin{bmatrix} \varphi_1(x_1^{(k)}, x_2^{(k)}) \\ \varphi_2(x_1^{(k)}, x_2^{(k)}) \end{bmatrix} + \left\{ \begin{bmatrix} \dfrac{\partial \varphi_1}{\partial x_1} & \dfrac{\partial \varphi_1}{\partial x_2} \\ \dfrac{\partial \varphi_2}{\partial x_1} & \dfrac{\partial \varphi_2}{\partial x_2} \end{bmatrix} \right.
$$

$$
\left. + \frac{1}{2} \begin{bmatrix} \dfrac{\partial^2 \varphi_1}{\partial x_1^2}(\tilde{x}_1^{(k+1)} - x_1^{(k)}) + \dfrac{\partial^2 \varphi_1}{\partial x_1 \partial x_2}(\tilde{x}_2^{(k+1)} - x_2^{(k)}), & \dfrac{\partial^2 \varphi_1}{\partial x_2^2}(\tilde{x}_2^{(k+1)} - x_2^{(k)}) + \dfrac{\partial^2 \varphi_1}{\partial x_1 \partial x_2}(\tilde{x}_1^{(k+1)} - x_1^{(k)}) \\ \dfrac{\partial^2 \varphi_2}{\partial x_1^2}(\tilde{x}_1^{(k+1)} - x_1^{(k)}) + \dfrac{\partial^2 \varphi_2}{\partial x_1 \partial x_2}(\tilde{x}_2^{(k+1)} - x_2^{(k)}), & \dfrac{\partial^2 \varphi_2}{\partial x_2^2}(\tilde{x}_2^{(k+1)} - x_2^{(k)}) + \dfrac{\partial^2 \varphi_2}{\partial x_1 \partial x_2}(\tilde{x}_1^{(k+1)} - x_1^{(k)}) \end{bmatrix} \right\}
$$

$$
\times \begin{bmatrix} x_1^{(k+1)} - x_1^{(k)} \\ x_2^{(k+1)} - x_2^{(k)} \end{bmatrix} = \begin{bmatrix} 0 \\ 0 \end{bmatrix}. \tag{9.3.8}
$$

Solving for $x_1^{(k+1)}, x_2^{(k+1)}$ gives the alternative form:

$$\begin{bmatrix} x_1^{(k+1)} \\ x_2^{(k+1)} \end{bmatrix} = \begin{bmatrix} x_1^{(k)} \\ x_2^{(k)} \end{bmatrix} - \left\{ \begin{bmatrix} \dfrac{\partial \varphi_1}{\partial x_1} & \dfrac{\partial \varphi_1}{\partial x_2} \\ \dfrac{\partial \varphi_2}{\partial x_1} & \dfrac{\partial \varphi_2}{\partial x_2} \end{bmatrix} \right.$$

$$\left. + \frac{1}{2} \begin{bmatrix} \dfrac{\partial^2 \varphi_1}{\partial x_1^2}(\tilde{x}_1^{(k+1)} - x_1^{(k)}) + \dfrac{\partial^2 \varphi_1}{\partial x_1 \partial x_2}(\tilde{x}_2^{(k+1)} - x_2^{(k)}), & \dfrac{\partial^2 \varphi_1}{\partial x_2^2}(\tilde{x}_2^{(k+1)} - x_2^{(k)}) + \dfrac{\partial^2 \varphi_1}{\partial x_1 \partial x_2}(\tilde{x}_1^{(k+1)} - x_1^{(k)}) \\ \dfrac{\partial^2 \varphi_2}{\partial x_1^2}(\tilde{x}_1^{(k+1)} - x_1^{(k)}) + \dfrac{\partial^2 \varphi_2}{\partial x_1 \partial x_2}(\tilde{x}_2^{(k+1)} - x_2^{(k)}), & \dfrac{\partial^2 \varphi_2}{\partial x_2^2}(\tilde{x}_2^{(k+1)} - x_2^{(k)}) + \dfrac{\partial^2 \varphi_2}{\partial x_1 \partial x_2}(\tilde{x}_1^{(k+1)} - x_1^{(k)}) \end{bmatrix} \right\}^{-1}$$

$$\times \begin{bmatrix} \varphi_1(x_1^{(k)}, & x_2^{(k)}) \\ \varphi_2(x_1^{(k)}, & x_2^{(k)}) \end{bmatrix}. \qquad (9.3.9)$$

It is by now clear that, for a system of n equations, the second-order Newton-Raphson method in the tangent parabola form appears as

$$\begin{bmatrix} x_1^{(k+1)} \\ x_2^{(k+1)} \\ \vdots \\ x_n^{(k+1)} \end{bmatrix} = \begin{bmatrix} x_1^{(k)} \\ x_2^{(k)} \\ \vdots \\ x_n^{(k)} \end{bmatrix} - \begin{bmatrix} \frac{\partial \varphi_1}{\partial x_1} & \frac{\partial \varphi_1}{\partial x_2} & \cdots & \frac{\partial \varphi_1}{\partial x_n} \\ \frac{\partial \varphi_2}{\partial x_1} & \frac{\partial \varphi_2}{\partial x_2} & \cdots & \frac{\partial \varphi_2}{\partial x_n} \\ \vdots & \vdots & & \vdots \\ \frac{\partial \varphi_n}{\partial x_1} & \frac{\partial \varphi_n}{\partial x_2} & \cdots & \frac{\partial \varphi_n}{\partial x_n} \end{bmatrix}^{-1} \begin{bmatrix} \varphi_1(x_1^{(k)}, \ldots, x_n^{(k)}) + \frac{1}{2} \sum_{i,j=1}^{n} \frac{\partial^2 \varphi_1}{\partial x_i \, \partial x_j} (\tilde{x}_i^{(k+1)} - x_j^{(k)})^2 \\ \varphi_2(x_1^{(k)}, \ldots, x_n^{(k)}) + \frac{1}{2} \sum_{i,j=1}^{n} \frac{\partial^2 \varphi_2}{\partial x_i \, \partial x_j} (\tilde{x}_i^{(k+1)} - x_j^{(k)})^2 \\ \vdots \\ \varphi_n(x_1^{(k)}, \ldots, x_n^{(k)}) + \frac{1}{2} \sum_{i,j=1}^{n} \frac{\partial^2 \varphi_n}{\partial x_i \, \partial x_j} (\tilde{x}_i^{(k+1)} - x_j^{(k)})^2 \end{bmatrix}, \quad (9.3.10)$$

where $\tilde{x}_i^{(k+1)}$ is used in both members of the quadratic term.

The other form, the method of tangent hyperbolas, appears as

$$\begin{bmatrix} x_1^{(k+1)} \\ x_2^{(k+1)} \\ \vdots \\ x_n^{(k+1)} \end{bmatrix} = \begin{bmatrix} x_1^{(k)} \\ x_2^{(k)} \\ \vdots \\ x_n^{(k)} \end{bmatrix} - \begin{bmatrix} \dfrac{\partial\varphi_1}{\partial x_1} + \dfrac{1}{2}\sum_{j=1}^{n} \dfrac{\partial^2\varphi_1}{\partial x_1\,\partial x_j}(\tilde{x}_j^{(k+1)} - x_j^{(k)}), \ldots, \dfrac{\partial\varphi_1}{\partial x_n} + \dfrac{1}{2}\sum_{j=1}^{n} \dfrac{\partial^2\varphi_1}{\partial x_n\,\partial x_j}(\tilde{x}_j^{(k+1)} - x_j^{(k)}) \\ \dfrac{\partial\varphi_2}{\partial x_1} + \dfrac{1}{2}\sum_{j=1}^{n} \dfrac{\partial^2\varphi_2}{\partial x_1\,\partial x_j}(\tilde{x}_j^{(k+1)} - x_j^{(k)}), \ldots, \dfrac{\partial\varphi_2}{\partial x_n} + \dfrac{1}{2}\sum_{j=1}^{n} \dfrac{\partial^2\varphi_2}{\partial x_n\,\partial x_j}(\tilde{x}_j^{(k+1)} - x_j^{(k)}) \\ \vdots \qquad\qquad \vdots \qquad\qquad \vdots \\ \dfrac{\partial\varphi_n}{\partial x_1} + \dfrac{1}{2}\sum_{j=1}^{n} \dfrac{\partial^2\varphi_n}{\partial x_1\,\partial x_j}(\tilde{x}_j^{(k+1)} - x_j^{(k)}), \ldots, \dfrac{\partial\varphi_n}{\partial x_n} + \dfrac{1}{2}\sum_{j=1}^{n} \dfrac{\partial^2\varphi_n}{\partial x_n\,\partial x_j}(\tilde{x}_j^{(k+1)} - x_j^{(k)}) \end{bmatrix}^{-1}$$

$$\times \begin{bmatrix} \varphi_1(x_1^{(k)}, \ldots, x_n^{(k)}) \\ \varphi_2(x_1^{(k)}, \ldots, x_n^{(k)}) \\ \vdots \\ \varphi_n(x_1^{(k)}, \ldots, x_n^{(k)}) \end{bmatrix}. \qquad (9.3.11)$$

So far the treatment has been purely formal. In Section 9.4 we give the theorems corresponding to Kantorovich's theorem for the two forms of the second-order Newton-Raphson method we have been discussing. From the theoretical point of view, the method of tangent parabolas has nothing to recommend it, since the available convergence theorem requires a better initial guess (reflected in a smaller h) and yet furnishes a rate of convergence estimate which is *slower* than the first-order Newton-Raphson method. The estimate of the rate of convergence of the method of tangent hyperbolas is faster than that for the first-order Newton-Raphson method, and it permits as large a value of h for its validity. From the practical point of view, however, the method of tangent hyperbolas requires the taking of an additional inverse, and this may discourage its application.

9.4. SECOND ORDER METHODS FOR P(x) = 0

With the backround of Sections 9.2 and 9.3, it is now clear how we proceed heuristically to develop second-order Newton-Raphson methods for the problem we are interested in, namely, the solution of the operator equation

$$P(x) = 0, \tag{9.4.1}$$

where P is a nonlinear mapping of the open set Ω of a Banach space X into X. That is, the second-order approximation to (9.4.1) is

$$P(x^{(k)})+P'(x^{(k)})\,\Delta x^{(k)}+\tfrac{1}{2}P''(x^{(k)})\,\Delta x^{(k)}\,\Delta x^{(k)} = 0. \tag{9.4.2}$$

Taking $\Delta\tilde{x}^{(k)}$ as the solution of the first-order approximation to (9.4.1),

$$\Delta\tilde{x}^{(k)} = -[P'(x^{(k)})]^{-1}\,P(x^{(k)}), \tag{9.4.3}$$

Equation (9.4.2) may be solved in two ways. In the method of "tangent parabolas", $\Delta\tilde{x}^{(k)}$ is substituted in both members of the quadratic term of (9.4.2). The resulting iteration formula is

$$\begin{aligned}\Delta x^{(k)} = -[P'(x^{(k)})]^{-1}[P(x^{(k)})+\tfrac{1}{2}P''(x^{(k)})\,&\{[P'(x^{(k)})]^{-1}P(x^{(k)})\}\\ &\times\{[P'(x^{(k)})]^{-1}P(x^{(k)})\}].\end{aligned} \tag{9.4.4}$$

The convergence properties of (9.4.4) are given by the following theorem due to Nečepurenko (see Moore [1]).

THEOREM 9.1. *Let the conditions corresponding to* (6.4.5), (6.4.6), *and* (6.4.7) *of Theorem* 6.3 *be fulfilled in the sphere* $S(x^{(0)}, r_0)$.† *Suppose that, instead of*

† Recall that the conditions of Theorem 6.3, which applies to sets of nonlinear equations, also apply to the operator equation $P(x) = 0$.

(6.4.8) *of Theorem* 6.3, *we have*

$$h_0 = B_0\eta_0 K \leq \sqrt{\tfrac{3}{2}} - 1 \sim 0.22. \qquad (9.4.5)$$

Then (9.4.4) *converges to the solution* x^* *of* (9.4.1) *in the sphere* $S(x^{(0)}, r_0)$, *where* $r_0 = (1-\sqrt{1-2h_0})\,\eta_0/h_0$. *The rate of convergence is given by*

$$\|x^{(n)} - x^*\| \leq (3/4)^n (\theta g_0)^{2^n - 1}\left(1 + \frac{h_0}{2}\right)\eta_0, \qquad n = 0, 1, \ldots, \qquad (9.4.6)$$

where $g_0 \leq h_0(1+h_0/2)$ *and* $\theta = 256/81$.

Comparison with the corresponding result for Kantorovich's theorem is not favorable for the method of tangent parabolas, which requires a better initial guess (smaller h_0) and exhibits slower convergence.

The alternative technique to (9.4.4) is the method of tangent hyperbolas, in which (9.4.3) is substituted in only one member of the quadratic term of (9.4.2). This yields

$$\Delta x^{(k)} = -[P'(x^{(k)}) - \tfrac{1}{2}P''(x^{(k)})\,\{[P'(x^{(k)})]^{-1}\,P(x^{(k)})\}]^{-1}\,P(x^{(k)}). \qquad (9.4.7)$$

The corresponding convergence theorem has been given by Mertvetsova (see Moore [1]).

THEOREM 9.2. *Let the conditions corresponding to* (6.4.4) *through* (6.4.8) *of Theorem* 6.3 *be satisfied, but in the sphere* $S(x^{(0)}, 2\eta_0)$ *instead of* $S(x^{(0)}, r_0)$ (*in general,* $S(x^{(0)}, 2\eta_0) \supset S(x^{(0)}, r_0)$). *Let P be three times differentiable with* $\|P'''(x)\| \leq L$ *in* $S(x^{(0)}, 2\eta_0)$ *and let*

$$\mu_0 = \left[\frac{L}{K^2 B_0}(2+h_0)+3\right](1+h_0) \leq 9. \qquad (9.4.8)$$

Then (9.4.7) *converges to the solution* x^* *of* (9.4.1) (*which lies in* $S(x^{(0)}, 2\eta_0)$), *and the rate of convergence is given by*

$$\|x^{(n)} - x^*\| \leq \frac{1}{2^{n-1}}(2h_0)^{3^n - 1}\eta_0, \qquad n = 0, 1, \ldots. \qquad (9.4.9)$$

This second-order Newton-Raphson method converges faster than the first-order method, as the exponent of $2h_0$ is $3^n - 1$ rather than $2^n - 1$, but of course requires more computation per step, in addition to requiring the existence of a third derivative.

Equations (9.4.4) and (9.4.7) are thus the operator equations for the two versions of the second-order Newton-Raphson method. Of course for finite sets of equations they reduce to their counterparts (9.3.10) and (9.3.11), which are the forms we are particularly interested in. By (9.4.4) and (9.4.7) the method of adjoints and the method of complementary functions can be extended to second-order methods. In addition, (9.4.4) and (9.4.7) could, in principle, be used to develop a second-order analog of quasilinearization.

Recall that in Section 6.7, for the two-point boundary value problem we defined

Y_m = mth specified terminal condition, $m = 1, 2, \ldots, n-r$,

$y_{im}^{(k)}(t_f)$ = the terminal values calculated for $m = 1, 2, \ldots, n-r$ by integrating (6.7.1) with the initial conditions $y_1(t_0), \ldots, y_r(t_0), y_{r+1}^{(k)}(t_0), \ldots, y_n^{(k)}(t_0)$,

$\varphi_m(\mathbf{y}^{(k)}(t_0)) = Y_m - y_{im}^{(k)}(t_f)$ = the miss distance between the specified and calculated terminal conditions, $m = 1, 2, \ldots, n-r$; $k = 0, 1, \ldots,$

$\varphi_m(\mathbf{y}(t_0))$ = 0 = the desired set of miss distances, which is approximated by $\varphi_m(\mathbf{y}^{(k)}(t_0))$, $m = 1, 2, \ldots, n-r$.

Recall further that we drive the miss distance to zero under appropriate conditions by generating a sequence of trial missing initial conditions.

If we expand $\varphi_m(\mathbf{y}(t_0))$ by a Taylor's series up to second-order terms, we have

$$\varphi_m(\mathbf{y}^{(k)}(t_0)) + \sum_{j=r+1}^{n} \frac{\partial \varphi_m}{\partial y_j(t_0)} \delta y_j^{(k)}(t_0) + \frac{1}{2} \sum_{s,j=r+1}^{n} \sum^{n} \frac{\partial^2 \varphi_m}{\partial y_s(t_0)\, \partial y_j(t_0)} \delta y_s^{(k)}(t_0)\, \delta y_j^{(k)}(t_0) = 0, \quad m = 1, 2, \ldots, r, \tag{9.4.10}$$

where $\varphi_m(\mathbf{y}(t_0))$ is set equal to zero.

If X is finite dimensional and if $P_m(x^{(k)})$, $P_m'(x^{(k)})\, \Delta x^{(k)}$, and $P_m''(x^{(k)})\, \Delta x^{(k)} \Delta x^{(k)}$ are the components of $P(x^{(k)})$, $P'(x^{(k)}) \Delta x^{(k)}$, and $P''(x^{(k)}) \Delta x^{(k)} \Delta x^{(k)}$, respectively, then we observe the correspondence between (9.4.2) and (9.4.10):

$$P_m(x^{(k)}) = \varphi_m(\mathbf{y}^{(k)}(t_0)), \tag{9.4.11}$$

$$P_m'(x^{(k)})\, \Delta x^{(k)} = \sum_{j=r+1}^{n} \frac{\partial \varphi_m}{\partial y_j(t_0)} \delta y_j^{(k)}(t_0), \tag{9.4.12}$$

$$P_m''(x^{(k)})\, \Delta x^{(k)} \Delta x^{(k)} = \sum_{s,j=r+1}^{n} \sum^{n} \frac{\partial^2 \varphi_m}{\partial y_s(t_0)\, \partial y_j(t_0)} \delta y_s^{(k)}(t_0)\, \delta y_j^{(k)}(t_0). \tag{9.4.13}$$

By the definition of the miss distance $\varphi_m(\mathbf{y}(t_0))$ we have

$$\frac{\partial \varphi_m}{\partial y_j(t_0)} = -\frac{\partial y_{i_m}(t_f)}{\partial y_j(t_0)}, \qquad m = 1,2,\ldots,n-r, \qquad j = r+1,\ldots,n, \tag{9.4.14}$$

and

$$\frac{\partial^2 \varphi_m}{\partial y_s(t_0)\,\partial y_j(t_0)} = -\frac{\partial^2 y_{i_m}(t_f)}{\partial y_s(t_0)\,\partial y_j(t_0)}. \tag{9.4.15}$$

Carrying out the computation thus requires the first and second partial derivatives. Since the calculation of these partial derivatives is such an important part in the solution by second-order Newton-Raphson methods, we devote the next section to it. The reader interested only in a report of numerical experience with a second-order Newton-Raphson method may turn to Section 9.6.

9.5. CALCULATION OF THE PARTIAL DERIVATIVES

In our treatment of the method of adjoints and the method of complementary functions, we first considered each of the methods as a complete scheme and showed how each would be used to solve a typical two-point boundary value problem. Then we examined the methods more closely, saw that each was a realization of the abstract first-order Newton-Raphson method, and that a certain part of the overall calculation actually computed the first order partial derivatives (9.4.14). For example, in the method of adjoints, the backward integration of the adjoint equations in effect determine the first-order partial derivatives (9.4.14).

Since in second-order Newton-Raphson methods we need not only the first-order partial derivatives (9.4.14) but also the second order partial derivatives (9.4.15), we take a new look at methods of computing these derivatives, keeping their use in second-order methods in mind.

There are at least three methods by which the derivatives (9.4.14), (9.4.15) can be found numerically: the finite difference method (sometimes called the secant method), variational equations, and adjoint equations.

9.5.1. Finite Difference Method

In this method the equations

$$\dot{y}_i = g_i(y_1, y_2, \ldots, y_n, t), \qquad i = 1,2,\ldots,n, \tag{9.5.1}$$

are solved with nominal trial values of the missing initial conditions $y_{r+1}(t_0), \ldots, y_n(t_0)$. Then the missing initial conditions are systematically perturbed in turn and (9.5.1) solved an additional $n-r$ times. The required partial derivatives (9.4.14) are then approximated by appropriate finite difference formulas. For example, denoting the solution of (9.5.1) by $y_i(y_1(t_0), \ldots, y_n(t_0), t)$ to indicate the dependence on the initial conditions,

$$\frac{\partial y_{i_m}(t_f)}{\partial y_j(t_0)} \approx \frac{y_{i_m}(\ldots, y_j(t_0)+\Delta_j, \ldots, t_f) - y_{i_m}(\ldots\ldots, t_f)}{\Delta_j}. \tag{9.5.2}$$

The second partial derivatives (9.4.15) would be obtained by perturbing *pairs* of initial conditions, integrating (9.5.1) an additional $(n-r+1)(n-r)/2$ times, and approximating (9.4.15), when $j \neq k$, by a formula such as

$$\frac{\partial^2 y_{i_m}(t_f)}{\partial y_j(t_0)\,\partial y_k(t_0)} \approx \frac{y_{i_m}(\ldots, y_j(t_0)+\Delta_j, \ldots, y_k(t_0)+\Delta_k, \ldots, t_f) - 2y_{i_m}(\ldots, y_j(t_0)+\Delta_j, \ldots, t_f) + y_{i_m}(\ldots, t_f)}{\Delta_j \Delta_k} \tag{9.5.3}$$

and, when $j = k$, by a formula such as

$$\frac{\partial^2 y_{i_m}(t_f)}{\partial y_j(t_0)^2} \approx \frac{y_{i_m}(\ldots, y_j(t_0)+2\Delta_j, \ldots, t_f) - 2y_{i_m}(\ldots, y_j(t_0)+\Delta_j, \ldots, t_f) + y_{i_m}(\ldots, t_f)}{\Delta_j^2}. \tag{9.5.4}$$

In the formulas above we have used dots (...) to indicate *unperturbed* initial conditions.

The finite difference method is relatively simple to apply, but care must be taken that the perturbations Δ_j in the initial conditions are neither too large (in which case the finite difference formula is not accurate enough) or too small (in which case the result will be just "noise" around the nominal solution).

9.5.2. Variational Equations

A second way to obtain the partial derivatives (9.4.14), (9.4.15) is the numerical integration of the variational equations. We have already met the variational equations and the closely related adjoint equations in connection with the methods of complementary functions and adjoints, respectively.

For the convenience of the reader we give a brief, self-contained discussion of variational equations here.

One approach to the variational equations corresponding to (9.5.1) is to consider a solution $\mathbf{y}(t)+\delta\mathbf{y}(t)$ which is close to the solution $\mathbf{y}(t)$. In this case, (9.5.1) may be written

$$\dot{y}_i+\delta\dot{y}_i = g_i(y_1(t)+\delta y_1(t),\ y_2(t)+\delta y_2(t),\ldots, y_n(t)+\delta y_n(t),\ t),$$
$$i = 1,2,\ldots,n. \qquad (9.5.5)$$

Expanding the right-hand side of (9.5.5) in a Taylor's series and retaining terms up through first order, we have

$$\dot{y}_i+\delta\dot{y}_i = g_i(y_1, y_2, \ldots, y_n, t) + \sum_{k=1}^{n} \frac{\partial g_i}{\partial y_k}\,\delta y_k(t), \qquad i = 1,2,\ldots,n. \qquad (9.5.6)$$

On subtracting (9.5.1) from (9.5.6) we obtain

$$\delta\dot{y}_i = \sum_{k=1}^{n} \frac{\partial g_i}{\partial y_k}\,\delta y_k(t), \qquad i = 1,2,\ldots,n. \qquad (9.5.7)$$

Equation (9.5.7) is called the variational equation associated with (9.5.1), and it may be thought of as the differential equation approximately satisfied by the difference between two solutions of (9.5.1) close to one another. Note that (9.5.7) is linear and homogeneous.

Another approach to the variational equations is to differentiate both sides of (9.5.1) with respect to the initial conditions $y_j(t_0)$; this gives the equations

$$\frac{\partial \dot{y}_i(t)}{\partial y_j(t_0)} = \sum_{k=1}^{n} \frac{\partial g_i}{\partial y_k}\frac{\partial y_k(t)}{\partial y_j(t_0)}, \qquad i,j = 1,2,\ldots,n. \qquad (9.5.8)$$

Now, if the functions $g_i(y_1, \ldots, y_n, t)$ in (9.5.1) have continuous second partial derivatives with respect to the y_j, then the theorem establishing the equality of mixed partial derivatives (see, for example, Courant [4]) can be applied to show that

$$\frac{\partial}{\partial y_j(t_0)}\frac{dy_i}{dt} = \frac{d}{dt}\frac{\partial y_i}{\partial y_j(t_0)} \qquad (9.5.9)$$

since the solutions of (9.5.1) satisfy the conditions of the theorem. Thus (9.5.8) can be written

$$\frac{d}{dt}\frac{\partial y_i(t)}{\partial y_j(t_0)} = \sum_{k=1}^{n} \frac{\partial g_i}{\partial y_k}\frac{\partial y_k(t)}{\partial y_j(t_0)}, \qquad i,j = 1,2,\ldots,n, \qquad (9.5.10)$$

which is seen to be of the same form as (9.5.7) with $\partial y_i(t)/\partial y_j(t_0)$ playing the role of δy_i. Therefore the variational equations can also be interpreted as the equations satisfied by the partial derivatives of the solution at time t with respect to initial conditions at time t_0, and this of course is the interpretation we are interested in. To obtain the equations satisfied by the *second* partial derivatives of the solution at t with respect to initial conditions at t_0 we may similarly differentiate (9.5.10) with respect to $y_j(t_0)$. If the $g_i(y_1, \ldots, y_n, t)$ have continuous *third* partial derivatives with respect to y_j, we can infer that

$$\frac{\partial}{\partial y_s(t_0)} \frac{d}{dt} \frac{\partial y_i(t)}{\partial y_j(t_0)} = \frac{d}{dt} \frac{\partial^2 y_i}{\partial y_s(t_0)\,\partial y_j(t_0)} \tag{9.5.11}$$

and the desired equations are

$$\frac{d}{dt} \frac{\partial^2 y_i(t)}{\partial y_s(t_0)\,\partial y_j(t_0)} = \sum_{k=1}^{n} \frac{\partial g_i}{\partial y_k} \frac{\partial^2 y_k(t)}{\partial y_s(t_0)\,\partial y_j(t_0)}$$

$$+ \sum_{k=1}^{n} \left(\sum_{q=1}^{n} \frac{\partial^2 g_i}{\partial y_k \partial y_q} \frac{\partial y_q(t)}{\partial y_s(t_0)} \right) \frac{\partial y_k(t)}{\partial y_j(t_0)}, \qquad i, j, s = 1, 2, \ldots, n. \tag{9.5.12}$$

Note that the "variational" equations for the second-order partial derivatives, although still linear, are no longer homogeneous.

Since there are n^2 first-order partial derivatives $\partial y_i(t)/\partial y_j(t_0)$, there are n^2 scalar equations (9.5.8). It is probably best to think of these n^2 equations as one matrix differential equation

$$\frac{d}{dt} \mathbf{J}(t) = \mathbf{G}(t)\mathbf{J}(t), \tag{9.5.13}$$

where $\mathbf{J}(t) =$ the $n \times n$ matrix,

$$\mathbf{J}(t) = \begin{bmatrix} \dfrac{\partial y_1(t)}{\partial y_1(t_0)} & \dfrac{\partial y_1(t)}{\partial y_2(t_0)} & \cdots & \dfrac{\partial y_1(t)}{\partial y_n(t_0)} \\[2ex] \dfrac{\partial y_2(t)}{\partial y_1(t_0)} & \dfrac{\partial y_2(t)}{\partial y_2(t_0)} & \cdots & \dfrac{\partial y_2(t)}{\partial y_n(t_0)} \\ \vdots & \vdots & \vdots & \vdots \\ \dfrac{\partial y_n(t)}{\partial y_1(t_0)} & \dfrac{\partial y_n(t)}{\partial y_2(t_0)} & \cdots & \dfrac{\partial y_n(t)}{\partial y_n(t_0)} \end{bmatrix},$$

and $\mathbf{G}(t)$ = the $n \times n$ matrix,

$$\mathbf{G}(t) = \begin{bmatrix} \dfrac{\partial g_1}{\partial y_1} & \dfrac{\partial g_1}{\partial y_2} & \cdots & \dfrac{\partial g_1}{\partial y_n} \\ \dfrac{\partial g_2}{\partial y_1} & \dfrac{\partial g_2}{\partial y_2} & \cdots & \dfrac{\partial g_2}{\partial y_n} \\ \vdots & \vdots & \cdots & \vdots \\ \dfrac{\partial g_n}{\partial y_1} & \dfrac{\partial g_n}{\partial y_2} & \cdots & \dfrac{\partial g_n}{\partial y_n} \end{bmatrix},$$

and where the partial derivatives $\partial g_i/\partial y_j$ are to be evaluated along the solution $(y_1(t), y_2(t), \ldots, y_n(t), t)$ corresponding to the initial conditions $(y_1(t_0), y_2(t_0), \ldots, y_n(t_0))$. The definition of the first partials as solutions of (9.5.13) will be complete as soon as the initial conditions have been specified for the set of equations.

For the matrix $\mathbf{J}(t)$ denote the i,j element by $J_{ij}(t)$. It turns out that the appropriate initial conditions are the Kronecker delta conditions, that is

$$J_{ij}(t_0) = \delta_{ij} \tag{9.5.14}$$

or, equivalently,

$$\frac{\partial y_i(t_0)}{\partial y_j(t_0)} = \left.\begin{matrix} 1, & i = j \\ 0, & i \neq j. \end{matrix}\right\} \tag{9.5.14'}$$

This statement will be justified shortly.

Again, since there are n^3 second-order partial derivatives $\partial^2 y_i(t)/\partial y_j(t_0)\, \partial y_s(t_0)$, there are n^3 scalar equations (9.5.12). However, under conditions usually met in applications, the mixed partial derivatives are equal, that is, $\partial^2 y_i(t)/\partial y_j(t_0)\, \partial y_s(t_0) = \partial^2 y_i(t)/\partial y_s(t_0)\, \partial y_j(t_0)$, so that only $n^2(n+1)/2$ different derivatives need be computed. Thus (9.5.12) can be considered the matrix differential equation

$$\frac{d}{dt}\mathbf{H}(t) = \mathbf{G}(t)\mathbf{H}(t) + \mathbf{F}(t), \tag{9.5.15}$$

where now $\mathbf{H}(t)$ is an $n \times n(n+1)/2$ rectangular matrix whose ith row is the vector of partial derivatives $\partial^2 y_i(t)/\partial y_j(t_0)\, \partial y_s(t_0)$ ordered by the pairs (j, s), $j \leq s$, that is $(1,1), \ldots, (1,n)$, $(2,2), \ldots, (2,n), \ldots, (n,n)$. We exhibit $\mathbf{H}(t)$:

$$
\mathbf{H}(t) = \begin{bmatrix}
\frac{\partial^2 y_1(t)}{\partial y_1(t_0)^2} & \frac{\partial^2 y_1(t)}{\partial y_1(t_0)\,\partial y_2(t_0)} & \cdots & \frac{\partial^2 y_1(t)}{\partial y_1(t_0)\,\partial y_n(t_0)} & \frac{\partial^2 y_1(t)}{\partial y_2(t_0)^2} & \frac{\partial^2 y_1(t)}{\partial y_2(t_0)\,\partial y_3(t_0)} & \cdots & \frac{\partial^2 y_1(t)}{\partial y_2(t_0)\,\partial y_n(t_0)} & \cdots & \frac{\partial^2 y_1(t)}{\partial y_n(t_0)^2} \\
\frac{\partial^2 y_2(t)}{\partial y_1(t_0)^2} & \frac{\partial^2 y_2(t)}{\partial y_1(t_0)\,\partial y_2(t_0)} & \cdots & \frac{\partial^2 y_2(t)}{\partial y_1(t_0)\,\partial y_n(t_0)} & \frac{\partial^2 y_2(t)}{\partial y_2(t_0)^2} & \frac{\partial^2 y_2(t)}{\partial y_2(t_0)\,\partial y_3(t_0)} & \cdots & \frac{\partial^2 y_2(t)}{\partial y_2(t_0)\,\partial y_n(t_0)} & \cdots & \frac{\partial^2 y_2(t)}{\partial y_n(t_0)^2} \\
\cdots & \cdots & & \cdots & \cdots & \cdots & & \cdots & & \cdots \\
\frac{\partial^2 y_n(t)}{\partial y_1(t_0)^2} & \frac{\partial^2 y_n(t)}{\partial y_1(t_0)\,\partial y_2(t_0)} & \cdots & \frac{\partial^2 y_n(t)}{\partial y_1(t_0)\,\partial y_n(t_0)} & \frac{\partial^2 y_n(t)}{\partial y_2(t_0)^2} & \frac{\partial^2 y_n(t)}{\partial y_2(t_0)\,\partial y_3(t_0)} & \cdots & \frac{\partial^2 y_n(t_0)}{\partial y_2(t_0)\,\partial y_n(t_0)} & \cdots & \frac{\partial^2 y_n(t)}{\partial y_n(t_0)^2}
\end{bmatrix}.
$$

$\mathbf{G}(t)$ is the same $n \times n$ matrix that appeared in (9.5.13), and $\mathbf{F}(t)$ is the $n \times (n+1)/2$ matrix of the "forcing" or nonhomogeneous terms in (9.5.12) given by the double summation. Letting

$$g_{isj} = \sum_{k=1}^{n} \left(\sum_{q=1}^{n} \frac{\partial^2 g_i}{\partial y_k \partial y_q} \frac{\partial y_q(t)}{\partial y_s(t_0)} \right) \frac{\partial y_k(t)}{\partial y_j(t_0)},$$

$\mathbf{F}(t)$ appears as

$$\mathbf{F}(t) = \begin{bmatrix} g_{111} & g_{112} & \cdots & g_{11n} & g_{122} & g_{123} & \cdots & g_{12n} & \cdots & g_{1nn} \\ g_{211} & g_{212} & \cdots & g_{21n} & g_{222} & g_{223} & \cdots & g_{22n} & \cdots & g_{2nn} \\ \vdots & \vdots & & \vdots & \vdots & \vdots & & \vdots & \vdots & \vdots \\ g_{n11} & g_{n12} & \cdots & g_{n1n} & g_{n22} & g_{n23} & \cdots & g_{n2n} & \cdots & g_{nnn} \end{bmatrix}.$$

For the matrix $\mathbf{H}(t)$, denote the element in row r and column s by $\mathrm{H}_{rs}(t)$. The appropriate initial conditions for (9.5.15) are the zero or homogeneous conditions

$$H_{rs}(t) = 0, \qquad \text{for all } r, s. \tag{9.5.16}$$

Comparing the matrix differential equations (9.5.13) and (9.5.15), we see that (9.5.13) is an $n \times n$ system, while (9.5.15) is an $n \times [n(n+1)/2]$ system. Further, (9.5.13) is a homogeneous system, while (9.5.15) is inhomogeneous with the forcing term the matrix $\mathbf{F}(t)$. Finally, for the solutions we are interested in, (9.5.13) has the identity initial conditions $\mathbf{J}(t_0) = \mathbf{I}$, where $\mathbf{I}$ is the $n \times n$ identity matrix, while (9.5.15) has the zero initial conditions $\mathbf{H}(t_0) = \mathbf{O}$, where $\mathbf{O}$ is the $n \times [n(n+1)/2]$ matrix of zeros. However, in spite of these apparent differences, the two equations are intimately related because the $n \times n$ matrix which really defines the equations, namely, $\mathbf{G}(t)$, is the same in both equations. This means that, once the solution $\mathbf{J}(t)$ of (9.5.13) is known, the solution $\mathbf{H}(t)$ of (9.5.15) can be found by quadratures, as we shall show.

In contrast to the first-order variational equations where the Kronecker delta initial conditions distinguish one set of solutions from another, the initial conditions for the second-order variational equations do not play that role since $\partial^2 y_i(t_0)/\partial y_j(t_0)\partial y_s(t_0) = 0$, $i,j,s = 1,2,\ldots,n$. For these equations the forcing function, the double summation term in (9.5.12) or (9.5.15), recognizes the subscripts on the $y_j(t_0)$ and $y_s(t_0)$ since the partial derivatives $\partial y_k(t)/\partial y_j(t_0)$ and $\partial y_j(t)/\partial y_s(t_0)$ are acquired from the solutions of the variational equations. Thus the forcing function in (9.5.12) or (9.5.15) distinguishes one set of second-order variational solutions from another.

We may provide a somewhat formal justification for the initial conditions

(9.5.14) and (9.5.16) by the following argument. We observe that $y_i(t)$ is a function of the initial conditions and the time t:

$$y_i(t) = y_i(v_1, v_2, \ldots, v_n, t), \qquad i = 1,2,\ldots,n \tag{9.5.17}$$

where for convenience we set

$$y_j(t_0) = v_j, \qquad j = 1,2,\ldots,n. \tag{9.5.18}$$

Let us consider a neighboring solution, $y_i(t)+\delta y_i(t)$, to $y_i(t)$:

$$y_i(t)+\delta y_i(t) = y_i(v_1+\delta v_1, v_2+\delta v_2, \ldots, v_n+\delta v_n, t), \quad i = 1,2,\ldots,n \tag{9.5.19}$$

where

$$\delta v_j = \delta y_j(t_0), \qquad j = 1,2,\ldots,n. \tag{9.5.20}$$

The right-hand side of (9.5.19) may be expanded in a Taylor's series through second-order terms:

$$y_i(v_1+\delta v_1, v_2+\delta v_2, \ldots, v_n+\delta v_n, t) = y_i(v_1, v_2, \ldots, v_n, t)$$
$$+ \sum_{j=1}^{n} \frac{\partial y_i}{\partial v_k}\delta v_k + \tfrac{1}{2}\sum_{j,k=1}^{n}\sum^{n} \frac{\partial^2 y_i}{\partial v_j \partial v_k}\delta v_j \delta v_k, \qquad i = 1,2,\ldots,n. \tag{9.5.21}$$

Since at $t = t_0$

$$y_i(v_1+\delta v_1, v_2+\delta v_2, \ldots, v_n+\delta v_n, t_0) - y_i(v_1, v_2, \ldots, v_n, t_0) = \delta v_i,$$
$$i = 1,2,\ldots,n, \tag{9.5.22}$$

by the definition of δv_i in (9.5.20), we can express (9.5.21) as

$$\delta v_i = \sum_{k=1}^{n} \frac{\partial y_i(t_0)}{\partial v_k}\delta v_k + \tfrac{1}{2}\sum_{j=1}^{n}\sum_{k=1}^{n} \frac{\partial^2 y_i(t_0)}{\partial v_j \partial v_k}\delta v_j \delta v_k, \qquad i = 1,2,\ldots,n. \tag{9.5.23}$$

Since the variations are nonzero independent perturbations to the initial conditions, the coefficients of δv_k, $k = 1,2,\ldots,n$, on both sides of the equation must be equal. It follows that

$$\frac{\partial y_i(t_0)}{\partial v_k} = \left.\begin{matrix} 1, & i = k, \\ 0, & i \neq k, \end{matrix}\right\} \tag{9.5.24}$$

$$\frac{\partial^2 y_i(t_0)}{\partial v_j \partial v_k} = 0, \qquad i,j,k = 1,2,\ldots,n. \tag{9.5.25}$$

This establishes the initial conditions (9.5.14), (9.5.16) for the first- and second-order variation equations (9.5.13), (9.5.15), respectively.

To summarize, the solutions of the first- and second-order variational equations are carried out in conjunction with the solution of (9.5.1) with trial values chosen for the missing initial conditions. The partial derivatives $\partial g_i/\partial y_k$, $i,k = 1,2,\ldots,n$ and $\partial^2 g_i/\partial y_j \partial y_k$, $i,j,k = 1,2,\ldots,n$, are formed analytically from (9.5.1) and evaluated numerically along the solution of (9.5.1).

The nonlinear differential equations (9.5.1) with the specified initial conditions and the trial values of the missing initial conditions, the first-order variational equations (9.5.13) with the initial conditions (9.5.14), and the second-order variational equations (9.5.15) with the initial conditions (9.5.16) are integrated forward and in parallel. At each step of the integration the nonlinear differential equations (9.5.1) are integrated first, the first-order variational equations integrated next, and finally the second-order variational equations are integrated. This order is essential for parallel integration procedure since, for efficient machine computation, only the current solution vectors and the succeeding vectors are stored, and since (9.5.1) supplies information required for the integration of (9.5.13) and (9.5.15), and, in turn, (9.5.13) supplies information required by (9.5.15). To be more specific, at each integration step of (9.5.1) we generate values for $y_i(t)$, $i = 1,2,\ldots,n$, which in turn are used to evaluate $\partial g_i(t)/\partial y_j$, $i,j = 1,2,\ldots,n$, and $\partial^2 g_i(t)/\partial y_j \partial y_k$, $i,j,k = 1,2,\ldots,n$. The partial derivatives $\partial g_i(t)/\partial y_j$ appear in both (9.5.13) and (9.5.15), and the $\partial^2 g_i(t)/\partial y_j \partial y_k$ terms appear in (9.5.15). Once the $\partial g_i(t)/\partial y_j$ terms are available, the variational equations can be integrated one step to yield $\partial y_i(t)/\partial y_j(t_0)$, $i,j = 1,2,\ldots,n$. With the solutions of the first-order variational equations in hand, the second-order variational equations may then be integrated one step.

By virtue of the fact that (9.5.13) and (9.5.15) are linear, certain efficiencies can be effected in the parallel integration. See Riley *et al* [5].

Another approach takes advantage of the fact, already pointed out, that the column vector differential equations in (9.5.15) are the inhomogeneous forms of the column vector differential equations (9.5.13). It is a standard result (see, for example, Bellman [6]) that, if $\mathbf{Y}(t)$ is the $n \times n$ matrix solution of

$$\frac{d\mathbf{Y}}{dt} = \mathbf{A}(t)\mathbf{Y}, \qquad \mathbf{Y}(t_0) = \mathbf{I}, \tag{9.5.26}$$

then the solution of the $n \times 1$ vector equation

$$\frac{d\mathbf{z}}{dt} = \mathbf{A}(t)\mathbf{z} + \mathbf{w}(t), \qquad \mathbf{z}(t_0) = \mathbf{c} \tag{9.5.27}$$

is

$$\mathbf{z} = \mathbf{Y}(t)\mathbf{c} + \int_{t_0}^{t} \mathbf{Y}(t)\mathbf{Y}^{-1}(t_1)\mathbf{w}(t_1)\,dt_1. \tag{9.5.28}$$

In the present application, $\mathbf{c} = \mathbf{0}$, the role of $\mathbf{z}$ is played by the columns of $\mathbf{H}(t)$, that of $\mathbf{Y}(t)$ by $\mathbf{J}(t)$, and that of $\mathbf{w}(t)$ by the appropriate columns of $\mathbf{F}(t)$. Use of (9.5.28) exchanges numerical integration of (9.5.15) for numerical quadrature, which is usually a quicker and simpler computation, plus a matrix inversion. Which of the two methods is preferable would have to be decided after an analysis of the programming involved for the particular problems at hand. We shall see below, in the following discussion of the adjoint equations, how the matrix inversion can be avoided.

Finally, it should be noted that in the solution of a two-point boundary value problem corresponding to (9.5.1) in which r initial conditions are specified, only the appropriate $(n-r)\times(n-r)$ submatrix of $\mathbf{J}(t)$ need be computed, and the corresponding $(n-r)\times((n-r)(n-r+1)/2)$ submatrix of $\mathbf{H}(t)$.

9.5.3. Adjoint Equations

Closely related to variational equations are adjoint equations, which we have already met in our discussion of the method of adjoints (Chapter 3). To recapitulate, to the system of linear ordinary differential equations

$$\frac{d\mathbf{y}}{dt} = \mathbf{A}(t)\mathbf{y} + \mathbf{f}(t) \tag{9.5.29}$$

corresponds the adjoint system

$$\frac{d\mathbf{x}}{dt} = -\mathbf{A}^T(t)\mathbf{x}, \tag{9.5.30}$$

where $\mathbf{x}$ and $\mathbf{y}$ are n-vectors, $\mathbf{A}$ is an $n\times n$ matrix, and $\mathbf{A}^T$ is its transpose. It has been shown in Section 3.2, Chapter 3 that the solutions of (9.5.29) and (9.5.30) are related by the identity

$$\sum_{k=1}^{n} x_k(t)y_k(t) - \sum_{k=1}^{n} x_k(t_0)y_k(t_0) = \int_{t_0}^{t} \sum_{k=1}^{n} x_k(\tau)f_k(\tau)\,d\tau. \tag{9.5.31}$$

In particular, if $\mathbf{f}(t) = \mathbf{0}$ and if (9.5.30) is integrated backward n times from

t to t_0 with the Kronecker delta conditions at t, $x_k^{(i)}(t) = \delta_{ik}$, where the superscript denotes integration number, then (9.5.31) reduces to

$$\sum_{k=1}^{n} x_k^{(i)}(t_0) y_k(t_0) = y_i(t), \qquad i = 1, 2, \ldots, n. \tag{9.5.32}$$

In the application to the computation of partial derivatives the columns of (9.5.13) correspond to (9.5.29), where $\mathbf{f}(t) = \mathbf{0}$. In particular, for the jth column of (9.5.13) we may write (9.5.31) as

$$\sum_{k=1}^{n} x_k^{(i)}(t) \frac{\partial y_k(t)}{\partial y_j(t_0)} = \sum_{k=1}^{n} x_k^{(i)}(t_0) \frac{\partial y_k(t_0)}{\partial y_j(t_0)}. \tag{9.5.33}$$

With the Kronecker delta conditions on the adjoint equation at t, $x_k^{(i)}(t) = \delta_{ik}$, and with the Kronecker delta conditions at t_0 on the partial derivatives $\partial y_k(t_0)/\partial y_j(t_0) = \delta_{jk}$, as specified by (9.5.14), we have from (9.5.33)

$$\frac{\partial y_i(t)}{\partial y_j(t_0)} = x_j^{(i)}(t_0). \tag{9.5.34}$$

Thus the required partial derivatives at t can be obtained by integrating the equations adjoint to (9.5.13) backward from t to t_0 with the identity matrix "initial" conditions at t, or

$$\mathbf{J}(t) = \mathbf{X}^T(t_0), \tag{9.5.35}$$

where $\mathbf{X}(t_0)$ is the solution at t_0 of

$$\frac{d\mathbf{X}(t)}{dt} = -\mathbf{G}^T(t)\mathbf{X}(t), \qquad \mathbf{X}(t) = \mathbf{I} \tag{9.5.36}$$

and $\mathbf{J}$ and $\mathbf{G}$ are as in (9.5.13).

A discussion along similar lines can be given for the computation of the second-order partial derivatives, which are the solutions of (9.5.15) with homogeneous initial conditions (9.5.16), by means of a backward integration of the equation adjoint to (9.5.15). In this case the vector $\mathbf{f}(t)$ appearing in (9.5.31) is no longer zero, but the column of the matrix $\mathbf{F}(t)$ in (9.5.15) corresponding to the column of $\mathbf{H}(t)$. However, the vector corresponding to $\mathbf{y}(t_0)$ in (9.5.31) is the zero vector and, perhaps more significantly, the equation adjoint to a column of (9.5.15) is the same as the equation adjoint to a column of (9.5.13). Thus the solutions of the adjoint equation $x_j^{(i)}(t)$ used in the computation of the first-order partial derivatives in (9.5.32) can be used in the computation of the second-order partial derivatives according to a formula of the form

$$\frac{\partial^2 y_i(t)}{\partial y_j(t_0)\, \partial y_k(t_0)} = \int_{t_0}^{t} \sum_{m=1}^{n} x_m^{(i)}(t) F_{m\mu}(\tau)\, d\tau, \tag{9.5.37}$$

where the subscript $\mu = \mu(j, k)$ indicates that column of $\mathbf{F}(t)$ corresponding to the indices j and k in the particular second-order partial derivatives on the left-hand side of (9.5.37). In applications the integral in (9.5.37) would, of course, be computed by a numerical quadrature formula. Comparing (9.5.37) with (9.5.28), we see that taking the inverse of the matrix $\mathbf{J}(t)$ has, in effect, been replaced by the backward integration of the adjoint equation.

9.6. NUMERICAL EXPERIENCE

In this section we describe numerical experience with the solution of a two-point boundary value problem solved by first- and second-order Newton-Raphson methods using variational equations to obtain the required first- and second-order partial derivatives.

Consider the equations of motion

$$\ddot{x} = -\frac{kx(t)}{r^3}, \qquad \ddot{y} = -\frac{ky(t)}{r^3}, \qquad \ddot{z} = -\frac{kz(t)}{r^3},$$

where

$$r = (x^2(t)+y^2(t)+z^2(t))^{1/2}, \qquad k = 1.0.$$

The boundary conditions at $t_0 = 0.0$ and $t_f = 2.0$ are

$$\begin{aligned} x(0) &= 1.076000, & x(2) &= 0.000000, \\ y(0) &= 0.000000, & y(2) &= 0.576000, \\ z(0) &= 0.000000, & z(2) &= 0.997661. \end{aligned}$$

The problem was solved by the first-order Newton-Raphson method using the first-order variational equations and by the "tangent parabola" second-order Newton-Raphson method†. The set of missing initial conditions and the set of target terminal conditions are tabulated in Tables 9.1A and 9.1B for the first- and second-order Newton-Raphson methods, respectively. Since the 0-th iteration initial conditions are identical in Tables 9.1A and 9.1B, we observe in this case that the second-order method converges slightly faster than the first-order method. This is a situation, then, in which the theoretical results, namely Theorem 9.1, do not apply since in this numerical example the actual rate of convergence of the first-order

† This problem was solved by the method of adjoints in sections 6.8 and 6.9, Chapter 6.

Newton-Raphson method is slower than that for the "tangent parabola" second-order Newton-Raphson method. This does not mean that the theory

Table 9.1A. First-Order Newton-Raphson Method

It. no.	$\dot{x}(0)$	$\dot{y}(0)$	$\dot{z}(0)$
0	$3.236826(10^{-1})$	$5.779974(10^{-1})$	1.001114
1	$8.710471(10^{-2})$	$4.488858(10^{-1})$	$7.774918(10^{-1})$
2	$1.017913(10^{-1})$	$4.709962(10^{-1})$	$8.157893(10^{-1})$
3	$1.016576(10^{-1})$	$4.722824(10^{-1})$	$8.180170(10^{-1})$
4	$1.016591(10^{-1})$	$4.722832(10^{-1})$	$8.180183(10^{-1})$
	$x(2)$	$y(2)$	$z(2)$
0	$7.165439(10^{-1})$	$9.196271(10^{-1})$	1.592830
1	$-8.621901(10^{-2})$	$5.011763(10^{-1})$	$8.680614(10^{-1})$
2	$-2.191255(10^{-1})$	$5.727143(10^{-1})$	$9.919702(10^{-1})$
3	$-5.296808(10^{-5})$	$5.759969(10^{-1})$	$9.976556(10^{-1})$
4	$-3.287300(10^{-12})$	$5.759999(10^{-1})$	$9.976609(10^{-1})$

Table 9.1B. Second-Order Newton-Raphson Method

It. no.	$x(0)$	$\dot{y}(0)$	$\dot{z}(0)$
0	$3.236826(10^{-1})$	$5.779974(10^{-1})$	1.001114
1	$1.075524(10^{-1})$	$4.642229(10^{-1})$	$8.040570(10^{-1})$
2	$1.016596(10^{-1})$	$4.722827(10^{-1})$	$8.180174(10^{-1})$
3	$1.016591(10^{-1})$	$4.722832(10^{-1})$	$8.180183(10^{-1})$
4	$1.016591(10^{-1})$	$4.722832(10^{-1})$	$8.180183(10^{-1})$
	$x(2)$	$y(2)$	$z(2)$
0	$7.165439(10^{-1})$	$9.196271(10^{-1})$	1.592830
1	$-6.466135(10^{-4})$	$5.590517(10^{-1})$	$9.683050(10^{-1})$
2	$2.468827(10^{-7})$	$5.759988(10^{-1})$	$9.976590(10^{-1})$
3	$1.249000(10^{-16})$	$5.760000(10^{-1})$	$9.976610(10^{-1})$
4	$-1.665334(10^{-16})$	$5.760000(10^{-1})$	$9.976610(10^{-1})$

is wrong. The point is that the theory only gives an estimate for the rate of convergence, and in the present case the second-order method converges faster than the estimate would predict.

We pointed out in Section 9.5 that the partial derivatives needed in the application of the Newton-Raphson method may be obtained in several ways, among them variational equations, finite difference methods, and the method of adjoints. For these three methods, respectively, in Tables 9.2A, 9.2B, and 9.2C are listed the matrix of first-order partial derivatives, developed for the 0th and 1st iterations in Table 9.1A. We observe that in this case all three methods produce approximately similar results. The matrix of second-order partials for 0th iteration of Table 9.1B are tabulated in Tables 9.3A and 9.3B for the second-order variational equation method and for the finite difference method.

Table 9.2A. Matrix of First-Order Partials for Table 9.1A by Variational Equations[a]

	0th Iteration	
2.445536	2.671861 (10^{-1})	4.627767 (10^{-1})
3.110074 (10^{-1})	1.716195	2.167441 (10^{-1})
5.386770 (10^{-1})	2.167441 (10^{-1})	1.966466
	1st Iteration	
2.618987	5.399491 (10^{-1})	9.352179 (10^{-1})
8.443554 (10^{-1})	1.543167	7.390275 (10^{-1})
1.462464	7.390275 (10^{-1})	2.396520

[a] Matrix appears as

$$\begin{bmatrix} \frac{\partial x(2)}{\partial \dot{x}(0)} & \frac{\partial x(2)}{\partial \dot{y}(0)} & \frac{\partial x(2)}{\partial \dot{z}(0)} \\ \frac{\partial y(2)}{\partial \dot{x}(0)} & \frac{\partial y(2)}{\partial \dot{y}(0)} & \frac{\partial y(2)}{\partial \dot{z}(0)} \\ \frac{\partial z(2)}{\partial \dot{x}(0)} & \frac{\partial z(2)}{\partial \dot{y}(0)} & \frac{\partial z(2)}{\partial \dot{z}(0)} \end{bmatrix}.$$

Table 9.2B. Matrix of First-Order Partials for Table 9.1A by Finite Difference Equations [a]

	0 th Iteration	
2.445536	2.671861 (10^{-1})	4.627767 (10^{-1})
3.110074 (10^{-1})	1.717195	2.167441 (10^{-1})
5.386770 (10^{-1})	2.167441 (10^{-1})	1.966466
	1st Iteration	
2.619137	5.398682 (10^{-1})	9.350777 (10^{-1})
8.440249 (10^{-1})	1.543198	7.386092 (10^{-1})
1.461891	7.386092 (10^{-1})	2.396068

Table 9.2C. Matrix of First-Order Partials for Table 9.1A by Method of Adjoints [a]

	0 th Iteration	
2.446159	2.674206 (10^{-1})	4.631829 (10^{-1})
3.113324 (10^{-1})	1.715916	2.169072 (10^{-1})
5.392397 (10^{-1})	2.169072 (10^{-1})	1.966375
	1st Iteration	
2.620629	5.408284 (10^{-1})	9.367409 (10^{-1})
8.462604	1.542658	7.401813 (10^{-1})
1.465763	7.401813 (10^{-1})	2.397343

[a] Note of Table 9.2A applies.

The partial derivatives obtained by the finite difference method were obtained by integrating the equations of motion 17 times with the nominal trial vector of velocities incremented as shown in Table 9.4. In Tables 9.2B and 9.3B, where the finite difference method was used to determine the partials, the $\Delta\dot{x}_0$, $\Delta\dot{y}_0$, $\Delta\dot{z}_0$ increments were set equal to 10^{-4} times $\dot{x}_0, \dot{y}_0, \dot{z}_0$, respectively.

Table 9.3A. Matrix of Second Partial Derivatives for Table 9.1B, 0th Iteration, by Variational Equations [a]

−4.366568 (10⁻²)	−7.501038 (10⁻¹)	−1.299209	2.015108 (10⁻¹)	−9.030534 (10⁻¹)	−3.198001 (10⁻¹)
−5.223034 (10⁻¹)	4.160592 (10⁻¹)	−1.143311	4.876258 (10⁻¹)	1.892078 (10⁻¹)	−2.691392 (10⁻¹)
−9.046499 (10⁻¹)	−1.143311	−9.04153 (10⁻¹)	9.460390 (10⁻²)	−5.382785 (10⁻¹)	2.838232 (10⁻¹)

Table 9.3B. Matrix of Second Partial Derivatives for Table 9.1B, 0th Iteration, by Finite Difference Method [a]

−4.366683 (10⁻²)	−7.501038 (10⁻¹)	−1.299209	2.015708 (10⁻¹)	−9.030535 (10⁻¹)	−3.198002 (10⁻¹)
−5.223037 (10⁻¹)	4.160592 (10⁻¹)	−1.143311	4.876257 (10⁻¹)	1.892078 (10⁻¹)	−2.691393 (10⁻¹)
−9.046473 (10⁻¹)	−1.143311	−9.041053 (10⁻¹)	9.460506 (10⁻²)	−5.382785 (10⁻¹)	2.838235 (10⁻¹)

[a] Matrix appears as

$$\begin{bmatrix} \frac{\partial^2 x(2)}{\partial \dot{x}(0)^2} & 2\frac{\partial^2 x(2)}{\partial \dot{x}(0)\partial \dot{y}(0)} & 2\frac{\partial^2 x(2)}{\partial \dot{x}(0)\partial \dot{z}(0)} & \frac{\partial^2 x(2)}{\partial \dot{y}(0)^2} & 2\frac{\partial^2 x(2)}{\partial \dot{y}(0)\partial \dot{z}(0)} & \frac{\partial^2 x(2)}{\partial \dot{z}(0)^2} \\ \frac{\partial^2 y(2)}{\partial \dot{x}(0)^2} & 2\frac{\partial^2 y(2)}{\partial \dot{x}(0)\partial \dot{y}(0)} & 2\frac{\partial^2 y(2)}{\partial \dot{x}(0)\partial \dot{z}(0)} & \frac{\partial^2 y(2)}{\partial \dot{y}(0)^2} & 2\frac{\partial^2 y(2)}{\partial \dot{y}(0)\partial \dot{z}(0)} & \frac{\partial^2 y(2)}{\partial \dot{z}(0)^2} \\ \frac{\partial^2 z(2)}{\partial \dot{x}(0)^2} & 2\frac{\partial^2 z(2)}{\partial \dot{x}(0)\partial \dot{y}(0)} & 2\frac{\partial^2 z(2)}{\partial \dot{x}(0)\partial \dot{z}(0)} & \frac{\partial^2 z(2)}{\partial \dot{y}(0)^2} & 2\frac{\partial^2 z(2)}{\partial \dot{y}(0)\partial \dot{z}(0)} & \frac{\partial^2 z(2)}{\partial \dot{z}(0)^2} \end{bmatrix}.$$

Table 9.4

Run no.	$\dot{x}_0$	$\dot{x}_0+\Delta\dot{x}_0$	$\dot{x}_0-\Delta\dot{x}_0$	$\dot{y}_0$	$\dot{y}_0+\Delta\dot{y}_0$	$\dot{y}_0-\Delta\dot{y}_0$	$\dot{z}_0$	$\dot{z}_0+\Delta\dot{z}_0$	$\dot{z}_0-\Delta\dot{z}_0$
1	×			×			×		
2	×			×				×	
3	×			×					×
4	×				×		×		
5	×					×	×		
6	×					×		×	
7	×					×			×
8		×		×			×		
9		×		×				×	
10		×		×					×
11		×			×		×		
12		×				×	×		
13			×	×			×		
14			×	×				×	
15			×	×					×
16			×		×		×		
17			×			×	×		

The first-order partials were determined numerically from the formula

$$\frac{\partial f_{0,0}}{\partial x} = \frac{1}{2h}(f_{1,0} - f_{-1,0}),$$

where

$f_{0,0}$ = value of the function at the point (0, 0),

$f_{1,0}$ = value of the function at the point (1, 0),

$f_{-1,0}$ = value of the function at the point (−1, 0),

h = increment in the independent variable x.

The second-order partials were determined by the formula

$$\frac{\partial f_{0,0}}{\partial x\,\partial y} = \frac{1}{4h^2}(f_{1,1} - f_{1,-1} - f_{-1,1} + f_{-1,-1}).$$

To be specific, if we let $w(2)$ stand for any one of $x(2), y(2)$, and $z(2)$, the first-order partial derivatives were determined numerically by

$$\frac{\partial w(2)}{\partial \dot{x}_0} = \frac{w(2)_{\dot{x}_0+\Delta\dot{x}_0,\dot{y}_0,\dot{z}_0} - w(2)_{\dot{x}_0-\Delta\dot{x}_0,\dot{y}_0,\dot{z}_0}}{2\Delta\dot{x}_0},$$

$$\frac{\partial w(2)}{\partial \dot{y}_0} = \frac{w(2)_{\dot{x}_0,\dot{y}_0+\Delta\dot{y}_0,\dot{z}_0} - w(2)_{\dot{x}_0,\dot{y}_0-\Delta\dot{y}_0,\dot{z}_0}}{2\Delta\dot{y}_0},$$

$$\frac{\partial w(2)}{\partial \dot{z}_0} = \frac{w(2)_{\dot{x}_0,\dot{y}_0,\dot{z}_0+\Delta\dot{z}_0} - w(2)_{\dot{x}_0,\dot{y}_0,\dot{z}_0-\Delta\dot{z}_0}}{2\Delta\dot{z}_0},$$

where

$w(2)_{\dot{x}_0+\Delta\dot{x}_0,\dot{y},\dot{z}}$ = the value of $w(2)$ on integrating with the trial velocity vector $\dot{x}_0+\Delta\dot{x}_0, \dot{y}_0, \dot{z}_0$.

The second-order partials were found numerically by

$$\frac{\partial^2 w(2)}{\partial \dot{x}_0^{\,2}} = \frac{w(2)_{\dot{x}_0+\Delta\dot{x}_0,\dot{y}_0,\dot{z}_0} - 2w(2)_{\dot{x}_0,\dot{y}_0,\dot{z}_0} + w(2)_{\dot{x}_0-\Delta\dot{x}_0,\dot{y}_0,\dot{z}_0}}{4\Delta\dot{x}_0^{\,2}},$$

$$\frac{\partial^2 w(2)}{\partial \dot{x}_0\,\partial \dot{y}_0} = \frac{w(2)_{\dot{x}_0+\Delta\dot{x}_0,\dot{y}_0+\Delta\dot{y}_0,\dot{z}_0} - w(2)_{\dot{x}_0-\Delta\dot{x}_0,\dot{y}_0+\Delta\dot{y}_0,\dot{z}_0} - w(2)_{\dot{x}_0+\Delta\dot{x}_0,\dot{y}_0-\Delta\dot{y}_0,\dot{z}_0} + w(2)_{\dot{x}_0-\Delta\dot{x}_0,\dot{y}_0-\Delta\dot{y}_0,\dot{z}_0}}{4\Delta\dot{x}_0\,\Delta\dot{y}_0}$$

$$\frac{\partial^2 w(2)}{\partial \dot{x}_0\,\partial \dot{z}_0} = \frac{w(2)_{\dot{x}_0+\Delta\dot{x}_0,\dot{y}_0,\dot{z}_0+\Delta\dot{z}_0} - w(2)_{\dot{x}_0-\Delta\dot{x}_0,\dot{y}_0,\dot{z}_0+\Delta\dot{z}_0} - w(2)_{\dot{x}_0+\Delta\dot{x}_0,\dot{y}_0,\dot{z}_0-\Delta\dot{z}_0} + w(2)_{\dot{x}_0-\Delta\dot{x}_0,\dot{y}_0,\dot{z}_0-\Delta\dot{z}_0}}{4\Delta\dot{x}_0\,\Delta\dot{z}_0}$$

While we do not intend to imply that the finite difference method is necessarily a recommended way to develop first- and second-order partials it nevertheless is a simple way to check the partial derivatives by an alternative means. We observe in this example good agreement between the second-order partials obtained from the variational equations and those obtained by the finite difference method.

The example in Tables 9.1, 9.2, and 9.3 deals with a case in which the utilization of second-order terms in the Newton-Raphson method applied to the two-point boundary value problem accelerates the convergence. For the same problem but with different trial initial velocities in Tables 9.5A and 9.5B we have a situation where the first-order Newton-Raphson method does not converge and where the second-order Newton-Raphson method produces on alternate iterations overwhelmingly large corrections to the missing initial conditions. The trajectory alternates from elliptic, where $v = (\dot{x}(0)^2 + \dot{y}(0)^2 + \dot{z}(0)^2)^{1/2} < 1.86$, to hyperbolic, where $v > 1.86$.

Table 9.5A. First-Order Newton-Raphson Method

It. no.	$\dot{x}(0)$	$\dot{y}(0)$	$\dot{z}(0)$
0	$-5.380000(10^{-1})$	$2.880000(10^{-1})$	$4.988300(10^{-1})$
1	2.978011	$6.094221(10^{-1})$	1.055548
2	$-1.808087(10^{-1})$	$3.197238(10^{-1})$	$5.537777(10^{-1})$
3	1.812083	$1.160494(10^{-1})$	$2.010034(10^{-1})$
4	$-2.482617(10^{-2})$	$3.151209(10^{-1})$	$5.458053(10^{-1})$
	$x(2)$	$y(2)$	$z(2)$
0	$6.086333(10^{-1})$	$-3.733741(10^{-1})$	$-6.467022(10^{-1})$
1	6.627845	1.189119	2.059611
2	$1.910401(10^{-1})$	$-3.119410(10^{-1})$	$-5.402974(10^{-1})$
3	4.094501	$2.191597(10^{-1})$	$3.795954(10^{-1})$
4	$-3.271626(10^{-3})$	1.872328	$5.149876(10^{-5})$

Table 9.5B. Second-Order Newton-Raphson Method

It. no.	$\dot{x}(0)$	$\dot{y}(0)$	$\dot{z}(0)$
0	$-5.380000(10^{-1})$	$2.880000(10^{-1})$	$4.988300(10^{-1})$
1	$3.371241(10^{2})$	$1.850288(10^{1})$	$3.204789(10^{1})$
2	$-5.220322(10^{-1})$	$2.880804(10^{-1})$	$4.989698(10^{-1})$
3	$3.111629(10^{3})$	$1.200540(10^{1})$	$2.079395(10^{1})$
4	$-5.235829(10^{-1})$	$2.880000(10^{-1})$	$4.988306(10^{-1})$
	$x(2)$	$y(2)$	$z(2)$
0	$6.086333(10^{-1})$	$-3.733741(10^{-1})$	$-6.467022(10^{-1})$
1	$6.752949(10^{2})$	$3.700574(10^{1})$	$6.409573(10^{1})$
2	$6.136449(10^{-1})$	$-3.647452(10^{-1})$	$-6.317570(10^{-1})$
3	$6.224305(10^{3})$	$2.401080(10^{1})$	$4.158791(10^{1})$
4	$6.136919(10^{-1})$	$-3.649470(10^{-1})$	$-6.321067(10^{-1})$

This illustrates the point that second-order corrections do not necessarily insure convergence to a solution.

REFERENCES

1. R. H. Moore, Newton's Method and Variations, in *Nonlinear Integral Equations*, P. M. Anselone, ed., University of Wisconsin Press, Madison, Wisc. 1964.
2. V. S. Grebenjuk, Application of the Principle of Majorants to a Class of Iteration Processes (in Russian), *Ukrain. Mat. Ž.* **18** (1966), 102–106.
3. R. McGill and P. Kenneth, Solution of Variational Problems by Means of a Generalized Newton-Raphson Operator, *AIAA J.*, (10), **2** (1964), 1761–1766.
4. R. Courant, *Differential and Integral Calculus*, Vol. II, 1st ed., Interscience, New York, 1936, p. 550.
5. J. Riley, M. Bennett, and E. McCormick, Numerical Integration of Variational Equations, *Math. Comp.*, (97) **21** (Jan. 1967), 12–17.
6. R. Bellman, *Stability Theory of Differential Equations*, McGraw-Hill, New York, 1953.

General References

T. R. Goodman and G. N. Lance, The Numerical Solution of Two Point Boundary Value Problems, *MTAC*, **10** (1956), 82–86.

APPENDIX

At various places in this book we have employed without discussion some of the concepts and theorems of functional analysis and point set topology. For the reader without background in these topics we collect here some of the ideas we have found useful and several theorems we have used, stated here without proof. We then list references with more detailed information for the interested reader.

We assume that the reader has some familiarity with the idea of a set or a collection of "points", in which the "points" may be the usual geometric points in space, or such objects as the continuous real-valued functions on the interval $[0, 1]$.

A *metric space* is a nonempty set X together with a *metric* or "distance" d defined between any two members x and y of X such that

(i) $d(x,y) \geqq 0$,

(ii) $d(x,y) = 0$, if and only if $x = y$,

(iii) $d(x,y) = d(y,x)$,

(iv) $d(x,y) \leqq d(x,z)+d(z,y)$, $x,y,z \in X$, the triangle inequality.

Some examples of metric spaces are

1. The set of all real numbers R_1 with $d(x,y) = |x-y|$.

2. The set of all n-tuples $(x_1, \ldots, x_n)$ of real numbers R_n with

$$d(x, y) = \sqrt{\sum_{i=1}^{n} (x_i-y_i)^2}.$$

3. The set of all continuous functions on the closed interval $[0, 1]$ with $d(x,y) = \max_{0 \leq t \leq 1} |x(t)-y(t)|$.

The set of points x in X with $d(x, x_0) < r$ is the *open sphere* of center x_0 and radius $r > 0$. A set $A \subset X$ is open if every point x in A is the center of some open sphere in X.

A sequence $\{x_n\}$ of points in X *converges* to a point x in X if $d(x_n, x) \to 0$ as $n \to \infty$. A sequence $\{x_n\}$ is a *Cauchy sequence* if $d(x_n, x_m) \to 0$ as $n, m \to \infty$.

A metric space X is *complete* if every Cauchy sequence converges to a point in the space. Example 1 is a complete metric space.

A metric space X is *compact* if, from every covering by a family of open sets, a finite subfamily can be extracted which also covers X. (A covering is a collection of sets such that every x in X is in at least one member of the collection.) Compactness is the generalization of the concept on the real line of closed and bounded sets.

The *Cartesian product* $X_1 \times X_2$ of two metric spaces X_1 and X_2 is the set of all ordered pairs (x_1, x_2) with x_1 in X_1 and x_2 in X_2. With the metric $d(x,y) = \max(d_1(x_1,y_1), d_2(x_2,y_2))$, where d_i is the metric in X_i, $X_1 \times X_2$ becomes a metric space. (The idea of the Cartesian product can be extended to any number of spaces X_α.) *Tychonoff's theorem* says that the Cartesian product of compact spaces is compact. We used Tychonoff's theorem in Sections 7.5 and 7.11, Chapter 7 to justify the continuation method.

If X and Y are two metric spaces and $y = f(x)$ is a single-valued function whose domain D is included in X and whose range R is included in Y, then f is called an *operator* or *mapping* from X to Y. If each y in Y mapped by f comes from a unique x in X, then f is said to be one-to-one, and there exists a unique single-valued function $f^{-1}(y)$ called the *inverse* of f with domain R and range D. Furthermore $f^{-1}(f(x)) = x$ for every x in D.

The operator $f(x)$ is *continuous* at x_0 in X if for every real $\varepsilon > 0$ there exists a $\delta > 0$ such that $d_X(x, x_0) < \delta$ implies that $d_Y(f(x), f(x_0)) < \varepsilon$, where d_X and d_Y are the metrics in X and Y, respectively. A continuous operator maps compact spaces into compact spaces.

If f maps X into X and x^* in X has the property that $x^* = f(x^*)$, then x^* is said to be a *fixed point* of f. If a real number $0 < \alpha < 1$ exists with the property that $d(f(x), f(y)) \leq \alpha d(x,y)$, then f is said to be a *contraction mapping*. The important *contraction mapping theorem* says that if f is a contraction mapping, then it has a unique fixed point x^*. The integral operator in Section 5.5, Chapter 5 was shown to be a contraction mapping.

A *linear space* L over the real numbers R_1 has an operation of addition defined between any two of its member x and y such that L is an Abelian group:

1. $x+y = y+x$.

2. $x+(y+z) = (x+y)+z$.

3. There is a unique element 0 with the property $x+0 = x$.

4. For every x in L there is an element denoted $-x$ with the property $x+(-x) = 0$.

Furthermore scalar multiplication by elements of R_1 is defined with the properties:

1. $\alpha(x+y) = \alpha x+\alpha y$, where α is in R_1, x and y are in L.
2. $(\alpha+\beta)x = \alpha x+\beta y$, where α and β are in R_1.
3. $(\alpha\beta)x = \alpha(\beta x)$.
4. $1x = x$.

Some examples of linear spaces are:

1. n-tuples of real numbers $(x_1, \ldots, x_n)$ with addition defined as $(x_1, \ldots, x_n)+(y_1, \ldots, y_n) = (x_1+y_1, \ldots, x_n+y_n)$ and scalar multiplication by $\alpha(x_1, \ldots, x_n) = (\alpha x_1, \ldots, \alpha x_n)$.
2. All continuous functions on the closed interval [0, 1].

The elements or, points of L are often called *vectors*, and L is often called a *vector space*.

A set of n elements x_i in L are said to be *linearly independent* if $\sum_{i=1}^{n} \alpha_i x_i = 0$ means that $\alpha_i = 0$. If not, the x_i are said to be *linearly dependent*. If L has a set of n linearly independent vectors, but every set of $n+1$ vectors is linearly dependent, then L has *dimension n*. If n is not finite, L is said to be *infinite dimensional*.

A linear space L is a *normed linear space* if, for every x in L, there is a real number $\|x\|$, the *norm* of x, such that:

1. $\|x\| \geqq 0$, x in L.
2. $\|x\| = 0$ if and only if $x = 0$.
3. $\|x+y\| \leqq \|x\|+\|y\|$, x, y in L.
4. $\|\alpha x\| = |\alpha|\ \|x\|$, α in R_1, x in L.

Since a metric in L can be defined by the relation $d(x,y) = \|x-y\|$, every normed linear space is also a metric space, and therefore the concepts of sequences, convergence, and completeness can be defined in L. A complete normed linear space is called a *Banach space*.

Operators $y = T(x)$ from a linear space L to a linear space M can be defined in the same manner as operators on metric spaces. T is said to be a *linear operator* if $T(\alpha x+\beta y) = \alpha T(x)+\beta T(y)$ for α, β in R_1 and x, y in L. If M is the set of real numbers R_1, T is often called a *linear functional*. If L and M are normed linear spaces, then T is said to be *bounded* if there is a number $\mu > 0$ such that $\|T(x)\| \leqq \mu\|x\|$ for all x in L. The number $\|T\| = \sup_{\|x\|=1} \|Tx\|$ is the *norm of T*.

A linear operator is continuous if and only if it is bounded.

The linear operator T has an *inverse* T^{-1} provided $T(x) = 0$ if and only if $x = 0$. Then T^{-1} is also a linear operator. If $\|T\| < 1$, then $(I-T)^{-1}$ exists and $\|(I-T)^{-1}\| \leqq (1-\|T\|)^{-1}$, where I is the *identity operator.*

If L and M are normed linear spaces and $y = T(x)$ is an operator (which may be nonlinear) from L to M with nonempty domain $D \subset L$ and range $R \subset M$, then, for $h \in L$, $x_0 \in D$, $x_0 + th \in D$, and t a scalar, the limit (if it exists)

$$\delta T(x_0, h) = \lim_{t\to 0} \frac{1}{t} [T(x_0+th)-T(x_0)]$$

is the *Gateaux differential* of T at x_0. The Gateaux differential is not necessarily linear or continuous in h. The operator T is said to have a *Fréchet differential* if its Gateaux differential is linear and continuous in h and if

$$\lim_{h\to 0} \frac{1}{\|h\|} \|T(x_0+h)-T(x_0)-\delta T(x_0, h)\| = 0.$$

Thus $\delta T(x_0, h)$ is the Fréchet differential of T at x_0 and is often written $dT(x_0, h)$ to distinguish it from the Gateaux differential. The Gateaux differential leads to the generalization of the directional derivative, while the Fréchet differential leads to the generalization of the ordinary derivative. In fact, the mapping $dT(x_0,.) = T'_{x_0}(.)$, which is itself a bounded linear operator, is the *Fréchet derivative* of T at x_0, which we take to be the derivative of the operator T. As an example of the Fréchet derivative, let L and M be spaces of dimension l and m, respectively. Then $y = T(x)$, where $y = (y_1, \ldots, y_l)$ and $x = (x_1, \ldots, x_m)$, is specified by the l functions $y_i = \varphi_i(x_1, \ldots, x_m)$, and $T'_{x_0}(.)$ is the $l \times m$ matrix of partial derivatives

$$\frac{\partial \varphi_i(x_1^{(0)}, \ldots, x_m^{(0)})}{\partial x_k}, \quad i = 1,2,\ldots,l, \quad k = 1,2,\ldots,m,$$

which is a linear operator. A *mean value theorem for* Fréchet derivatives can be established. Further the integral of an operator $T(x)$ can be defined, and the *fundamental theorem of integral calculus,*

$$\int_{x_0}^{x_0+\Delta x} T'(x)\,dx = T(x_0+\Delta x)-T(x_0),$$

can be proved.

In our exposition we have followed Antosiewicz and Rheinboldt [1]. An excellent expository article on derivatives of operators is by Nashed [2]. Several other useful sources are also listed in the General References.

For the convenience of the reader we list the various symbols used in the book and give the symbol definition.

Symbol	Symbol Definition
$\exists x$	there exists an x
$\forall x$	for each or for every x
$A \Rightarrow B$	A implies B
$\ni$	such that
$x \in A$	x is a member of the set A
$A \subset B$	inclusion symbol, A is a subset of B
$T: \ A \to B$	set A is mapped into set B by operator T
$x_n \to x^*$	x_n converges to x^*, $n = 0, 1, 2, \ldots$
$\|x\|$	norm of x

REFERENCES

1. H. A. Antosiewicz and W. C. Rheinboldt, Numerical Analysis and Functional Analysis, in *A Survey of Numerical Analysis*, John Todd, ed., McGraw-Hill, New York, 1962.
2. M. Z. Nashed, Some Remarks on Variations and Differentials, *Amer. Math. Monthly*, (4) 73 (April 1966), Pt. II, 63–76.

General References

L. Collatz, *Functional Analysis and Numerical Mathematics*, Academic, New York, 1966.

W. Fulks, *Advanced Calculus*, Wiley, New York, 1964.

C. Goffman, *Real Functions*, Holt, Rinehart, and Winston, New York, 1953.

C. Goffman and G. Pedrick, *First Course in Functional Analysis*, Prentice-Hall, Englewood Cliffs, N.J., 1965.

L. V. Kantorovich and G. P. Akilov, *Functional Analysis in Normed Spaces*, Macmillan, New York, 1964.

A. N. Kolomogorov and S. V. Fomin, *Elements of the Theory of Functions and Functional Analysis*, Vol. 1, Graylock, Rochester, 1957.

L. A. Liusternik and V. J. Sobolev, *Elements of Functional Analysis*, Ungar, New York, 1961.

G. F. Simmons, *Topology and Modern Analysis*, McGraw-Hill, New York, 1963.

A. E. Taylor, *Advanced Calculus*, Blaisdell, Waltham, Mass., 1955.

A. E. Taylor, *Introduction to Functional Analysis*, Wiley, New York, 1958.

B. Z. Vulikh, *Introduction to Functional Analysis for Scientists and Technologists*, Pergamon Press, New York, and Addison-Wesley, Reading, Mass., 1953.

AUTHOR INDEX

SUBJECT INDEX